Multidisciplinary Applications of Computer–Mediated Communication

Hung Phu Bui
University of Economics Ho Chi Minh City, Vietnam

Raghvendra Kumar
GIET University, India

A volume in the Advances in
Wireless Technologies and
Telecommunication (AWTT) Book
Series

Published in the United States of America by
    IGI Global
    Information Science Reference (an imprint of IGI Global)
    701 E. Chocolate Avenue
    Hershey PA, USA 17033
    Tel: 717-533-8845
    Fax: 717-533-8661
    E-mail: cust@igi-global.com
    Web site: http://www.igi-global.com

Library of Congress Cataloging-in-Publication Data

Names: Bui, Hung Phu, editor. | Kumar, Raghvendra, 1987- editor.
Title: Multidisciplinary applications of computer-mediated communication /
    edited by: Hung Phu Bui, and Raghvendra Kumar.
Description: Hershey PA : Information Science Reference, [2023] | Includes
    bibliographical references. | Summary: "Multidisciplinary Applications
    of Computer-Mediated Communication considers the future use of CMC and
    recent applications of CMC in different contexts in the world with
    implications for further development. Covering key topics such as
    learning environments, business communication, and social media, this
    reference work is ideal for industry professionals, researchers,
    scholars, academicians, practitioners, instructors, and students"--
    Provided by publisher.
Identifiers: LCCN 2023008195 (print) | LCCN 2023008196 (ebook) | ISBN
    9781668470343 (hardcover) | ISBN 9781668470350 (paperback) | ISBN
    9781668470367 (ebook)
Subjects: LCSH: Computer-assisted instruction--Vietnam--Case studies. |
    Telematics--Vietnam--Case studies.
Classification: LCC LB1028.5 .M83 2023 (print) | LCC LB1028.5 (ebook) |
    DDC 371.33/4--dc23/eng/20230324
LC record available at https://lccn.loc.gov/2023008195
LC ebook record available at https://lccn.loc.gov/2023008196

This book is published in the IGI Global book series Advances in Wireless Technologies and
Telecommunication (AWTT) (ISSN: 2327-3305; eISSN: 2327-3313)

British Cataloguing in Publication Data
A Cataloguing in Publication record for this book is available from the British Library.

For electronic access to this publication, please contact: eresources@igi-global.com.

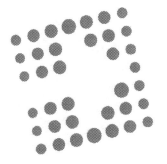

Advances in Wireless Technologies and Telecommunication (AWTT) Book Series

ISSN:2327-3305
EISSN:2327-3313

Editor-in-Chief: Xiaoge Xu University of Nottingham Ningbo China, China

MISSION

The wireless computing industry is constantly evolving, redesigning the ways in which individuals share information. Wireless technology and telecommunication remain one of the most important technologies in business organizations. The utilization of these technologies has enhanced business efficiency by enabling dynamic resources in all aspects of society.

The **Advances in Wireless Technologies and Telecommunication Book Series** aims to provide researchers and academic communities with quality research on the concepts and developments in the wireless technology fields. Developers, engineers, students, research strategists, and IT managers will find this series useful to gain insight into next generation wireless technologies and telecommunication.

COVERAGE

- Mobile Communications
- Wireless Sensor Networks
- Wireless Technologies
- Cellular Networks
- Virtual Network Operations
- Digital Communication
- Global Telecommunications
- Mobile Web Services
- Radio Communication
- Wireless Broadband

IGI Global is currently accepting manuscripts for publication within this series. To submit a proposal for a volume in this series, please contact our Acquisition Editors at Acquisitions@igi-global.com or visit: http://www.igi-global.com/publish/.

Titles in this Series

For a list of additional titles in this series, please visit:
www.igi-global.com/book-series/advances-wireless-technologies-telecommunication/73684

Economic and Social Implications of Information and Communication Technologies
Yilmaz Bayar (Bandirma Onyedi Eylul University, Turkey) and Lina Karabetyan (Independent Researcher, Turkey)
Information Science Reference • © 2023 • 318pp • H/C (ISBN: 9781668466209) • US $235.00

Modelling and Simulation of Fast-Moving Ad-Hoc Networks (FANETs and VANETs)
T.S. Pradeep Kumar (Vellore Institute of Technology, India) and M. Alamelu (Kumaraguru College of Technology, India)
Information Science Reference • © 2023 • 251pp • H/C (ISBN: 9781668436103) • US $240.00

Challenges and Risks Involved in Deploying 6G and NextGen Networks
A.M. Viswa Bharathy (GITAM University, Bengaluru, India) and Basim Alhadidi (Al-Balqa Applied University, Jordan)
Information Science Reference • © 2022 • 258pp • H/C (ISBN: 9781668438046) • US $250.00

Achieving Full Realization and Mitigating the Challenges of the Internet of Things
Marcel Ohanga Odhiambo (Mangosuthu University of Technology, South Africa) and Weston Mwashita (Vaal University of Technology, South Africa)
Engineering Science Reference • © 2022 • 263pp • H/C (ISBN: 9781799893127) • US $240.00

Handbook of Research on Design, Deployment, Automation, and Testing Strategies for 6G Mobile Core Network
D. Satish Kumar (Nehru Institute of Engineering and Technology , India) G. Prabhakar (Thiagarajar College of Engineering, India) and R. Anand (Nehru Institute of Engineering and Technology, India)
Engineering Science Reference • © 2022 • 490pp • H/C (ISBN: 9781799896364) • US $360.00

701 East Chocolate Avenue, Hershey, PA 17033, USA
Tel: 717-533-8845 x100 • Fax: 717-533-8661
E-Mail: cust@igi-global.com • www.igi-global.com

Table of Contents

Preface... xv

Chapter 1
Asynchronous and Synchronous Learning and Teaching.....................................1
   Nguyen Tien Dung, Thainguyen University of Medicine and Pharmacy,
      Vietnam
   Pham Chien Thang, Thai Nguyen University of Sciences, Vietnam
   Ta Thi Nguyet Trang, International School, Thai Nguyen University,
      Vietnam

Chapter 2
Language Teachers' Investment in Digital Multimodal Composing (DMC) as
a Manifold Application of Computer-Mediated Communication.......................17
   Nurdan Kavaklı Ulutaş, Izmir Demokrasi University, Turkey
   Aleyna Abuşka, Izmir Demokrasi University, Turkey

Chapter 3
Technology-Assisted Self-Regulated Learning in EMI Courses: A Case Study
With Economics Students in a Vietnamese Higher Education Setting................31
   Tho Doan Vo, University of Economics Ho Chi Minh City, Vietnam

Chapter 4
Boredom in Online Language Classrooms: Vietnamese EFL Students'
Perspectives..51
   Sieu Khai Luong, University of Economics Ho Chi Minh City, Vietnam
   Chau Thi Hoang Hoa, Tra Vinh University, Vietnam

Chapter 5
The Impact of 4English Mobile App of EFL Students' Reading Performance
in a Secondary Education Context 67
> Nguyen Ngoc Vu, Ho Chi Minh City University of Foreign Languages
>   and Information Technology, Vietnam
> Dang Thanh Tam, Chu Van An High School, Vietnam
> Le Nguyen Nhu Anh, Ho Chi Minh City University of Education,
>   Vietnam
> Nguyen Thi Hong Lien, Hoa Sen University, Vietnam

Chapter 6
EFL Students' Perceptions and Practices Regarding Online Language
Learning: A Case in Vietnam 88
> Thanh Nguyet Anh Le, Dong Thap University, Vietnam

Chapter 7
Facebook-Based Language Learning: Vietnamese University EFL Students'
Attitudes and Practices..110
> Tham My Duong, Ho Chi Minh City University of Economics and
>   Finance, Vietnam
> Thao Quoc Tran, HUTECH University, Vietnam

Chapter 8
Explicit Lexical Collocation Instruction in Online Teaching 133
> Cao Yen Ngoc, Victory Language Center, Tra Vinh University, Vietnam
> Chau Thi Hoang Hoa, International Collabouration Office, Tra Vinh
>   University, Vietnam

Chapter 9
Video-Mediated Dialogic Reflection for Teacher Professional Development:
A Case in Vietnam 155
> Khoa Do, University of Warwick, UK

Chapter 10
CMC Users' Positive and Negative Emotions: Features of Social Media
Platforms and Users' Strategies 188
> Hong Quan Bui, Ho Chi Minh City University of Education, Vietnam
> Thanh Tra Tran, Ho Chi Minh City Open University, Vietnam

Chapter 11

Computer-Mediated Communication and the Business World.........................211
   Ta Thi Nguyet Trang, International School, Thai Nguyen University,
       Vietnam
   Pham Chien Thang, Thai Nguyen University of Sciences, Vietnam

Chapter 12

Electronic Word of Mouth (eWOM) in Consumer Communication 225
   Ayushi Gupta, Indian Institute of Foreign Trade, New Delhi, India

Chapter 13

Customer Satisfaction With a Named Entity Recognition (NER) Store-Based
Management System Using Computer-Mediated Communication 256
   Le Thi Hong Vo, University of Economics Ho Chi Minh City, Vietnam
   Thien Tuan Hang, Mobile World Investment Corporation, Vietnam
   Ayman Youssef Nassif, University of Portsmouth, UK

Compilation of References 277

About the Contributors 315

Index...320

Detailed Table of Contents

Preface xv

Chapter 1
Asynchronous and Synchronous Learning and Teaching 1
    Nguyen Tien Dung, Thainguyen University of Medicine and Pharmacy,
       Vietnam
    Pham Chien Thang, Thai Nguyen University of Sciences, Vietnam
    Ta Thi Nguyet Trang, International School, Thai Nguyen University,
       Vietnam

Traditional teaching, in which teachers and students need to meet face to face, have been widely applied for centuries. In the traditional classroom, the teacher plays the primary function, trying to provide students with knowledge. However, the needs for distance learning has provoked ways to provide remote learners with desired knowledge and skills. Assisted by technology, online classes have been pervasively pervasively offered, especially in higher education Students can enroll in online courses, saving time for travel and other difficulties. While online learning is computer-based, traditional teaching and learning involves in-person meetings. This chapter first critically reviews aspects and traditional and online teaching before arguing for the advantages and disadvantages of the online education mode. It then presents the employment of synchronous and asynchronous computer-mediated communication in online education. Finally, directions for the future of online teaching and learning are discussed.

Chapter 2
Language Teachers' Investment in Digital Multimodal Composing (DMC) as
a Manifold Application of Computer-Mediated Communication 17
    Nurdan Kavaklı Ulutaş, Izmir Demokrasi University, Turkey
    Aleyna Abuşka, Izmir Demokrasi University, Turkey

As an instructional potential for language learning purposes to integrate digital technologies, digital multimodal composing (DMC) has mushroomed as a textual practice which involves the exploitation of digital tools in order to produce texts. In doing this, multiple semiotic modes (e.g., word, image, soundtrack etc.) are combined, and involved in the process of text production. This chapter is assumed to envision the changes in the educational landscape as a result of the age of digitization, and to understand the potential contributions of digital technologies and novel literacies to language learning and teaching. Specifically, DMC-oriented language learning and teaching will be scrutinized in order to maximize the potentials of language teachers, and thereof language learners, by investing in a post-pandemic virtual technology as an application of computer-mediated communication.

Chapter 3
Technology-Assisted Self-Regulated Learning in EMI Courses: A Case Study
With Economics Students in a Vietnamese Higher Education Setting.................31
    Tho Doan Vo, University of Economics Ho Chi Minh City, Vietnam

English-medium instruction (EMI) has challenged university learners in many Asian countries including Vietnam. The students in these contexts who are not English native speakers express concerns associated with learning both content knowledge and English. In the digital age, one question is whether and how they are adapting to the emerging context of EMI. This paper addresses the query by reporting results from a qualitative study on the self-regulated learning strategies of 24 students in EMI economics courses at one Vietnamese university. The data collected from classroom observations and focus group discussions were thematically analysed. Results revealed that the students deployed technology-assisted self-regulated strategies in their learning activities both inside and outside class, which are in line with three phases of forethought, performance, and self-reflection in the self-regulation model. This reflects special characteristics of students in a digital age and raises implications for the learning support and EMI teaching practices integrated with digital technologies.

Chapter 4
Boredom in Online Language Classrooms: Vietnamese EFL Students'
Perspectives...51
    Sieu Khai Luong, University of Economics Ho Chi Minh City, Vietnam
    Chau Thi Hoang Hoa, Tra Vinh University, Vietnam

This study investigates the causes of and suggested solutions to students' boredom in online EFL learning in Vietnam during Covid - 19. The study follows descriptive qualitative research design with the use of semi-structured interview as the sole instrument. Due to the social distance, online interviews with 38 student participants were conducted via Google Meet. Findings show that among the three factors (teacher-related, IT-related, and task-related), the teacher-related factors were the leading causes of students' boredom. Likewise, most of the students' suggestions to mitigate boredom are teacher-related: teachers' IT competency, interactive classrooms, real-life task types and authentic materials, games, bonus points, teachers' and students' relationships. On that basis, the study recommends teachers' training to improve online instruction, not only while but also post-Covid – 19 area because remote teaching using technology is an unavoidably rising trend in modern society.

Chapter 5
The Impact of 4English Mobile App of EFL Students' Reading Performance

in a Secondary Education Context 67

*Nguyen Ngoc Vu, Ho Chi Minh City University of Foreign Languages
and Information Technology, Vietnam*
Dang Thanh Tam, Chu Van An High School, Vietnam
*Le Nguyen Nhu Anh, Ho Chi Minh City University of Education,
Vietnam*
Nguyen Thi Hong Lien, Hoa Sen University, Vietnam

The main goal of this study was to examine how the 4English mobile app affected 10th-grade students' reading abilities and how they felt about using it to learn to read. At a high school in the province of An Giang, Vietnam, 90 10th graders are chosen and split equally into two groups. The curriculum, materials, school resources, and classroom instruction are the same for both groups. The experimental group is instructed to use the 4English mobile app to improve their reading skills, while the control group is given traditional reading assignments from the teacher. Four research tools employed in the study were a questionnaire, an interview, a pretest, and a posttest. The results demonstrate that the experimental participants performed better in the posttest and had favorable sentiments toward using the 4English mobile app to teach reading.

Chapter 6
EFL Students' Perceptions and Practices Regarding Online Language

Learning: A Case in Vietnam 88

Thanh Nguyet Anh Le, Dong Thap University, Vietnam

Virtual teaching and learning have become a hot topic in the Covid-19 pandemic. This chapter will present an investigation of learners' perceptions and practices of learning online. The current research was conducted with 161 EFL students at Dong Thap University, Vietnam. Questionnaires and observations were set to collect data. The findings showed that most students indicated benefits of online learning, such as saving time and money, protecting their health from Coronavirus, and being a suitable way of learning during lockdown time. However, in practice, they felt stress in long online classes. They could not fix technological problems as well as understand rules, technology culture, and attitudes when learning online. Furthermore, freshmen met several difficulties in learning English major online rather than seniors. Participants also showed their wishes and suggestions to improve virtual language teaching and learning platforms in the distant future.

Chapter 7

Facebook-Based Language Learning: Vietnamese University EFL Students' Attitudes and Practices...110
Tham My Duong, Ho Chi Minh City University of Economics and
*   Finance, Vietnam*
Thao Quoc Tran, HUTECH University, Vietnam

The emergence of Facebook has benefited language educators and learners in different ESL/EFL contexts as Facebook can function as a learning management system (LMS), facilitating the English language teaching and learning (ELTL) process. Nonetheless, research on the Facebook-based language learning (FBLL) activities is scarce. This book chapter presents a study delving into tertiary EFL English majors' attitudes toward FBLL activities, their FBLL strategy use, and the correlation between the two variables mentioned above at an institution of higher education in Vietnam. This study adopted the postpositivist perspective for quantitative data collection from a cohort of 126 English majors answering closed-ended questionnaires. The SPSS software processed the data in terms of descriptive and inferential statistics. The findings revealed that tertiary English majors showed positive attitudes towards FBLL activities and employed FBLL strategies at a high level. Furthermore, the English majors' attitudes towards FBLL activities did not affect how they utilized their FBLL strategies. This book chapter suggests some practical pedagogical implications for both teachers and students aiming to leverage the quality of ELTL concerning the use of FBLL activities.

Chapter 8

Explicit Lexical Collocation Instruction in Online Teaching 133
Cao Yen Ngoc, Victory Language Center, Tra Vinh University, Vietnam
Chau Thi Hoang Hoa, International Collabouration Office, Tra Vinh
*   University, Vietnam*

This present study aims to investigate the impact of explicit lexical collocation instruction on Vietnamese students' vocabulary use and their attitudes on the explicit lexical collocation instruction during online teaching due to COVID-19. Participants were 47 EFL students, divided into the experimental group and the control group learning online via Google Meet. Results collected from two writing tests (pre and posttest) and semi-structured interview demonstrated that the vocabulary use in students' writing performance in the experimental group had a significant improvement after receiving the explicit lexical collocation instruction online and students had positive attitude toward the explicit lexical collocation instruction. It is recommended that the application of explicit lexical collocation instruction at the beginning of a writing period in EFL teaching. It is recommended that EFL teachers can integrate lexical collocation instruction into teaching writing online to improve students' English proficiency in terms of quality of vocabulary use and writing performance.

Chapter 9

Video-Mediated Dialogic Reflection for Teacher Professional Development:
A Case in Vietnam 155
Khoa Do, University of Warwick, UK

This study investigates the nature of discourse, the advantages and disadvantages of Video-mediated Cooperative Development (VMCD) and the potential of the technique being applied as a means of Continuing Professional Development in Vietnam. The study is a qualitative case study, and the data encompass recordings of the VMCD sessions, recordings of the interviews, recordings of the online lessons, and the participants' drawings. Overall, the nature of discourse in VMCD meetings is similar to that of face-to-face Group Development (GD), with Attending suffering the most due to problems which emerge from online interaction. Additionally, even though they experienced some difficulties being the Understanders, the participants generally enjoyed the VMCD sessions, acknowledging its novelty, its relevance, and the fact that it conveys a sense of community. It is also reported that solutions or the 'moments of enlightenment' can come during the VMCD sessions, or later on when the teachers have their own time to reflect. The results have implications for the possible modifications of future versions of VMCD, or any VMCD-integrated teacher training programmes.

Chapter 10

CMC Users' Positive and Negative Emotions: Features of Social Media
Platforms and Users' Strategies 188
Hong Quan Bui, Ho Chi Minh City University of Education, Vietnam
Thanh Tra Tran, Ho Chi Minh City Open University, Vietnam

The relationship between social media and users' emotions is a prevalent research topic in computer-mediated communication. This study aims to explore the features of social media platforms that affect users' emotional self-expression and the strategies employed by users to express their emotions in text-based communication. Semi-structured interviews with ten regular users of major social media platforms, namely Facebook and Zalo, were conducted to collect data. Results showed that online network density and size, visual properties, and content display were three features that influenced users' emotional expressions. To convey their emotions on social media platforms, users employed both verbal strategies (e.g., using affect terms and verbosity) and non-verbal strategies (e.g., using punctuations and emoticons). Some participants used punctuation marks, especially question marks, and emoticons to express positive emotions, while others used verbosity to express negative feelings. Implications for using social media platforms are discussed.

Chapter 11

Computer-Mediated Communication and the Business World..........................211
*Ta Thi Nguyet Trang, International School, Thai Nguyen University,
    Vietnam*
Pham Chien Thang, Thai Nguyen University of Sciences, Vietnam

The importance of the Internet and online communication in business has significantly increased in recent years as more companies have been turning to digital tools to conduct their operations. This chapter presents how the Internet and online communication have impacted the business world, focusing on the B2B (business-to-business) sector. The chapter first will present the advantages and disadvantages of the Internet and online communication in the B2B sector, including the benefits of connecting and collaborating with other businesses in real time and the challenges of maintaining cybersecurity and building personal connections. It will then address the current state of the Internet and online communication in the business world, including updated facts and figures on its usage and impact. Also, strategies for businesses to effectively navigate the rapidly evolving landscape of the Internet and online communication, including best practices for utilizing these tools and overcoming challenges. By understanding and utilizing the internet and online communication, businesses can significantly enhance their operations and reach new levels of success.

Chapter 12

Electronic Word of Mouth (eWOM) in Consumer Communication 225
Ayushi Gupta, Indian Institute of Foreign Trade, New Delhi, India

Electronic word of mouth (eWOM) has garnered substantial interest from academic and market practitioners due to its considerable influence on consumer behavior. In the virtual world, consumer interactions have strengthened owing to the high use of digital technology. Although extant literature is available in this area, the corpus of academic literature is expanding due to fragmented published studies, which increases the complexity of the current research. In this study, the author attempts to integrate findings on the meaning of eWOM; theories used to study the area and its impact on influencing consumer behavior to synthesize existing literature. Finally, the paper explicates the scope for future research as identified from previous literature.

Chapter 13
Customer Satisfaction With a Named Entity Recognition (NER) Store-Based
Management System Using Computer-Mediated Communication 256
 Le Thi Hong Vo, University of Economics Ho Chi Minh City, Vietnam
 Thien Tuan Hang, Mobile World Investment Corporation, Vietnam
 Ayman Youssef Nassif, University of Portsmouth, UK

With the rise of the popularity of e-commerce, it is evident that the service retail
industries aim to reduce inventory and increase sales and profit margins. To achieve
this, it is of paramount importance to establish excellent and effective interaction
between customers and customer support. When a customer orders a product online,
it is essential that the store demonstrates whether the products are in stock and the
nearest stores to where the customers are. Currently, the needs of the customers are
unlikely to be effectively met. Hence, the stores are unlikely to provide desirable
products to customers even with high inventory. This paper investigates this issue
at a typical and popular retail store in Vietnam. The authors present an investigation
of this issue through two main stages. Corpus analysis for a set of collected text
messages posted on the stores' websites for customer support was first carried out to
explore the lexical patterns that indicate the customers' needs. This analysis revealed
the frequency of customers' requests for the stores' locations where they can buy the
goods and/or whether they are in stock. In the second stage of the investigation, the
valuable findings from the corpus analysis were used for data extraction based on
Named Entity Recognition (NER) software. The NER recognizes entities, including
locations and names.

Compilation of References 277

About the Contributors 315

Index 320

Preface

It is a great opportunity for two of us to collaborate on this book and is our honor to coedit endeavors by many contributors based in different parts of the world. The first editor Dr Hung Phu Bui works as a lecturer and researcher at School of Foreign Languages, University of Economics Ho Chi Minh City, Vietnam. Hung also serves as an editor for several journals indexed in Scopus and Web of Science. His broad research interests have stretched across different aspects of second/foreign language (L2) education and the use of computer-mediated communication (CMC) in L2 teaching and learning. His recently published works have mainly concentrated on applications of cognitive linguistics in L2 acquisition, sociocultural theory in L2 acquisition, teacher and student cognition, social interaction in L2 classrooms, L2 classroom assessment, teaching English for specific purposes, and computer-assisted language teaching and learning. Influenced by educational, linguistic, and psychological perspectives, his endeavors, mainly published in leading journals in the fields of language education, applied linguistics, and educational psychology, have been stimulating interesting discussions. Serving as the keynote and plenary speaker in many national and international conferences in the world, Hung has had opportunities to spread his knowledge and research interests to students, colleagues, and novice researchers.

The second editor, Dr Raghvendra Kumar, is an Associate Professor in Computer Science and Engineering Department at GIET University, India. He received a bachelor's degree and then master's degree in Technology and Ph.D. in Computer Science and Engineering, India, and postdoctoral fellowship from Institute of Information Technology, Virtual Reality and Multimedia, Vietnam. He serves as a series editor in Internet of Everything (IOE): Security and Privacy Paradigm, Green Engineering and Technology, and Concepts and Applications, published by CRC Press, Taylor & Francis Group, USA, and Bio-Medical Engineering: Techniques and Applications, Published by Apple Academic Press, CRC Press, Taylor & Francis Group, USA. He also serves as acquisition editor for Computer Science by Apple Academic Press, CRC Press, Taylor & Francis Group, USA. He has published a number of research papers in international journal (SCI/SCIE/ESCI/Scopus) and conferences including IEEE and Springer as well as served as organizing chair (RICE-2019, 2020), volume Editor (RICE-2018), Keynote speaker, session chair, Co-chair, publicity chair, publication chair, advisory board, Technical program Committee members in many international and national conferences and served as guest editors in many special issues from reputed journals (Indexed by: Scopus, ESCI, SCI). He also published 13 chapters in edited book published by IGI Global, Springer and Elsevier. His research areas are Computer Networks, Data Mining, cloud computing and Secure Multiparty Computations, Theory of Computer Science and Design of Algorithms. He authored and edited 23 computer science books in the fields of Internet of Things, Data Mining, Biomedical Engineering, Big Data, and Robotics for IGI Global Publication (USA), IOS Press Netherland, Springer, Elsevier, CRC Press (USA).

This edited book is composed of concept-based and research-based chapters which present the current interests in applying CMC in different disciplines, including general education, language education, and business. These chapters are theoretical and research-based, reflecting applications of CMC across the globe.

Computer-mediated communication or CMC is traditionally defined as humans' use of computers and Internet-based social networking to communicate with one another (Hung et al., 2022). Technological advances have expanded the use of this term, and it now generally refers to the use of electronic devices, therefore including mobile devices. CMC is of two main types: synchronous and asynchronous.

Synchronous CMC is referred to as real-time communication in which all parties are involved in text-based interaction simultaneously. They can use audio or video calls to communicate even though they are far away from one another (Hung & Nguyen, 2022). Asynchronous CMC occurs when the recipient does not receive messages from the sender immediately because they do not need to be engaged in networked communication simultaneously. As such, asynchronous CMC includes the use of email and text messages (Hung et at., 2022). In short, in CMC, social interaction can be in the two-way mode. The message sender and recipient can take turns to give and receive message.

The contemporary literature in CMC shows its potential applications in many fields. In education, CMC has provided opportunities for teacher-student and student-student interaction in distance learning programs. Teachers can deliver synchronous instruction using video calls or create online platforms for collaborative learning. Students do not need to meet in person to discuss their issues of concern but use CMC to exchange their knowledge and experiences to negotiate arising issues, such as task sharing, to reach the mutual and individual goals. In applied linguistics, CMC gives opportunities to retest Interaction Hypothesis. Originally, Interaction Hypothesis is interested in the relationship between social interaction (face-to-face) and learners' language development (Huong & Hung, 2021). The introduction of CMC facilitates real-time interaction in which parties can meet online. Language development depends not only on the learners' interaction time but also the characteristics of their interaction (Vu et al., 2021). In other words, the complexity of their language use, attention, and time investment may contribute to the success of their language learning.

During the outbreak of the COVID-19 pandemic, many educational institutions worldwide have taken advantage of CMC to deliver online instruction and communicate information to students. Although this unprecedented abrupt use indicates potential shortcomings mainly due to the inadequate preparation of the institutions and stakeholders, CMC shows its usefulness for making learning uninterrupted. Students can voice up their problems and ask for immediate support in need.

In business, CMC mediates communication between sellers and buyers. Companies may understand what their customer need and expect by sending surveys and providing means of communications. Consumers can contact people in charge when they request assistance by sending emails or text messages in the company's website or fan page. In short, CMC can bridge the gap between producers, intermediaries, and consumers.

This book consists of two main sections. Section One spans the first ten chapters mainly reflecting the use of CMC in education. The first two chapters are conceptual in nature, providing necessary background knowledge of CMC, and the following chapters make insightful applications. Section Two, with three chapters, straddle theoretical and practical aspects of CMC.

Chapter One, as mentioned earlier, is theoretical, outlining the current literature in CMC. It also gives implications of CMC for online learning and teaching. Several aspects of online learning and teaching are critically reviewed and compared to the face-to-face education mode.

Chapter Two reviews the use of digital multimodal composing (DMC) in language education. It first describes the current education landscape and then argues for the essence of DMC in language education. The author also suggests the necessity for including digital and multimodal knowledge in language teacher training.

Chapter Three reports a case study on technology-assisted self-regulated learning in English-medium instruction (EMI) courses. Results from classroom observations and focus group discussions indicated that the students deployed a variety of technology-assisted self-regulated strategies in their learning activities both inside and outside the classroom. The study argues for EMI teaching practices integrated with digital technologies.

Chapter Four explores language students' boredom in online learning. Qualitative data collected from interviews with 38 secondary school students showed the main causes of boredom and their coping strategies. The authors argue for the importance of teacher training to improve their online teaching practices, reducing students' boredom in online learning.

Chapter Five reports the use of a mobile app to improve the reading performance of secondary school students of English as a foreign language. Experimental results showed that the students benefited from using the app, resulting in their significant reading score gain. The authors argue for the use of mobile apps in teaching reading to students of English as a foreign language.

Chapter Six reports students' perceptions and practices regarding online teaching and learning of English as a foreign language. Recruiting 161 students at a Vietnamese university, this study reports that the students were aware of the benefits of online language learning. They also revealed challenges which they dealt with from online learning. From the students' difficulty learning online and suggestions to improve this learning mode, the author argues for the necessity for training students for online learning.

Chapter Seven argues for the use of Facebook, a social media platform, to facilitate foreign language learning. The quantitative data collected from questionnaires administered to 126 students revealed that the students in general held positive attitudes toward the practice. They employed a variety of strategies to be benefited from Facebook-based language learning. The authors give several implications for using Facebook in foreign language education.

Chapter Eight reports experimental results from explicit instruction on lexical collocations in online teaching. The vocabulary use in students' writing performance improved significantly after receiving the explicit lexical collocation instruction online. Implications for teaching collocations are discussed.

Chapter Nine investigates the use of video-mediated dialogic reflection to enhance language teachers' professional development. Although the teachers involved in the study revealed several difficulties, they acknowledged the benefits of video-mediated cooperative development (VMCD). The results have implications for the possible modifications of future versions of VMCD, or any VMCD-integrated teacher training programs.

Chapter Ten presents findings about CMC users' emotions regarding features of social media platforms and users' strategies in an educational setting. Results from in-depth interviews with ten social media users showed that they communicated on social media quite regularly. There was found a relationship between features of social media platforms, users' strategies and emotions, and their social networks. In general, they experienced both positive and negative feelings from communicating with people on social media.

Chapter Eleven outlines key issues in CMC and settles arguments for the use of CMC in the business world. Acknowledging the characteristics of CMC and results of recent studies on CMC, the chapter first theorizes the use of CMC in communication between parties inside the company and between the company and the outside parties. The chapter also makes several suggestions for improving the use of CMC in the business world.

Chapter Twelve acknowledges the impacts of technological advances on customer communication. This theoretical chapter first critically reviews aspects of customers' online communication. Several communication models and theories are reviewed to pave the way for theorizing customer communication regarding importance, characteristics, and benefits.

Chapter Thirteen reports a Vietnamese investigation into applications of CMC in companies to improve customer satisfaction. Reasoning the importance of communication between companies and their customers, the study used corpora of text messages retrieved from communication between customers and companies' employees. Results showed that the characteristics and responsiveness of companies as well as the features of their communication platforms applied by companies could affect customer satisfaction.

Interested in multidisciplinary applications of CMC, this edited book is a collection of meaningful and trendy endeavors in the fields of education and business. These applications attempt to advance the communication between teachers and students and between companies and customers. Providing insightful findings, they both theorize the use of CMC, visualize its future, and provide significant implications. It will be of great interest to researchers, teachers, students, and organizations. With great contributions to the development of CMC, this book provides insightful information about CMC and gives implications for improving communication between related parties (teachers, students, school administrations, companies, and office staffs) in the field of education and business.

Hung Phu Bui
University of Economics Ho Chi Minh City, Vietnam

Raghvendra Kumar
GIET University, India

REFERENCES

Hung, B. P., & Nguyen, L. T. (2022). Scaffolding language learning in the online classroom. In R. Sharma & D. Sharma (Eds.), New trends and applications in Internet of Things (IoT) and Big Data Analytics (pp. 109-122). Intelligent System Reference, Vol. 221. Springer. doi:10.1007/978-3-030-99329-0_8

Hung, B. P., Pham, A. T. D., & Purohit, P. (2022). Computer mediated communication in second language education. In R. Sharma & D. Sharma (Eds.), New trends and applications in Internet of Things (IoT) and Big Data Analytics (pp. 45-60). Intelligent System Reference, Vol. 221. Springer. doi:10.1007/978-3-030-99329-0_4

Huong, L. P. H., & Hung, B. P. (2021). Mediation of digital tools in English learning. *LEARN Journal: Language Education and Acquisition Research Network, 14*(2), 512-528. https://so04.tci-thaijo.org/index.php/LEARN/article/view/253278

Vu, N. N., Hung, B. P., Van, N. T. T., & Lien, N. T. H. (2021). Theoretical and instructional aspects of using multimedia resources in language education: A cognitive view. In R. Kumar, R. Sharma, & P. K. Pattnaik (Eds.), Multimedia technologies in the Internet of Things environment (pp. 165-194), Studies in Big Data, vol. 93. Springer. doi:10.1007/978-981-16-3828-2_9

Chapter 1
Asynchronous and Synchronous Learning and Teaching

Nguyen Tien Dung
Thainguyen University of Medicine and Pharmacy, Vietnam

Pham Chien Thang
https://orcid.org/0000-0002-5982-4173
Thai Nguyen University of Sciences, Vietnam

Ta Thi Nguyet Trang
International School, Thai Nguyen University, Vietnam

ABSTRACT

Traditional teaching, in which teachers and students need to meet face to face, have been widely applied for centuries. In the traditional classroom, the teacher plays the primary function, trying to provide students with knowledge. However, the needs for distance learning has provoked ways to provide remote learners with desired knowledge and skills. Assisted by technology, online classes have been pervasively pervasively offered, especially in higher education Students can enroll in online courses, saving time for travel and other difficulties. While online learning is computer-based, traditional teaching and learning involves in-person meetings. This chapter first critically reviews aspects and traditional and online teaching before arguing for the advantages and disadvantages of the online education mode. It then presents the employment of synchronous and asynchronous computer-mediated communication in online education. Finally, directions for the future of online teaching and learning are discussed.

DOI: 10.4018/978-1-6684-7034-3.ch001

INTRODUCTION

New methods for 21st-century education have provoked educational change. The emergence of online learning in the 1990s, driven by societal changes and technological advancements, has changed the importance of traditional education. Traditionally, acquiring academic skills is the primary motivation for learning; nevertheless, the practicality of knowledge is now the primary component that determines learning style (Wardani et al., 2018). Traditional and online education can be compared by examining their characteristics. In short, online education requires students to fully take on the responsibility for their learning, while in traditional education, this task is shared by both students and teachers (Vroeginday, 2005). Hence, related to the resources of learning, exams, tests, quizzes and assignments, the role of the teacher in both traditional and online education can be distinguished (Li, 2020). All curricula, including the assignments and quizzes for assessing students' comprehension of the content covered in class, and printed or electronic reference materials are supposed to be utilized by the students to the scope of their knowledge of various ideas. However, there are variations between the two systems in terms of capability, interaction, communication, cognitive experiences, and professional growth. While online learning is computer-based, traditional education involves a classroom arrangement (Sun & Chen, 2016). A new viewpoint on the effectiveness and utility of the two educational systems seems to have emerged because of the transition from a 19th century to a 21st century educational system. The rapid advancement of information technology, whose applications have spread from industrial production to practically every aspect of global life, is responsible for this transition (Li, 2020).

There has been a significant increase in the use of online education in recent years. Online education, also known as distance learning or e-learning, refers to delivering educational content and instruction via the Internet. On the other hand, traditional teaching refers to in-person education delivery in a classroom setting. Several advantages to online education make it an attractive option for students. Additionally, online learning is frequently more affordable than traditional learning. Students can save money on transportation and housing costs as they do not need to attend school (Cammarano et al., 2019) physically. Additionally, online courses and degree programs may have lower tuition fees than in-person ones.

Online education does, however, have some potential downsides. The lack of in-person connection with peers and instructors is one of the biggest problems with online learning. Because of this, it could be more challenging for students to stay motivated and seek the assistance they need. Another concern with online education is the quality of the instruction (Sun and Chen, 2016). While many high-quality online programs and courses are available, some are less rigorous or do not provide the same level of support as traditional programs. Traditional teaching and online education each have their own benefits and drawbacks. Traditional education gives face-to-face engagement but may be more expensive and inflexible (Parapi et al., 2020). Online education offers flexibility and cost savings but may lack personal interaction and support in the traditional mode. Ultimately, deciding which type of education to pursue will depend on the individual student's needs, preferences, and goals.

Traditional Teaching

A typical classroom was said to as a space with rows of desks and chairs facing a podium and clean, hued walls. People discovered that classrooms could affect learning because education has always focused on daily lesson attendance. In a traditional classroom, learning primarily relies on the teaching methodology, which frequently emphasizes the content more than the learners or differences in talents and learning styles. Additionally, it is only natural for learners to coordinate their learning styles and methods. On the other hand, the traditional classroom does not stimulate creativity or the intellect encouraging rote learning (Park & Choi, 2014).

Going to school traditionally entails attending lectures in a classroom setting with other students and a teacher present for in-person interaction. Students are required to show up for the classes at the scheduled times and locations. They might need to go to lectures and laboratories. While traditional schooling gives you a more practical education, it may take a long time. Students might not have time to attend classes where they must arrive on time if they juggle employment, school, and a family (Ni, 2013). Additionally, traditional education usually costs more money. Students must additionally pay for travel expenses in addition to their tuition. The enormous range of needs, interests, past knowledge, and support that students bring to the classroom makes it difficult for teachers to diversify instruction to meet these needs while raising the comprehension of every student. The concept of education is evolving away from the one-size-fits-all study guide and task, urging teachers to push new ideas and leave their comfort zones in order to enhance the learning of every one of their students (Gasevic et al., 2016).

Teacher-centered and learner-centered classrooms are the two categories under which traditional classrooms fall. The teacher plays the primary function in the traditional academic classroom, which strives to convey knowledge to students. It is well-established in the education literature that traditional teaching methods, such as hands-on activities, can be effective in helping students learn and retain information. However, the use of technology in the classroom, such as interactive whiteboards and online resources, has also become more common in recent years (Ahmed, 2013).

According to Scherus and Dumbraveanu (2014), the quality of online education can vary, with some programs and courses being more rigorous and supportive than others. Research on the comparative effectiveness of traditional teaching and online education is ongoing, with studies published as recently as 2021 and 2022. Some studies have found that traditional teaching can be more effective for specific subjects or student populations. In contrast, others have found that online education can be just as effective or even more effective in certain situations. Educators and policymakers must continue to study this issue and consider both traditional and online education's strengths and limitations to provide students the best possible learning opportunities.

Traditional strategies are a variety of methods teachers have used in the past, many of which are not generally reported to be successful, while others are still employed in one form or another. According to Tularam (2018), Traditional teaching methods are a collection of formerly popular approaches but are now widely viewed as outdated. During the period from 1800 to the 1930s, when teachers were not required to have any lesson planning, these strategies were the standard teaching practices used in the early stages of the modern education system. Until the past century, most countries had minimal government involvement or influence in the education system. Parents would only provide payment to the teacher if they found the service to be worthwhile.

Throughout that time, many teaching methods focused less on rational reflection, problem-solving skills, conceptual understanding, and social interaction and more on monotonous assignments that prioritized memorization. According to Bano et al. (2015), The teacher spoke a lot, the students did not talk much, and by today's standards, positive discipline was frequently harsh. But there is another side to traditional teaching approaches that is generally ignored. Many of the techniques used in the past by teachers are still used today, albeit in different ways. They have been altered, nonetheless, to consider our current ideas and perceptions of what makes a high-quality educational system. The "look-cover-write-check" method is a modern way to improve one's vocabulary and spelling skills. The most successful method of learning, according to teachers in the past, was to continually write out words (Kolesnikova, 2016).

Alzube (2013) states that drills and rote exercises are frequently the first things that come to mind when discussing traditional teaching methods. Drilling involves repeatedly completing a short, tedious task, such as reciting a spelling word or solving a math problem, to "get it right." Rote essentially involves repeating something until it is memorized. However, the issue with rote learning is that students often struggle to apply what they have learned in a practical setting. For example, students may memorize a dozen or more vocabulary words through rote learning but have difficulty using them in context. Another problem with rote learning is that information is mostly maintained in short-term memory; long-term memory is not committed to information until it is applied in a practical setting. As a result, other tasks must come after rote and drill teaching (Johnson, 2015). The teacher-centered approach is another characteristic of traditional education. Teacher-centered activities include demonstrating, discussions, shared learning, and working cases. Student-centered strategies involve the teacher assisting more and 'teaching' little, which is the opposite of teacher-centered strategies. Well-known cases of student-centered strategies include teamwork, scientific breakthrough learning, and self-directed learning

The distinction between the two strategies is quite hazy and, in many ways, more intellectual than practical. Teachers incorporate elements of several approaches. For instance, work examples frequently include questions inviting involvement and dialogue from the students. High-performing teachers mix teacher- and student-centered exercises into most of their classes. The issue is not which technique to employ but which is most likely to enable students to achieve modern educational objectives (Rufii, 2015).

ONLINE TEACHING AND LEARNING

Large portions of the modern world have undergone a technological revolution in recent decades. The industrial era's living conditions no longer define society as it does now, described by a knowledge-based economy stimulated and propelled by innovation and creativity. A knowledge society is essentially a learning society. Knowledge societies manage information and knowledge in ways that foster learning, stimulate creativity and innovation, and enhance people's capacity for initiating and adapting to change (Palvia et al., 2018). In the past, societies were largely characterized by an educational system that involved teachers and students physically interacting in the classroom. How we approach education has evolved because of significant technological advancements that have transformed civilization, particularly in the last 20 years and partly because of the Internet. Today, for instance, we are discussing ideas like "instructional strategies". The use of various information and communication technologies in classroom activities by teachers and students in contemporary educational systems is one example of what is meant by this notion (Kuo et al., 2013).

In this era of information technology, many people, particularly university students, rely on computers to complete their job. In addition, many institutions of higher education are aware that network technology may be utilized to grow, foster, encourage, and support learning as well as enhance students' experiences and comprehension. According to Stimpson and Cummings (2014), higher education has been significantly impacted by the quick advancements and growth of information and communication technologies. This is referred to as e-learning, which means that the teachers and students complete course assignments online, in contrast to a traditional classroom. The idea that the traditional teaching method is always the best way to facilitate learning in a university has been contested by supporters of computer-based learning over the last 20 years. Through comparison studies about the differences between e-learning and traditional classroom, people discovered that e-learning has its advantages on student learning results.

According to Cheok and Wang (2015), in online education, students can access course materials and complete tasks on their own time and from any location with an internet connection. This might be especially advantageous for students with hectic schedules or who live in remote areas. Online education provides access to a wider range of educational opportunities, including courses and programs that may not be available locally. Online education provides access to a vast amount of information, including course materials, lectures, and other resources. This information can be easily shared with other students and instructors, facilitating collaboration and learning. Online education can create a sense of community among students and instructors, even when they are physically distant. This can be achieved through discussion forums, group projects, and other interactive activities. Online education can help to promote equality by providing access to education for students who may not have the means to attend a traditional school. This can include students with disabilities, students in rural areas, and students who are unable to afford the costs of traditional education. Online education can be delivered using either synchronous or asynchronous methods. Synchronous methods involve real-time interactions, such as live lectures or webinars. Asynchronous methods involve self-paced learning and do not require real-time interactions.

Asynchronous teaching, as defined by Benta et al. (2015), is giving students access to course materials and tasks on their own schedules so they can finish them at their own pace. Online tools like learning management systems and textbooks can be used for this. For example, students and instructors can engage in real-time while learning through live lectures or webinars. This can be accomplished using online resources like video conferencing. Some asynchronous and synchronous methods are combined in online education programs to offer a balance of freedom and interactivity. Self-paced instruction combined with live lectures or debates can be used for this (Bui et al., 2022). There are several ways to give education online, such as through video lectures, podcasts, online readings, and interactive exercises. The preferred techniques will vary depending on the course and the instructor. The role of the teacher in online education is like that in traditional education, with the added responsibilities of creating and managing course materials and facilitating online interactions with students (Bui & Nguyen, 2022). Some potential disadvantages of online education include the lack of face-to-face interaction, the potential for technical issues, and the need for self-motivation. It is important for students to carefully consider these factors before enrolling in an online program.

Online Education and Computer-Mediated Communication

Communication is an essential part of any educational activity. To get achievement in an educational course, students should have to cooperate with the teachers as well as the classmates. The lack of a proper setting that is physical in nature, would promote more dialogue. This is one of an online education platform's most basic and inherent characteristics. A modernist method of learning requires that students develop communities where they can confront one another, raise questions, and create new knowledge. An online class gives students this opportunity (Edwards and Mercer, 2013). Therefore, the teacher's duty is selecting the mixed technologies that could be used for communication in online education. The various difficulties in choosing and structuring the interactive aspect of online education are reviewed in this study section. Asynchronous or synchronous communication can both be possible in terms of online education. Synchronous communication denotes real-time communication relative to one that would occur in a conventional classroom (Ogbonna et al., 2019).

According to Motteram (2013), Facetime or digital conferences would be necessary for similar interactions in an online course. Blackboard Collaborate, ooVoo, and Google Hangouts are some tools that make synchronous communication easier. This is known as asynchronous communication when there is a lag between a teacher's first communication and the students' following reactions. Email and discussion boards are two instances of asynchronous communication techniques. It is crucial to be better aware of the obstacles students face in relation to communication and collaboration because both the synchronous and asynchronous means of activities make the students share ideas with their teachers and classmates. The hardest and most important components of meaningful online learning are seen to be coordination and communication. Owing to the difficulty, some academics focus primarily on synchronous techniques, some just on asynchronous techniques, and there is a scarcity of studies on the consequences of combining different tools.

Several strategies are required to keep students motivated and interested when working with other students. One tool may be overly restrictive when employing collaborative tools to gauge participation. Many students today use social networking sites more for leisure or consumption than for collaborative learning with other students. So that students can experience editing, uploading, and posting the information through any communication channel being used, teachers should set up a space and exercises for them (Abramenka, 2015).

Delivery of Information in Online Education

Online education refers to delivering educational content and instruction via the Internet. There are a few different ways in which information is delivered in online education:

Asynchronous Delivery refers to pre-recorded lectures, readings, and other course materials that students can access at their convenience. This allows students to learn at their own pace and fit their studies into their own schedules.

Synchronous Delivery refers to the use of real-time video conferencing and other interactive tools, such as webinars, to deliver course content. Synchronous delivery allows more direct interaction between students and instructors but requires students to be available at a specific time.

Self-Paced Delivery is a combination of asynchronous and synchronous delivery. Students can complete coursework at their own pace but are also given the option to participate in real-time discussions or other interactive activities.

Overall, the delivery of information in online education highly depends on the specific program or course. Some programs rely heavily on pre-recorded lectures and readings, while others use a mix of synchronous and asynchronous activities. Students need to consider the delivery methods when choosing an online program or course to ensure that it aligns with their learning style and needs (Hass and Joseph, 2018).

ROLE OF THE TEACHERS IN ONLINE EDUCATION

Learning is most effective when the teacher centrally manages the setting. By teaching students how to correctly use the equipment, this control minimizes many of the technical issues that can arise in online education courses (e.g., checking audio and video). In these structured learning environments, students still have many opportunities to design their own learning. They can experiment with more student-centered controllable environments as their technological skills advance. According to Abramenka (2015), the effectiveness of an online course ultimately depends on its teacher. Even though students have been using computer and internet-based innovations, teachers cannot assume they are familiar with formal online educational technologies. Also, those doing their degrees in IT need proper direction and training in using these teaching tools. Any tool chosen in the classroom must be used correctly, which is the teacher's duty. However, a teacher should not just pick or select an online tool without the proper plan for conducting the course content. When designing educational programs, it is essential to have a thorough understanding of the learning environment, considering all of the students' needs and how they would access the information. Additionally, implementation of digital innovations, other options for learning spaces, and pertinent policy advancement should be given special consideration.

Means et al. (2013) suggest that one strategy teachers can employ to assist students in navigating formal learning innovations is to create a dedicated space and exercises for them to experience interpersonal editing, writing, and sharing content across any combination of platforms. Another tactic is to require students to take instruction or complete a prerequisite course that tests their familiarity with all available materials. It is not realistic to assume that after just a few online sessions, every student will be able to understand technical jargon and pick up all technical abilities (Nguyen, 2015). Setting up the collaborative space, creating and distributing clear instructions for task completion, and encouraging learning activities are all part of the teacher's responsibilities in a virtual classroom. The teacher's first responsibility is to give clear instructions and explain the duties so that students will understand their expectations and be prepared for the knowledge they will be expected to learn. In order to react, students must take responsibility for their education and use their existing knowledge to complete assessment activities (Palloff & Pratt, 2013).

REMAINING PROBLEMS AND FUTURE OF ONLINE TEACHING AND LEARNING

By utilizing articles, videos, sounds, cooperative communication, and multimedia presentations, e-learning instructors can ensure that students are fully engaged in the learning process. E-learning may increase student access to education and training in a globalized environment, highlight the need for higher education institutions to maintain a competitive advantage, and improve the quality of teaching and learning. The use of e-learning and IT has reduced the cost of education for students while raising the level of instruction. Another key benefit of online education is flexibility, which allows students to attend classes whenever and whenever they choose. According to Urokova (2020), e-learning supports a wide range of learning styles by employing dynamic features readily available online. The necessity for internet research and education has increased as a result of technology's availability and the Internet's vastness. Users can work without regard to time or physical limitations thanks to the continually evolving environment of distance learning. Online education is characterized as acquiring new knowledge entirely or in part over the Internet in higher education. Online education, which is growing in acceptance in settings ranging from primary schools to university education.

CONCLUSION

Online education can be adapted to meet the needs of individual pupils. For example, knowledge can be effectively conveyed by focusing on the needs of individual students rather than the needs of educational institutions or teachers. Online learning allows for the most efficient and convenient achievement of objectives (Vu et al., 2021). When administering the online learning environment, equitable access to information is ensured independent of the location, race, age, or racial background of the users (Urokova, 2020). There has been a significant increase in the use of online education in recent years, and research in this field has grown accordingly. According to Alshamrani (2019), the advantages of online education, flexibility is often cited as a major benefit. Students can access course materials and complete assignments on their own time if they meet the deadlines. This can be especially useful for students who have full-time jobs or other time commitments. However, there has also been research on the importance of online assessment and testing in maintaining academic rigor and ensuring student learning outcomes.

In addition to flexibility, cost-effectiveness is another advantage of online education. Students can save money on transportation and housing costs, as they do not need to attend school physically. Additionally, online courses and degree programs may have lower tuition fees than in-person ones. Dumford and Miller (2018) have shown that student learner autonomy, or the ability to manage one's own learning independently, is crucial in online education.

Traditional teaching and online education each have their own benefits and drawbacks. Traditional education offers the benefits of face-to-face interaction and support but may be more expensive and inflexible. As said that Paul and Jefferson (2019), recent research in the last 5 years has focused on the role of online assessment and testing, student learner autonomy, and the overall quality of online instruction.

Everyone is aware that with the aid of a dependable internet connection, online learning and courses are more flexible and conveniently available from anywhere. Buffering could cause you to miss out on a lot of crucial information if your phone or computer isn't linked to a reliable internet connection. Additionally, families in lower- and middle-class economic levels are at a disadvantage since parents must pay monthly or yearly fees for a dependable internet connection. Digital online classes make it difficult for teachers to evaluate their pupils during tests, in contrast to traditional education where they can monitor students closely. The teachers won't be able to tell if the pupils cheated because they won't be there in person to evaluate the kids (Arkorful & Abaidoo, 2015).

In addition to flexibility and the ability to learn at one's own pace, online education also offers the opportunity for students to customize their learning experience. Many online programs and courses offer a range of elective options, allowing students to choose the coursework that aligns with their interests and goals. This customization can help make learning more engaging and meaningful for students. Online education also allows students to access educational resources from anywhere in the world. This can be especially advantageous for students who reside in rural, or who may not have the means to physically attend a school.

While online education offers several opportunities for students, it is important to note that it also has its own challenges. One of the primary challenges is the absence of face-to-face interaction with teachers, which can make it difficult for students to obtain assistance when needed and may impact motivation (Dumford & Miller, 2018). Additionally, there is a concern about the quality of instruction in online education. While many high-quality online programs and courses are available, some are less rigorous or do not provide the same level of support as traditional programs. Students need to do their research and choose a reputable program or course. Online education offers flexibility, the ability to learn at one's own pace, customization, and global access to resources, but may lack the personal interaction and support of traditional education. Ultimately, the decision of which type of education to pursue will depend on the individual student's needs, preferences, and goals.

Attending school traditionally involves studying in a classroom with other students and a teacher for face-to-face interaction. In contrast, this is not true with online classes or online education. Technology has made it possible for this learning to take place online. The goals of all instructions are to encourage learners to acquire their own responsibility for learning by gaining the necessary tools and abilities. From students' perspectives, numerous studies have demonstrated the beneficial effects of online education. But, the cons of online education are also there, which push back the students from achieving their academic goals. As a result, we can say that traditional classroom setup and traditional teaching strategies can be proven more beneficial in many scenarios.

REFERENCES

Abramenka, V. (2015). *Students' motivations and barriers to online education.* [Masters Thesis, Grand Valley State University]. https://scholarworks.gvsu.edu/theses/776

Ahmed, A. K. (2013). Teacher-centered versus learner-centered teaching style. *Journal of Global Business Management*, 9(1), 22.

Al-Zu'be, A. F. M. (2013). The difference between the learner-centred approach and the teacher-centred approach in teaching English as a foreign language. *Education Research International*, 2(2), 24–31.

Alshamrani, M. (2019). *An investigation of the advantages and disadvantages of online education* [Doctoral dissertation, Auckland University of Technology].

Arkorful, V., & Abaidoo, N. (2015). The role of e-learning, advantages and disadvantages of its adoption in higher education. *International journal of instructional technology and distance learning, 12*(1), 29-42.

Bano, N., Arshad, F., Khan, S., & Aqeel Safdar, C. (2015). Case based learning and traditional teaching strategies: Where lies the future? *Pakistan Armed Forces Medical Journal, 65*(1).

Benta, D., Bologa, G., Dzitac, S., & Dzitac, I. (2015). University level learning and teaching via e-learning platforms. *Procedia Computer Science, 55*, 1366–1373. doi:10.1016/j.procs.2015.07.123

Bui, H. P., & Anh, D. P. T. (2022). Computer-mediated communication and second language education. In R. Sharma & D. Sharma (Eds.), *New trends and applications in Internet of things (IoT) and big data analytics* (pp. 109–122). Springer.

Bui, H. P., & Nguyen, L. T. (2022). Scaffolding language learning in the online classroom. In R. Sharma & D. Sharma (Eds.), *New trends and applications in Internet of things (IoT) and big data analytics* (pp. 45–60). Springer., doi:10.1007/978-3-030-99329-0_8

Cammarano, A., Michelino, F., & Caputo, M. (2019). Open innovation practices for knowledge acquisition and their effects on innovation output. *Technology Analysis and Strategic Management, 31*(11), 1297–1313. doi:10.1080/09537325.2019.1606420

Cheok, M. L., & Wong, S. L. (2015). Predictors of e-learning satisfaction in teaching and learning for school teachers: A literature review. *International Journal of Instruction, 8*(1), 75–90. doi:10.12973/iji.2015.816a

Dumford, A. D., & Miller, A. L. (2018). Online learning in higher education: Exploring advantages and disadvantages for engagement. *Journal of Computing in Higher Education, 30*(3), 452–465. doi:10.100712528-018-9179-z

Edwards, D., & Mercer, N. (2013). *Common Knowledge (Routledge Revivals): The Development of Understanding in the Classroom.* Routledge. doi:10.4324/9780203095287

Gašević, D., Dawson, S., Rogers, T., & Gasevic, D. (2016). Learning analytics should not promote one size fits all: The effects of instructional conditions in predicting academic success. *The Internet and Higher Education, 28*, 68–84. doi:10.1016/j.iheduc.2015.10.002

Hass, A., & Joseph, M. (2018). Investigating different options in course delivery–traditional vs online: Is there another option? *The International Journal of Information and Learning Technology, 35*(4), 230–239. doi:10.1108/IJILT-09-2017-0096

Johnson, W. A. (2015). Learning to read and write. *A Companion to Ancient Education, 120,* 137.

Kolesnikova, I. V. (2016). Combined Teaching Method: An Experimental Study. *World Journal of Education, 6*(6), 51–59. doi:10.5430/wje.v6n6p51

Kuo, Y. C., Walker, A. E., Belland, B. R., & Schroder, K. E. (2013). A predictive study of student satisfaction in online education programs. *International Review of Research in Open and Distributed Learning, 14*(1), 16–39. doi:10.19173/irrodl.v14i1.1338

Li, L. (2020). Education supply chain in the era of Industry 4.0. *Systems Research and Behavioral Science, 37*(4), 579–592. doi:10.1002res.2702

Means, B., Toyama, Y., Murphy, R., & Baki, M. (2013). The effectiveness of online and blended learning: A meta-analysis of the empirical literature. *Teachers College Record, 115*(3), 1–47. doi:10.1177/016146811311500307

Motteram, G. (2013). *Innovations in learning technologies for English language teaching.* British Council.

Nguyen, T. (2015). The effectiveness of online learning: Beyond no significant difference and future horizons. *Journal of Online Learning and Teaching, 11*(2), 309–319.

Ni, A. Y. (2013). Comparing the effectiveness of classroom and online learning: Teaching research methods. *Journal of Public Affairs Education, 19*(2), 199–215. doi:10.1080/15236803.2013.12001730

Ogbonna, C. G., Ibezim, N. E., & Obi, C. A. (2019). Synchronous versus asynchronous e-learning in teaching word processing: An experimental approach. *South African Journal of Education, 39*(2), 1–15. https://hdl.handle.net/10520/EJC-168a98cd12. doi:10.15700aje.v39n2a1383

Palloff, R. M., & Pratt, K. (2013). *Lessons from the virtual classroom: The realities of online teaching.* John Wiley & Sons.

Palvia, S., Aeron, P., Gupta, P., Mahapatra, D., Parida, R., Rosner, R., & Sindhi, S. (2018). Online education: Worldwide status, challenges, trends, and implications. *Journal of Global Information Technology Management, 21*(4), 233–241. doi:10.1080/1097198X.2018.1542262

Parapi, J. M. O., Maesaroh, L. I., Basuki, B., & Masykuri, E. S. (2020). Virtual education: A brief overview of its role in the current educational system. *Scripta: English Department Journal, 7*(1), 8–11. doi:10.37729cripta.v7i1.632

Park, E. L., & Choi, B. K. (2014). Transformation of classroom spaces: Traditional versus active learning classroom in colleges. *Higher Education, 68*(5), 749–771. doi:10.100710734-014-9742-0

Paul, J., & Jefferson, F. (2019). A comparative analysis of student performance in an online vs. face-to-face environmental science course from 2009 to 2016. *Frontiers of Computer Science, 7*, 7. doi:10.3389/fcomp.2019.00007

Rufii, R. (2015). Developing module on constructivist learning strategies to promote students' independence and performance. *International Journal of Education, 7*(1), 18. doi:10.5296/ije.v7i1.6675

Scheurs, J., & Dumbraveanu, R. (2014). A shift from teacher centered to learner centered approach. *learning, 1*(2).

Stimpson, A. J., & Cummings, M. L. (2014). Assessing intervention timing in computer-based education using machine learning algorithms. *IEEE Access: Practical Innovations, Open Solutions, 2*, 78–87. doi:10.1109/ACCESS.2014.2303071

Sun, A., & Chen, X. (2016). Online education and its effective practice: A research review. *Journal of Information Technology Education, 15*, 157–190. doi:10.28945/3502

Tularam, G. A. (2018). Traditional vs Non-traditional Teaching and Learning Strategies-the case of E-learning! *International Journal for Mathematics Teaching and Learning, 19*(1), 129–158.

Urokova, S. B. (2020). Advantages and disadvantages of online education. *ISJ Theoretical & Applied Science, 9*(89), 34–37. doi:10.15863/TAS.2020.09.89.9

Vroeginday, B. J. (2005). *Traditional vs. online education: A comparative analysis of learner outcomes.* Fielding Graduate University.

Vu, N. N., Hung, B. P., Van, N. T. T., & Lien, N. T. H. (2021) Theoretical and Instructional Aspects of Using Multimedia Resources in Language Education: A Cognitive View. In: Kumar R., Sharma R., Pattnaik P.K. (eds) Multimedia Technologies in the Internet of Things Environment, (pp. 165-194). Springer. doi:10.1007/978-981-16-3828-2_9

Wardani, D. K., Martono, T., Pratomo, L. C., Rusydi, D. S., & Kusuma, D. H. (2018). Online learning in higher education to encourage critical thinking skills in the 21st century. *International Journal of Educational Research Review*, *4*(2), 146–153. doi:10.24331/ijere.517973

ADDITIONAL READINGS

Dejnaronk, A. (2000). A Preliminary Investigation of Online Education, Online Students, and their Learning Outcomes. *AMCIS 2000 Proceedings, 6*. AIS. https://aisel.aisnet.org/amcis2000/6

Hurlbut, A. R. (2018). Online vs. traditional learning in teacher education: A comparison of student progress. *American Journal of Distance Education*, *32*(4), 248–266. doi:10.1080/08923647.2018.1509265

Mortagy, Y., & Boghikian-Whitby, S. (2010). A longitudinal comparative study of student perceptions in online education. *Interdisciplinary Journal of E-Learning and Learning Objects, 6*(1), 23-44. https://www.learntechlib.org/p/44772/

Rawat, R., & Singh, P. (2020). A Comparative Study between Traditional and Online Teaching-Learning: Medical Students' Perspective in the Wake of Corona Pandemic. *National Journal of Community Medicine*, *11*(09), 341–345. doi:10.5455/njcm.20200902070715

Salter, G. (2003). Comparing online and traditional teaching–a different approach. *Campus-Wide Information Systems*, *20*(4), 137–145. doi:10.1108/10650740310491306

Shen, Q., Chung, J. K., Challis, D. I., & Cheung, R. C. (2007). A comparative study of student performance in traditional mode and online mode of learning. *Computer Applications in Engineering Education*, *15*(1), 30–40. doi:10.1002/cae.20092

Smith, S. B., Smith, S. J., & Boone, R. (2000). Increasing access to teacher preparation: The effectiveness of traditional instructional methods in an online learning environment. *Journal of Special Education Technology*, *15*(2), 37–46. doi:10.1177/016264340001500204

Stone, M. T., & Perumean-Chaney, S. (2011). The benefits of online teaching for traditional classroom pedagogy: A case study for improving face-to-face instruction. *Journal of Online Learning and Teaching*, *7*(3), 393–400.

Umeh, A., & Salatian, A. (2011, November). A comparison of traditional teaching versus integrated Internet—Based learning in an American style university in Nigeria. In *3rd IEEE International Conference on Adaptive Science and Technology (ICAST 2011)* (pp. 132-136). IEEE. 10.1109/ICASTech.2011.6145165

Wang, M., & Wang, F. (2021, August). Comparative Analysis of University Education Effect under the Traditional Teaching and Online Teaching Mode. In *The Sixth International Conference on Information Management and Technology* (pp. 1-6). 10.1145/3465631.3465864

KEY TERMS AND DEFINITIONS

Asynchronous CMC: Communication mediated by computers and networks in which the sender and recipient are not required to be online at the same time.

Computer-Mediated Communication (CMC): Communication mediated by computers and networks.

Online Education: Online education, also known as distance learning or e-learning, refers to a mode of instruction in which students receive course materials and interact with instructors and peers remotely, typically via the Internet.

Synchronous CMC: Communication mediated by computers and networks in which the sender and recipient need to be online at the same time.

Traditional Teaching: Traditional teaching, also known as the lecture-based or didactic approach, refers to a method of instruction in which the teacher transmits knowledge to students through lectures, readings, or other forms of presentation.

Chapter 2

Language Teachers' Investment in Digital Multimodal Composing (DMC) as a Manifold Application of Computer–Mediated Communication

Nurdan Kavaklı Ulutaş

🆔 https://orcid.org/0000-0001-9572-9491
Izmir Demokrasi University, Turkey

Aleyna Abuşka
Izmir Demokrasi University, Turkey

ABSTRACT

As an instructional potential for language learning purposes to integrate digital technologies, digital multimodal composing (DMC) has mushroomed as a textual practice which involves the exploitation of digital tools in order to produce texts. In doing this, multiple semiotic modes (e.g., word, image, soundtrack etc.) are combined, and involved in the process of text production. This chapter is assumed to envision the changes in the educational landscape as a result of the age of digitization, and to understand the potential contributions of digital technologies and novel literacies to language learning and teaching. Specifically, DMC-oriented language learning and teaching will be scrutinized in order to maximize the potentials of language teachers, and thereof language learners, by investing in a post-pandemic virtual technology as an application of computer-mediated communication.

DOI: 10.4018/978-1-6684-7034-3.ch002

INTRODUCTION

The age of digitization has proven to be a keystone for linking modern language learning practices with technological developments (Kavaklı Ulutaş & Abuşka, 2022), and these developments have led to a pedagogical shift among second language (L2) teachers since digital literacy is assumed to offer the exploitation of digital technologies with a given language construct to gain language ability (Dudeney et al., 2013). In this sense, L2 teachers have been presented with new perspectives and frameworks to use in multidisciplinary and multimodal learning environments.

Confronted with these new perspectives and frameworks, educators and field practitioners are prompted to take advantage of new opportunities and presented them with new chllanges, as most courses were delivered online via synchronous and asynchronous learning environments, which is also confirmed by the current impact of the COVID-19 pandemic. As a result, courses were delivered using virtual technologies so that learners could continue learning and educators could continue teaching. Accordingly, for developing multiliteracies, digital multimodal composing (DMC) has emerged as a type of textual practice involving digital tools for text production.

Basically, DMC involves combining different semiotic modes and incorporating them into text production. On the grounds of learning, DMC is exemplified mainly through video composing, digital storytelling and podcasting (Early et al., 2015); and, thus, considered by L2 teachers as playing a significant role in language teaching. In this context, it is of utmost importance to develop L2 teachers' awareness of and engagement in DMC-oriented language learning and teaching opportunities. However, there is a dearth of research investigating how DMC can ease L2 learning.

What is apropos in this chapter is to envision the changes in the educational landscape as a result of the age of digitization and to understand the potential contributions of digital technologies and novel literacies to language learning and teaching. Specifically, DMC-oriented language learning and teaching will be scrutinized to maximize the potentials of language teachers, and language learners, by investing in a post-pandemic virtual technology as an application of computer-mediated communication (CMC).

CMC: TRENDS AND IMPACT IN THE CHANGING EDUCATIONAL LANDSCAPE

CMC basically targets the role of interactivity between stakeholders by means of mediated and multimodal communication channels (Rafaeli, 1988). CMC, along with its newer modes, are inclining towards dyadic and ephemeral environments

(Herring, 2002) instead of online interactions and interplay between different modes. Since it is not doomed to a single mode, some additional modes are employed to communicate and enhance the relationships between the stakeholders since it pays attention to not only the individual but the collective influences of multiple modes of CMC.

In essence, it is exploited as a generic term nestled by various communication forms through networked computers. Well-known examples of such systems may include electronic mail, computer conferencing, bulletin boards, discussion lists, collaborative work through interconnected computers, networking, and the like. The central focus is on the relationship of new messages being created with preceding messages rather than other variables (i.e., frequency, timing, content, number, etc.) of exchange. Herein, technology increases the plethora of interactivity in which it occurs, disbursing communication from structured nature of technology. In all veins, it promotes both mediated and face-to-face communication in order to provide responsiveness amidst conversational partners.

By creating a medium of multimodal communication, conversational partners can jointly construct social realities, putting aside their temporal and geographic barriers. In doing so, CMC supports processes of interaction in a dynamic and digital environment where feedback is limitless and not preprepared, albeit emotional overtones of the conversational partners are present. This also paves the way creating a more flexible and possibly richer interactional and multi-way environment.

In this environment, since any form of interpersonal communication is mediated through computer-assisted digital technologies, it is not restricted solely to online interaction. That said, CMC effectively transforms human relations, but this does not solve the complexity of these effects, albeit remain elusive since this interactional environment is also molded by linguistic symbols from images to metaphors. Yet, since technology evolves in time, definitions of the symbols do also evolve; thus, the notion of 'text' is becoming blur as a delimitation.

Triggered by the digital revolution brought by the utilization of CMC-oriented practices in language classrooms together with the nascent views on literacy practices, DMC has gained a heightened growth of interest in the last decade (Zhang, Akoto & Li, 2021) in order to reshape the future of language learning and teaching environments through multimodality (Hancı-Azizoğlu & Kavaklı, 2021a; Kavaklı & Hancı-Azizoğlu, 2021, Hancı-Azizoğlu & Kavaklı, 2021b), which also mushrooms as a potential future research topic to verify to what extent the use of DMC as a multimodal application of CMC can be effective as an alternative to traditional methods of teaching and learning.

THE ROLE OF DMC IN LANGUAGE EDUCATION: WHAT IT MEANS?

A tremendous experience has been witnessed at various levels of education over the past years, one of which is assumed to be the pandemic period since many different countries all over the world have shifted their systems from face-to-face to online and/or virtual ones. With no or little time to prepare for this unexpected situation, whoever is involved in the process of education (i.e., teachers, students, high stakes holders, practitioners, etc.) has experienced the adaptation process thoroughly, which then paves the way towards taking advantage of online and/or virtual education laced my digital learning materials in a digital world of learning in many ways. Not only was this tied to the case of the pandemic, but educational institutions continue to look for alternatives to address the damage resulting from the pandemic period. As digital tools mostly influence learning pathways, the lasting traces of the pandemic will continue to accelerate as the flexibility of time and place of learning in the essence of digital transformation.

Accordingly, a pedagogical shift has also been listed for language education, where digital literacy generates the use of digital technologies with a specific language construct to develop a prescribed language ability (Dudeney et al., 2013). With the introduction of multiliteracy development, investment and engagement in multimodality has gained importance, providing a path for engaging with various multimodal texts to make sense of them (Yi et al., 2020). And then, DMC has blossomed as a novel approach to meet the ever-changing needs of language learners (Hafner, 2014). Correlatively, DMC, as a product of the age of digitization and its relevance to language learning and teaching, has mushroomed as a nascent way to provide ubiquitous learning environments for teachers and students in a computer-mediated and -saturated world.

In conjunction with various theories (e.g., systemic functional grammar, sociocultural theory, social semiotic approach, etc.), DMC is essentially based on multimodality from a social semiotic perspective (Jewitt, 2008). Here, multimodality is considered as a "normal state of human communication" (Kress, 2010, p. 1), and individuals create meaning through socially shaped semiotic modes and resources. In doing so, different modes and resources of meaning are addressed through the complexity of renewed literacy practices.

In L2 education, DMC refers to "activities that engage learners in using digital tools to construct texts in a variety of semiotic modes, including writing, image, and sound (to name a few)" (Hafner, 2015, p. 487). Thus, DMC has often been used in L2 settings for a variety of purposes, as it has myriad of benefits for L2 learners (Bradley et al., 2017). For example, DMC is considered to have remarkable benefits for English language learners in the L2 when used properly in the classroom. In addition to using digital tools to engage learners in the creation of advanced texts, previous research has found that conference posters designed with DMC efficiently support learners' knowledge prior to the submission process, boost learners' confidence, and increase their self-efficacy in academic work (Lynch, 2017). Another crystal-clear advantage of DMC is that it promotes students' fluency and semantic complexity in English as a foreign language writing and improves their overall writing competence (Hafour & Al-Rashidy, 2020). DMC also helps learners express their valued identities, which leads them to enjoy investing in academic tasks (Wilson et al., 2012).

To understand why DMC is such a motivational tool in L2 classes, we can benefit from Malone and Lepper's (1987) taxonomy. According to them, 7 notable characteristics mark DMC as motivational. These are (a) challenge, (b) curiosity, (c) control, (d) imagination, (e) cooperation, (f) competition, and (g) recognition. First, it is known that L2 learners have more fun when they have tasks with an optimal level of difficulty. Second, when L2 learners are curious about what they will learn, they are more motivated. Environments that allow for the use of technology and multimodal aids, such as videos and/or audios, can also inspire curiosity (Miller, 2013). Third, as Malone and Lepper (1987) assert, motivation is best fostered when a learning activity provides "a sense of personal control over meaningful outcomes" (p. 238), and they offer that "empowering learning environments ... are those in which options are rich and dependent on the learner's response" (p. 238), as L2 learners have options to choose from that can promote their autonomy and thus increase their motivation. In addition, it is possible to support the emotional needs of L2 learners through fantasies on a digital platform. By allowing them to create their own worlds and/or avatars in digital environments, L2 teachers increase their motivation as they experience different ways of learning.

Obviously, it is required from future language teachers and students since they engage in various kinds of literacy practices supported by computer-mediated communication; they attempt to create meaning out of learning by creating multimodal messages, storing digital materials, taking photos, playing music, recording voice-overs, creating videos, and playing games. "the more exposure and practice students have with multiple genres and registers ..., the more likely they are to gain both competency and confidence in dealing with the 21st-century texts" (Rowsell et al., 2017, p. 158).

Because "meanings must be represented in a way that conforms to culturally accepted conventions of representations" (Mills, 2010, p. 232), integrating DMC into language instruction is known to be a subject matter in the growing body of research in literature (Williamson et al., 2019). The development of digital technologies fuels this growing interest to create meaning and communicate through texting as the multimodal nature of human communication has also changed. Therefore, following section will highlight the DMC-oriented affordances in language education for future directions.

THE ROLE OF DMC IN LANGUAGE EDUCATION: WHAT IT OFFERS FOR THE FUTURE OF LEARNING?

Since the age of digitization has emerged to be lacing learning practices with the developments in technology, what we try to mean by the word 'knowledge' has experienced a change since the new form of 'knowledge' is digitally formulated. As there is an abundance of information, one cannot thoroughly learn to know everything, albeit performance knowledge is becoming increasingly important.

This is also confirmed by the bringing of the 21st-century learning and teaching trends, through which students are investing more in outside-of-the-school learning environments (Gee, 2004) in order to prepare future students for the workplace of the 21st century (Partnership for 21st Century Skills, 2006). In this respect, integrating information and communication technologies into language learning environments to establish multimodal affordances is of crucial importance for language teachers, as well.

Prosumer is a newly coined term used to describe those who are both producers and consumers of digital multimodal resources to enable meaning-making (Duncum, 2011). There is now greater pressure on schools to provide learners with the digital access, knowledge, and skills needed for DMC-oriented practicesThere is now greater pressure on schools to provide learners with the digital access, knowledge, and skills required for DMC-oriented practices. On another the other hand, the 21st-century learning purports "more dynamic, interactive, generative, exploratory, visual and collaborative" (Conference on English Education [CEE], 2005). That said, learners are now considered prosumers of digital multimodal resources (Ritzer & Jurgenson, 2010).

That said, as a literacy practice that deploys digital tools to produce texts by multiple semiotic resources (Smith et al., 2020), DMC uses four essential phases (Hafner & Ho, 2020). The very first phase is pre-design. Here, learners are encouraged to be engaged in the planning process. The second phase is design. It includes improving plans and creating digital compositions. In this phase, technology use is promoted. The third phase is sharing, which expects a product. Learners are expected to present their work to the audience either synchronously or asynchronously using a technology. The fourth (final) stage is called reflection. This phase encourages students to reflect on their work (products) and the process of creating their digital texts using DMCs.

So, in designing the L2 environment, L2 teachers are assumed to nurture learners by developing a metalanguage to be used in multimodal writing. In this regard, metalanguage is considered a critical area because multimodal texts are introduced using a metalanguage to describe the semiotic choices made for multimodal meaning making. As learners are introduced to a metalanguage used to describe multimodal texts, they are advised to evaluate the possible influences of their choices that may contribute to the communicative purposes of text production (i.e., media production) through a creative domain (i.e., design thinking) and a technical domain (i.e., digital tools).

The creative domain is embellished by design thinking, which is a collaborative and inquiry-based approach to address the problems learners face in the creative learning process (Aflatoony et al., 2018). Since it is process-oriented, design thinking tends to expand learners' innovations and increase their engagement. Moreover, design thinking stimulates learners' creative outcomes, consistent with DMC's creative domain through deeper understanding. For example, learners may state their goals for their (video) productions; the assumption is that they brainstorm and evaluate several other ideas preferred by other group members. Then, they discuss, question, and reconsider their ideas together in order to select the content with which they will creatively design their product (video). After selecting the content, appropriate semiotic modes are defined and used to achieve the intended effect with DMC. Through designing and prototyping, learners can visualize their ideas and finally present the product (video) to the audience to get feedback for possible revision.

On the other hand, the technical domain is permeated by digital tools. Here, learners are supported by technology-oriented knowledge and skills required for DMC (e.g., editing, production, etc.). Thus, learners are guided to edit their work using digital tools and techniques (e.g., subtitles, filters, special effects, etc.). Rather than explaining every bit of information about how to use each digital tool and/or feature, learners are encouraged to discover these tools for themselves and be guided by online tutorials or peers.

Through these areas, learners can focus on form by distinguishing between different genres. Throughout the process, they can also learn about the importance of theme and setting, which are necessary for a particular event. Throughout the process, they can reflect on the semiotic modes used while building a relationship with the audience. Accordingly, they can draw meaning from the process using the scaffolding of metalanguage. By integrating all these elements, learners can connect their own learning to the DMC tools to achieve a final product at the end of the process that is both compelling and effective.

CONCLUSION

In today's digital world, it is believed that when students are exposed to multiple genres and registers, they would gain improved competency and confidence in comprehending the 21st-century texts (Rowsell et al., 2017, p. 158) and in writing more advanced ones. DMC, herein, has appeared with a need to focus on the learners' growing needs, especially in second language classrooms. There is no question that the use of DMC in L2 instruction is desirable in many ways (Jiang & Luk, 2016); however, there does not seem to be a common ideology for the use of DMC in L2 instruction, as teachers and learners have differing opinions (Tan & McWilliam, 2009). Although learners "generally" have a positive attitude toward DMC (Lotherington & Jenson, 2011), teachers "mostly" doubt the idea of DMC use in school (Tan & McWilliam, 2009) and have therefore removed it from the L2 curriculum (Early et al., 2015):

"Some teachers are ambivalent about or resistant to embracing multimodal composing in their L2/multilingual classrooms because they may feel multimodal composing is far removed from their classroom realities, and because they lack resources, time, and training for multimodal composing instruction" (Yi et al., 2020, p. 1).

Arguing on this side, L2 teachers might feel insecure because they lack digital and multimodal knowledge in their academic training (Siegel, 2012). Alternatively, they may reject DMC because of the lack of paper-based activities, thus labeling it as "less academic" (Yi & Choi, 2015, p. 844). Furthermore, because L2 teachers have expressed that they see DMC as a "side dish to the pedagogical roast of traditional code/print-based literacy and academic literacy" (Tan & McWilliam, 2009, p. 223), it has been assumed that they are the ones struggling with a new technology (Lotheringtan & Jenson, 2011), even though learners are familiar with and interested in the digital world.

To envision L2 teachers' engagement with DMC, Leander (2009) provides four examples, which he refers to as resistance, replacement, return, and reparation. Resistance is manifested in teachers' reluctance to engage in DMC while choosing to engage in traditional literacy instruction. In the replacement stance, teachers feel that print literacy is no longer practical, albeit outdated, and that they, therefore, need a more conventional form of literacy instruction. In the return stance, teachers tend to use a 'weak version of multimodality' (Smythe & Neufeld, 2010). Finally, in the remediation stance, teachers are likely to accept new literacies if they do not conflict with traditional ones; therefore, these teachers tend to use a 'strong version of multimodality' (Grapin, 2019). In related literature, teachers are usually found to have a resistant attitude when interacting with DMC-oriented practices (Tour, 2015).

Since DMC requires advanced skills to be operated in language classes in order to engage learners with it, L2 teachers should be very cautious, which requires an essential amount of workload. This, itself, is a shortcoming for L2 teachers. In literature, there is a bunch of research that highlights the use of DMC in L2 settings; however, a scarcity of research in terms of L2 teachers' engagement with DMC in order to help L2 learners to engage with digital literacies. As a prerequisite for L2 learners' engagement with DMC, L2 teachers are expected to add to the current theorizing of DMC in language classes, which may also elucidate teacher education in terms of multimodal changes in the digital age.

To shake this belief, recommendations in line with integrating DMC into L2 pedagogy can be listed as stimulating technology-oriented courses for teacher education, making future teachers more technologically talented and innovative. In addition, academic training is critical to integrating new tools and innovations into the classroom. Several studies have shown that teachers' anxiety about integrating new technologies into their teaching is due to the nature of their academic training. Although prospective teachers have had a contemporary and technology-oriented education, they seem unprepared and disqualified when asked to integrate DMC into their future teaching, as the use of DMC in language teaching may seem quite complex. To note more from the learners' perspective, DMC is an important motivation source. Being familiar with technology, they benefit more from the technological tools used in the classroom. As a functional tool, DMC supports multiple learning activities in the second language classroom, as language learning can be supported by videos, podcasts, and/or visual images. The use of DMC in the language classroom enriches instruction and encourages learners to complete their academic studies.

Ultimately, it is apparent that DMC is a powerful tool to use in L2 classes as it is supportive and promotive in language education by its conveniences. Thus, L2 teachers' engagement with DMC-oriented practices is significant, albeit largely absent in their professional experiences. That said, how DMC can be incorporated into language education to embellish L2 teachers with such practices should deal with enormous caution for future directions.

REFERENCES

Aflatoony, L., Wakkary, R., & Neustaedter, C. (2018). Becoming a design thinker: Assessing the learning process of students in a secondary level design thinking course. *International Journal of Art & Design Education, 37*(3), 438–453. doi:10.1111/jade.12139

Bradley, J. P., Cabell, C., Cole, D. R., Kennedy, D. H., & Poje, J. (2018). From which point do we begin?: On combining the multiliteral and multiperspectival. *Stem Journal*, *19*(2), 65–93. doi:10.16875tem.2018.19.2.65

Conference on English Education (CEE). (2005) *Beliefs about technology and the preparation of English teachers.* NCTE. http://www.ncte.org/cee/positions/beliefsontechnology

Dudeney, G., Hockly, N., & Pegrum, M. (2013). *Digital literacies.* Pearson Education.

Duncum, P. (2011). Youth on YouTube: Prosumers in a peer-to-peer participatory culture. *The International Journal of Arts Education*, *9*(2), 24–39.

Early, M., Kendrick, M., & Potts, D. (2015). Multimodality: Out from the margins of English language teaching. *TESOL Quarterly*, *49*(3), 447–460. doi:10.1002/tesq.246

Gee, J. P. (2004). *Situated language and learning: A critique of traditional schooling.* Routledge.

Grapin, S. (2019). Multimodality in the new content standards era: Implications for English learners. *TESOL Quarterly*, *53*(1), 30–55. doi:10.1002/tesq.443

Hafner, C. (2014). Embedding digital literacies in English language teaching: Students' digital video projects as multimodal ensembles. *TESOL Quarterly*, *48*(4), 655–685. doi:10.1002/tesq.138

Hafner, C. (2015). Remix culture and English language teaching: The expression of learner voice in digital multimodal compositions. *TESOL Quarterly*, *49*(3), 486–509. doi:10.1002/tesq.238

Hafner, C., & Ho, W. (2020). Assessing digital multimodal composing in second language writing: Towards a process-based model. *Journal of Second Language Writing*, *47*, 100710–100714. doi:10.1016/j.jslw.2020.100710

Hafour, M. F., & Al-Rashidy, A. S. M. (2020). Storyboarding-based collaborative narratives on Google Docs: Fostering EFL learners' writing fluency, syntactic complexity, and overall performance. *The JALT CALL Journal*, *16*(3), 123–146. doi:10.29140/jaltcall.v16n3.393

Hancı-Azizoğlu, E. B., & Kavaklı, N. (2021a). Rewriting the future through rhetorical technology. In E. B. Hancı-Azizoğlu & N. Kavaklı (Eds.), *Futuristic and linguistic perspectives on teaching writing to second language students* (pp. 1–15). IGI Global. doi:10.4018/978-1-7998-6508-7.ch001

Hancı-Azizoğlu, E. B., & Kavaklı, N. (2021b). Creative digital writing: A multilingual perspective. In M. Montebello (Ed.), Handbook of Research on Digital Language Pedagogies (pp. 250-266). IGI Global. doi:10.4018/978-1-7998-6745-6.ch013

Jewitt, C. (2008). Multimodality and literacy in school classrooms. *Review of Research in Education, 32*(1), 241–267. doi:10.3102/0091732X07310586

Jiang, L., & Luk, J. (2016). Multimodal composing as a learning activity in English classrooms: Inquiring into the sources of its motivational capacity. *System, 59*, 1–11. doi:10.1016/j.system.2016.04.001

Kavaklı, N., & Hancı-Azizoğlu, E. B. (2021). Digital storytelling: A futuristic-second-language-writing method. In E. B. Hancı-Azizoğlu & N. Kavaklı (Eds.), *Futuristic and linguistic perspectives on teaching writing to second language students* (pp. 66–83). IGI Global. doi:10.4018/978-1-7998-6508-7.ch005

Kavaklı Ulutaş, N., & Abuşka, A. (2022). Understanding L2 teachers engagement with digital multimodal composing (DMC) in the changing educational landscape. In E. Duruk (Ed.), *The new normal of online language education* (pp. 127–144). Eğiten Kitap.

Kress, G. (2010). *Multimodality: A social semiotic approach to contemporary commmunication.* Routledge.

Leander, K. M. (2009). Composing with old and new media: Toward a parallel pedagogy. In V. Carrington & M. Robinson (Eds.), *Digital literacies: Social learning and classroom practices* (pp. 147–165). Sage. doi:10.4135/9781446288238.n10

Lotherington, H., & Jenson, J. (2011). Teaching multimodal and digital literacy in L2 settings: New literacies, new basics, new pedagogies. *Annual Review of Applied Linguistics, 31*, 226–246. doi:10.1017/S0267190511000110

Lynch, M. W. (2018). Using conferences poster presentations as a tool for student learning and development. *Innovations in Education and Teaching International, 55*(6), 633–639.

Malone, T. W., & Lepper, M. R. (1987). Making learning fun: A taxonomy of intrinsic motivations for learning. In R. E. Snow & M. J. Farr (Eds.), *Aptitude, learning and Instruction III: Conative and affective process analysis* (pp. 223–253). Erlbaum.

Miller, S. M. (2013). A research metasynthesis on digital video composing in classrooms: An evidence-based framework towards a pedagogy for embodied learning. *Journal of Literacy Research, 45*(4), 385–430. doi:10.1177/1086296X13504867

Mills, K. A. (2010). What learners 'know' through digital media production: Learning by design. *E-Learning and Digital Media, 7*(3), 223–236. doi:10.2304/elea.2010.7.3.223

Partnership for 21st Century Skills. (2006). *Results that matter: 21st century skills and high school reform.* 21st Century Skills. http://www.21stcenturyskills.org/documents/RTM2006.pdf

Rafaeli, S. (1988). Interactivity: from new media to communication. In R. P. Hawkins, J. M. Wiemann, & S. Pingree (Eds.), *Advancing communication science: Merging mass and interpersonal process* (pp. 110–134). Sage.

Ritzer, G., & Jurgenson, N. (2010). Production, consumption, prosumption: The nature of capitalism in the age of the digital 'prosumer'. *Journal of Consumer Culture, 10*(1), 13–36. doi:10.1177/1469540509354673

Rowsell, J., Morrell, E., & Alvermann, D. E. (2017). Confronting the digital divide: Debunking brave new world discourses. *The Reading Teacher, 71*(2), 157–165. doi:10.1002/trtr.1603

Siegel, M. (2012). New times for multimodality? Confronting the accountability culture. *Journal of Adolescent & Adult Literacy, 55*(8), 671–681. doi:10.1002/JAAL.00082

Smythe, S., & Neufeld, P. (2010). 'Podcast time': Negotiating digital literacies and communities of learning in a middle years ELL classroom. *Journal of Adolescent & Adult Literacy, 53*(6), 488–496. doi:10.1598/JAAL.53.6.5

Tan, J. P. L., & McWilliam, E. (2009). From literacy to multiliteracies: Diverse learners and pedagogical practice. *Pedagogies, 4*(3), 213–225. doi:10.1080/15544800903076119

Tour, E. (2015). Digital mindsets: Teachers' technology use in personal life and teaching. *Language Learning & Technology, 19*, 124–139. http://llt.msu.edu/issues/october2015/tour.pdf

Williamson, B., Potter, J., & Eynon, R. (2019). New research problems and agendas in learning, media and technology: The editors' wishlist. *Learning, Media and Technology, 44*(2), 87–91. doi:10.1080/17439884.2019.1614953

Wilson, A. A., Chaves, K., & Anders, P. L. (2012). "From the Koran and Family Guy": Expressions of identity in English learners' digital podcasts. *Journal of Adolescent & Adult Literacy, 55*(5), 374–384. doi:10.1002/JAAL.00046

Yi, Y., & Choi, J. (2015). Teachers' views of multimodal practices in K-12 classrooms: Voices from teachers in the United States. *TESOL Quarterly*, *29*(4), 838–847. https://www.jstor.org/stable/43893789. doi:10.1002/tesq.219

Yi, Y., Shin, D., & Cimasko, T. (2020). Special issue: Multimodal composing in multilingual learning and teaching contexts. *Journal of Second Language Writing*, *47*, 100717–100716. doi:10.1016/j.jslw.2020.100717

Zhang, M., Akoto, M., & Li, M. (2021). Digital multimodal composing in post-secondary L2 settings: A review of the empirical landscape. [CALL]. *Computer Assisted Language Learning*, 1–28. doi:10.1080/09588221.2021.1942068

ADDITIONAL READING

Jiang, L. (2017). The affordances of digital multimodal composing for EFL learning. *ELT Journal*, *71*(4), 413–422. doi:10.1093/elt/ccw098

Kavaklı Ulutaş, N., & Abuşka, A. (2022). Understanding L2 teachers engagement with digital multimodal composing (DMC) in the changing educational landscape. In E. Duruk (Ed.), *The new normal of online language education* (pp. 127–144). Eğiten Kitap.

Liang, W. J., & Lim, F. V. (2021). A pedagogical framework for digital multimodal composing in the English Language classroom. *Innovation in Language Learning and Teaching*, *15*(4), 306–320. doi:10.1080/17501229.2020.1800709

Miller, S. M., & McVee, M. B. (2012) (Eds.). Multimodal composing in classrooms: Learning and teaching for the digital world. Taylor & Francis.

Zhang, M., Akoto, M., & Li, M. (2021). Digital multimodal composing in post-secondary L2 settings: A review of the empirical landscape. *Computer Assisted Language Learning*, 1–28. Advance online publication. doi:10.1080/09588221.2021.1942068

KEY TERMS AND DEFINITIONS

CMC: It is a generic term shaped by various communication forms through networked computers and types of computer technologies through different forms of texts.

Digital Literacy: It is an individual's ability to asset, evaluate, and disseminate information through media channels and typing by using digital platforms.

DMC: It refers to the "activities that engage learners in using digital tools to construct texts in a variety of semiotic modes, including writing, image, and sound (to name a few)" (Hafner, 2015, p. 487).

ICT: It is the acronym for information and communications technology (or technologies), assumed as the components and infrastructure that empower contemporary computing.

L2: It is an individual's second language that is different from his/her native language (L1, first language), albeit learnt later within the scope of another language of one's home country, a neighbouring language, or a foreign language other than L1.

Multimodality: It is known to be the "normal state of human communication" (Kress, 2010, p. 1), through and in which individuals create meaning through socially shaped semiotic modes and resources.

Prosumer: It is a newly coined term used to describe those who are both producers and consumers of digital multimodal resources to enable meaning-making (Duncum, 2011).

Chapter 3

Technology–Assisted Self–Regulated Learning in EMI Courses:
A Case Study With Economics Students in a Vietnamese Higher Education Setting

Tho Doan Vo
University of Economics Ho Chi Minh City, Vietnam

ABSTRACT

English-medium instruction (EMI) has challenged university learners in many Asian countries including Vietnam. The students in these contexts who are not English native speakers express concerns associated with learning both content knowledge and English. In the digital age, one question is whether and how they are adapting to the emerging context of EMI. This paper addresses the query by reporting results from a qualitative study on the self-regulated learning strategies of 24 students in EMI economics courses at one Vietnamese university. The data collected from classroom observations and focus group discussions were thematically analysed. Results revealed that the students deployed technology-assisted self-regulated strategies in their learning activities both inside and outside class, which are in line with three phases of forethought, performance, and self-reflection in the self-regulation model. This reflects special characteristics of students in a digital age and raises implications for the learning support and EMI teaching practices integrated with digital technologies.

DOI: 10.4018/978-1-6684-7034-3.ch003

INTRODUCTION

Hand in hand with globalisation, the process of internationalisation has also gained much attention in the higher education (HE) sectors. Universities have created strategies to "develop and strengthen relations across national borders" (Larsen, 2016, p. 398). These strategies are associated with increased student intake, recruitment of international students, promotion of institutional profiles and ranking, enhanced international collaboration in research and curriculum development, and a new focus on foreign language instruction (Delgado-Márquez et al., 2013). Accordingly, developing competency in English as a global language is viewed as an essential strategy in the process of internationalisation. More and more universities are implementing English-medium instruction (EMI) programmes, "an educational system where content is taught through English in contexts where English is not used as the primary, first, or official language" (Rose & McKinley, 2018, p. 114). Given that many HE institutions are using "Englishisation" (Dafouz & Smit, 2020) to internationalise their programmes, EMI has become intertwined with internationalisation (Kirkpatrick, 2011). This worldwide trend includes multilingual Asian countries (Kirkpatrick, 2011) such as Vietnam (Tran & Nguyen, 2018).

The promotion of EMI is crucial to the internationalisation of HE in Vietnam. Vietnamese universities now have the chance to improve the standard of their instruction and learning, market their institution as multilingual and globally minded, and draw in more local and foreign students as a result of the adoption of EMI. This was officially initiated in the National Foreign Language 2020 (NFL2020) project of promoting English capability promulgated by the Vietnamese government. However, there are challenges in adopting EMI in Vietnamese universities. For example, the insufficiency of EMI teaching and learning resources and professional learning for EMI teachers has also challenged the promotion of EMI programmes (Dang & Moskovsky, 2019; Vu & Burns, 2014). Moreover, there is little evidence in literature that EMI has a positive impact on students' language competence and academic performance (Tran & Nguyen, 2018). More importantly, the biggest issue is related to the level of English language competency of instructors and students enrolled in EMI courses (Vu & Burns, 2014). Limited English competence has the potential to inhibit teachers' ability to teach in English or students' ability to understand content taught in English.

Facing those problems, EMI students appear to find ways to adapt to the emerging context of EMI. Given insufficient support from teachers and universities, students are required to regulate their learning by looking for new learning strategies. While the extant literature shows an increasing number of studies on students' self-regulated learning (SRL), there has been a dearth of research exploring the students' use of SRL strategies in EMI settings. This paper addresses the issue by investigating how EMI students employ SRL strategies in learning economics subjects in their EMI classes.

Self-Regulated Learning

Generally, self-regulated learning (SRL) is described as "an active, constructive process whereby learners set goals for their learning and then attempt to monitor, regulate and control their cognition, intentions and behaviour, guided and constrained by their goals and the contextual features of the environment" (Pintrich, 2000, p. 453). In other words, learners conducting SRL actively engage in their learning metacognitively, motivationally, and behaviourally (Zimmerman, 2002). SRL involves three phases: the forethought, the performance, and the reflection phase. In each phase, learners conduct different strategies to self-regulate their learning. For example, in the first phase of forethought, learners' strategies include setting goals and making plan for their learning. In the second phase of performance, learners focus on conducting the task, monitoring their learning, seeking support, and staying focused. In the last phase of reflection, learners' typical strategies are associated with their reflecting on their own progress and the cognitive strategies they used (Puustinen & Pulkkinen, 2001; Zimmerman, 2002). The question is whether such strategies have influenced learners' academic performance.

Researchers have reported evidence on the influence of SRL on course outcomes and academic achievement. Typically, they found significant and positive correlation between SRL and academic achievement and course outcomes (Boer et al., 2013; Broadbent & Poon, 2015). In other words, learners are able to enhance their achievement when engaging more in SRL. For example, ChanLin (2012) explored the relationship between students' learning strategies and their academic outcomes and found that all students using self-regulation strategies of time and anxiety management successfully achieved their learning objectives and enhanced their interaction. Likewise, Pelikan et al. (2021) investigated the roles of SRL strategies and reported that students who perceived themselves as highly competent in using SRL strategies such as goal setting and planning, time management, metacognitive strategies tended to successfully deal with challenges even with little support in learning. This provides empirical evidence supporting the value of self-regulated learning strategies for student learning.

In EMI classes, studies on self-regulated learning albeit with limited numbers have revealed some positive findings. EMI students who have been facing a number of challenges in learning subject matter through an additional language are in need of new learning strategies to enhance their learning. In this case, SRL strategies appear to be of interests to many students. Zhou & Rose (2021) explored the listening strategies used by 412 students at an EMI university in China. They reported that "students engaged in a holistic self-regulatory cycle of learning to cope with the transition to listening to EMI classes" (p.1). The results also highlight the significance of those strategies associated with students' lesson preparation before class and their revising activities after class. This calls for more research examining the impacts of SRL strategies on EMI practices as there has been a dearth of literature in this area.

Technologies may help enhance the development of the self-regulation cycle. Despite the fact that there is still very little empirical study in this field, Kitsantas & Dabbagh (2011) have reported that 2.0 social software technologies offer a lot of promise to promote self-regulation (Hung et al., 2022). According to Kitsantas (2013), specific technologies help with all three of the aforementioned phases. This raises significant implications for teachers in designing activities to promote students' learning self-regulation. Digital technologies open up new learning possibilities and options that encourage the development of self-control skills (Bernacki et al., 2011). At the same time, the deployment of SRL appears to be crucial for students in their technology-based learning experiences (Winters et al., 2008). This explicitly indicates the casual technology and SRL tactics relationship in students' learning practices (Valentín et al., 2013).

A few studies have examined university students' use of technologies in both official and casual learning settings (Hung & Nguyen, 2022; Margaryan et al., 2011). The findings have shown a wide range of technologies employed for learning but failed to clarify specific learning purposes of using technologies. For instance, Lai & Gu (2011) have tried to understand how university students use technologies through the lens of SRL as the theoretical framework. They found that students have little knowledge on how to use technology for learning. This echoes previous studies and raises some concerns regarding learners' application of technology for self-regulating learning as learners nowadays are prone to increasingly use technology to self-regulate their learning.

Students' Learning Practices in EMI Classes

Students in EMI classes are required to use new strategies for learning new content in an additional language. In Yeh's (2014) study, the researcher reported how frequently the students adopted specific strategies in EMI courses. The most often employed techniques were paying attention in class, taking notes, asking for assistance from classmates, and spending extra time reviewing texts. It was discovered that few students were engaging in active learning techniques including creating study groups, asking questions in class, or previewing texts (Yeh, 2014), suggesting that learners were not aware of specific strategies available to them. In Chappel's (2015) research, Japanese students mentioned comparing notes with foreign peers, downloading lecture slides, recording and listening to the lectures again, keeping vocabulary logs and asking Japanese peers as activities to keep up with the EMI classes. This suggests a need for learning strategies to enhance students' language and content learning in EMI.

Students also adopted various strategies to overcome unfamiliar language structures in EMI classes. According to Hu et al. (2014), students in China devised a range of activities they used inside and outside classes including asking teachers to code-switch to Chinese for abstract concepts, referencing Chinese language books, looking up unknown words in the textbooks before class, preparing lessons at home by reading relevant sections, or reviewing slides in line with books written in Chinese after class. Some students even translated the content from English into Chinese, did relevant readings in Chinese, and took notes based on Chinese and English textbooks when preparing for tests (Hu et al., 2014). These findings showed a reliance on the L1 to deal with the new content language, a finding that aligns with Tarnopolsky and Goodman's (2014) study whereby both the teachers and students in Ukraine "consider the use of the L1 in the classroom to be a natural function of the need for comprehension" (p. 383).

Despite the dearth of research investigating teaching and learning strategies used in EMI settings, the few studies available point to the importance of strategies applicable to the dual objectives of EMI contexts. The increasing popularity of EMI and the evolution of digital technologies are considered to be key issues affecting HE in today's globalised academic world (Querol-Julián & Camiciottoli, 2019). In the process of teaching and learning, both EMI and technology integration appear to influence learning. Despite its rapid development, EMI has created significant challenges for teachers and students who are not native speakers of English and researchers have investigated whether and how digital technologies might contribute to EMI practices.

There is empirical evidence supporting the integration of digital technologies in EMI settings. For example, in their study, Paliwoda-Pękosz and Stal (2015) reported that most students were satisfied with the range of Moodle tools in the Virtual Learning Environment system and perceived it useful in providing students with more interactive resources for learning content and creating a collaborative environment where students can share their experience, find support, and interact with friends and teachers. The authors also suggested a framework of blended learning using technologies to enhance the effectiveness of EMI courses. The use of VLE in blended learning or mobile learning was presented as "an effective approach that improves lecture comprehension, encourages more class engagement, promotes collaborative learning, and achieves better learning outcomes" (Chuang, 2017, p. 640). These studies reported positive impacts from using various technologies on EMI courses. There are concerns, however, about the trustworthiness of assessing and evaluating the effectiveness of these technologies in EMI settings as there appears to be little evidence reporting the use of measurement tools for assessment and evaluation in these studies.

Recent research focuses on pedagogical approaches which combine a range of digital technologies in EMI classes, with mixed results. Flipped classroom pedagogy has been investigated to see whether it enhances teaching and learning practices. For instance, Choi et al. (2015) surveyed 75 students in an EMI nursing course using the flipped learning model and revealed that the students preferred the flipped learning strategy as this method helped enhance their understanding of lecture content. This finding appears to be inconsistent with what Karjanto and Simon (2018) reported in their study where the students perceived an improvement in communication and engagement in the course, but they still struggled to understand the materials and content (Karjanto & Simon, 2018). This points to a need for further investigation. There are limited studies researching the adoption of digital technologies in EMI teaching and learning.

RESEARCH QUESTION

The study aims to explore the undergraduate students' use of SRL strategies in the EMI economics courses at a higher education institution in Vietnam. As such, the study focusses on answering the following question:

How do students use learning strategies to regulate their own learning in EMI courses with the assistance of technology?

METHODOLOGY

This study examined undergraduate economics-related courses offered in English at a Vietnamese institution using a qualitative multiple case study approach.

Participants

The study investigated four individual cases, each of which had a different major. Each case had one subject teacher teaching a group of 40–50 students in an EMI course. Students were recruited for focus groups from the classes suggested by the subject teachers. After the classroom observations, each student was provided with information about the study and a form on which students were invited to share personal information including their names, ages, genders, contact information, their levels of confidence in learning subject content, literacy about digital technologies,

Table 1. Summary of student participants

MAJOR	Management	7
	Economics	6
	Finance	4
	Electronic commerce	7
YEAR	First year	6
	Third year	18
GENDER	Male	7
	Female	11

English competence, and information about providing consent to participate in a focus group interview. Students interested in focus group discussions were able to specify their available time on the sheets. Out of 169 students, 71 expressed interests in participating in a focus group discussion, so I used criteria to choose candidates in each class. In order to keep the focus groups diverse, I took into account their gender and professional backgrounds. To facilitate the students' engagement, I also considered their availability. There were 24 students in all that participated in focus groups. These students, who ranged in age from 18 to 20, were majoring in various aspects of economics. Six of them were first-year students, while 18 were third-year students (see Table 1).

Data Collection

Data were collected during the second semester of the school year, from August to December. Two major sources of data in this study included one classroom observation and one student focus group in each case. I scheduled student focus groups within the week of the classroom observations at a time and venue that suited the students. Interview questions were sent to students via email three days prior to the focus group discussions. At the beginning of each discussion, after we discussed their rights and responsibilities in the research, students gave written consent. Students agreed on certain ground rules for their focus group interview. All focus group discussions were audio-recorded and conducted in Vietnamese at the students' request. Discussions lasted 45 minutes to 60 minutes. I transcribed the data in Vietnamese and processed them using N-Vivo11. I only translated significant parts into English to report the findings. The scripts in both languages were easier to understand thanks to the back-translation approach, which involves translating from the source language into the target language (Chen & Boore, 2010).

Data Analysis

Data from observation field notes and students' focus group discussions were managed both manually and using computer software. Transcripts and field-notes of four cases were separated into four different document folders with identified labels. The process of data analysis included individual case descriptions and then a thematic cross case study analysis. Within-case analysis identified distinctive contextual features, and cross-case analysis enabled the building of "abstractions across cases" (Khoa et al., 2022; Merriam, 2014, p. 234). The coding process was conducted both manually and with N-Vivo11. I started the procedure by highlighting and labelling each data chunk with codes created using coloured sticky notes. Each study case used the same coding approach. I also coded the field notes and transcripts that were saved in an N-Vivo 11 project at the same time. I was able to compare the coding process outcomes using the code-recode method, which refers to the act of coding the same data twice over the course of a period of time (Krefting, 1991). The themes were developed in alignment with the three phases of the self-regulated learning procedure.

FINDINGS

The Forethought Phase

Most students in the focus groups shared the common challenges in learning economics subjects through English. Many of them possessed such limited English that they could not fully understand their lectures. Others with high levels of English proficiency had concerns about their teachers' English proficiency and noticed their grammatical mistakes, inaccurate pronunciation or intonation, and Vietnamese accents. They worried that these weaknesses might hinder their comprehension. Some complained about the amount of content they were expected to cover in a lecture. All the students said these issues prompted them to engage in self-study at home, as exemplified in the following representative quote:

My English is not good enough, and there is too much content knowledge to learn. I have to invest more time to study at home. I have to prepare the lessons and find ways to comprehend the content by myself because I cannot follow the lectures in class. (Student 4 - Focus group 1)

The students therefore prepared carefully at home before each lecture so that they would be able to understand what the teachers taught in class. Part of this preparation involved regularly reading textbooks in advance using two distinct strategies which neatly divided the participating students. The first group tried to translate reading texts into Vietnamese, with some participants using both Vietnamese and English textbooks at the same. After comprehending the main ideas in Vietnamese, these students read the English texts to compare and understand meanings in English. Meanwhile, the other group of students read the texts exclusively in English, as given in this quote:

I am using an English course book. The teacher has provided us with English slides and lectured in English, too, so I read the texts, take notes and understand business concepts in English. Only when I encounter difficult words do I have to look up their meaning in Vietnamese. (Student 6 – Focus group 2)

Despite using different strategies, all students searched for additional sources of reference to understand significant content. In this case, the students reported accessing multiple digital resources to assist their preparation. They searched on the Internet for relevant articles, eBooks or video clips on YouTube that could provide them with detailed explanations or practical examples. Some joined Facebook groups and received help from senior students. They believed that preparation at home was essential if they were to follow lectures easily in class, as evidenced in the quote below:

I often have difficulties understanding subject matter in the reading texts. I have to search on Google for relevant information, especially articles, online lectures or video clips. I can find useful advice or thorough explanations in some forums or Facebook groups. (Student 6 – Focus group 4)

Some students complained that there was too much subject content for them to prepare at home. This meant that they were unable to follow parts of the lectures. Another problem occurred when the teachers skipped some sections in the textbooks, which meant that students had difficulty in keeping up with the lectures. Students used particular strategies for learning business concepts in English. The most common was to follow the process of "definition – explanation – example". Many students explained how they read the definitions of business concepts, identified key words and checked their meanings to gain a general understanding. Next, they would go over the explanations and relevant examples to fully comprehend the concepts. These students believed that making a list of essential terms improved their ability to retain information. In other words, they focused on mastering important terminology, as described in this following quote:

I usually pay close attention to important words that can clarify concepts when I'm learning them. I try to utilize those words as prompts to restate the meanings by noting them down in my notes. Only when studying for tests or exams do I need to memorize specific terms. (Student 2 – Focus group 3)

One student who was confident in using computers said that he learned the concepts through video clips. He read teachers' slides and used business terms to search for relevant clips on YouTube. He felt that those clips with visual information helped him understand the concepts better. Students had their own ways of memorising the concepts. Most of them used key words as clues to review a business definition. One student said that she paraphrased the definitions of business concepts using key terms and her own understanding. Four other students looked for links between the concepts and subject matter and reorganised diagrammatically, as explained by this student:

I often use my notes and a mind map to display the business concepts. Each concept includes some key words supported with relevant examples or explanations. After each chapter, I try to link the concepts together. This has helped me to memorise them more easily. (Student 2 – Focus group 1)

In short, the students in EMI classes reported having to prepare their lessons before class using different strategies in which they accessed multiple digital resources and deployed a wide range of digital technologies to assist their learning.

The Performance Phase

In class, students responded differently to learning activities. While all first year-students tried to pay attention to what their teachers said, the third year-students had their own ways of learning. Some of them found it hard to understand the teacher's and their friends' English, so they decided to work on their own, as stated in the quote below:

I am not familiar with the teacher's pronunciation, so I find it hard to understand him. When my friends presented their solutions to assigned case studies, I could not follow their English either. I focused on what I had prepared at home, solved the case by myself and compared with the teacher's slides to know the final answers. (Student 3 – Focus group 1)

Students also made their own choices in managing their language learning. As seen in the classroom, Vietnamese and English were also used by certain learners who were not confident in their English competence. To ensure that they fully understood everything, they translated ideas, essential terms, and reading passages into Vietnamese. Then, in order to pass the examinations, they concentrated on memorization of English vocabulary. Some students followed the lectures in class using textbooks in both Vietnamese and English. Others attempted to study in English only, thinking that doing so would help them:

I make an effort to utilize as much English as I can in class so that I will feel more comfortable using business words that I might come across in publications like books, journals, magazines, or the news. Most significantly, I can practice thinking in English, which I believe to be highly beneficial for my studies and future employment. (Student 2 – Focus group 1)

All the students mentioned their group activities in class. Students in the first focus group reported working on case studies in groups. Those in the second and third focus groups joined in group discussions to solve problems. Those in the last focus group stated that they explored video cases by collaborating with group mates. Most of these students valued working with others, as indicated below:

I enjoy group work activities in class. The teacher often provides us a case study or shows us a video clip followed by some questions to discuss. We work in groups to find out the answers. We can choose to report our answers with or without slides. If we have good answers, our group will get one bonus for our mid-term paper. (Student 5 – Focus group 4)

This was evidenced in the classroom observations as all students were seen engaging in their group activities.

Some students found that discussing or working in groups was useful for their learning as they could share their opinions easily with their friends. The students conducted group work in different ways. Some selected a group leader who was responsible for allocating tasks to each member. They discussed how to complete the assignment the most effectively. Students were in charge of their own section and then shared their work with the whole group. Many students who were self-conscious about their command of the English language thought that group interaction and assistance would help them study more effectively, as evidenced in this quote:

Since I struggle with English, I frequently ask my group leader to translate for me. In a group assignment, the leader first instructs us to read all texts and gather the necessary materials, then assigns responsibilities to each participant. Finally, we share our understanding as a group. This encourages my self-assurance and aids in my comprehension of the entire task. (Student 4 – Focus group 1)

In addition, the students were seen using their mobile phones and laptops in class. Specifically, when they involved in groupwork discussions, some students seemed to search for information using their phones and exchange the information with friends. When reporting their work in front of the whole class, the students used different presenting tools such as PowerPoint, Canva, or Prezi to assist their presentation as well.

To sum up, the students in EMI classes appreciated the value of collaborative work in which they can interact with and get support from their peers. The students also showed their implementation of technology in their in-class activities such as searching information with mobile phones or laptops and presenting with presentation software.

The Reflection Phase

Outside class, the students reported using digital technologies for different purposes. They all spent a great deal of time using digital devices to support their learning. All the students reported using laptops and mobile phones to search for information. They said that their devices were connected to the Internet most of the time. They used wireless Internet on the university campus, at the university library, at coffee shops and at home. They had data on their phones as well which allowed them to use Google to search whenever they needed extra information or reference materials, as in the example below:

The Internet at university sometimes lags seriously due to too many users. However, I can get Wi-Fi access easily at coffee shops around the university. The connection may be slower at home, but I can use 3G sims or data packages on my mobile phone. (Student 5 – Focus group 2)

The students were able to access multiple online resources, so they needed to process the online information to select what was relevant to their learning. They were aware of strategies to look for information from different academic websites, electronic journals, forums or social networks like YouTube or Facebook, as indicated in this quote:

Using key words in the course books or the teacher's slides is the most effective way when searching for something. I often start with reading Q&A sections in a forum or Facebook groups. I also use the references provided in the course book or further reading materials from the teachers. (Student 3 – Focus group 4)

Many students used digital technologies to exchange information and interact with others. All of them had class Facebook accounts through which they updated information about their subjects and courses. They spoke about different groups on Facebook where they sought help or support in their learning, as given by this student:

My class has a common group on Facebook to update information related to the course. I am a member of a student association group where I can chat with senior students or those from different classes. I also joined a marketing group and a business administration group whose members are not only students but also employers, researchers, lecturers or experts in the field. (Student 5 – Focus group 1)

Three students said that digital technologies helped them learn both English and content by themselves. For example, they enrolled in complementary online courses in different areas:

I am taking one course in digital marketing on Facebook and another course in Finance on Coursera, which is a website providing online courses in various areas. These courses help me learn subjects related to economics, which is very useful for my major. (Student 1 – Focus group 3)

Two students stated that they used applications installed on their laptops and mobile phones to manage their learning, as this student explained:

I always transfer my notes into digital formats that I can store in Dropbox and Google Drive. I use a mind map application to summarise each lesson and arrange these lessons into different categories. I can retrieve them easily from my laptop and mobile phone when I need to review for exams. (Student 4 – Focus group 1)

All students believed that online learning or e-learning was helpful for them. They could contact the teachers, receive updates about the subject and instant feedback on assignments, review the lectures with quizzes, and receive extra links or materials from the teachers. However, they all felt that the LMS used at the university was still limited, as stated in this quote:

The teacher requires us to use the LMS, mainly for receiving and submitting assignments, getting notices and reference materials. However, the system runs slowly. It does not have a notification function, so I have to check it every day and feel disappointed when having no updates. I think it is boring and time-consuming, too. (Student 6 – Focus group 2)

In short, the students revealed their competent use of digital technologies for different purposes after class to assist their learning and comprehension of content knowledge through English.

DISCUSSION

The study findings show that the students conducted different strategies to regulate their own learning in the new contexts of EMI. Those strategies were in alignment with the self-regulatory cycle including the forethought, the performance, and the self-reflection phase. Typically, to deal with challenges in EMI classes, the students spent a considerable amount of time preparing their lesson before class. The study reveals different ways the students used to prepare themselves for learning EMI classes such as previewing reading texts, checking lesson slides, and. looking up new vocabulary. These strategies indicate the students' engagement in the first phase of forethought in which they planned thoroughly for their learning. This finding echoes Zhou & Rose's (2021) study, in which the authors reported that EMI students used different strategies in learning before and after class. This also implies the insufficient preparation of the students who participated in the EMI programmes (Yeh, 2014).

The study also provides evidence supporting the students' practices associated with the second phase of performance in the self-regulatory cycle. Specifically, the students reported their high levels of staying focused in class and their participation in problem solving tasks and collaborative learning. The students expressed strong appreciation for the teachers' use of case study in which they could actively engage in problem solving tasks and group work discussion. This is in line with previous studies by Vo (2021), Jiang et al. (2019), and Choi (2013) who have reported the value of engaging students in learning collaboratively to solve practical problems in their majors. Moreover, the students stated their difficulties in maintaining high level of attention during long lectures in English, which required them to make effort in staying focussed and learning effectively. This fits with the findings of Hua (2020) and Soruç & Griffiths's (2018), which raises some implications for EMI teachers in designing teaching activities to enhance students' learning in EMI classes.

The study findings also indicate how the students reviewed their lesson, retrieved knowledge, and reflected their own learning in EMI classes. Given facing difficulties associated with both content knowledge and language proficiency, the students in those EMI classes had to regulate their learning after class by finding ways to organise information for revision and consolidation, interacting with others to get advice, and conducting extra tasks to enhance their comprehension. This explicitly shows how the students used the learning strategies in the last phase of self-refection (Boer et al., 2013; Broadbent & Poon, 2015).

Finally, the study findings reveal the students' competent use of digital technology in conducting all the SRL strategies to accommodate EMI challenges. The students participating in EMI courses had to contend with new content learning, limited English proficiency, and lack of guidance and support. Fortunately, they had access to multiple resources on the Internet, which enabled them to utilise their technical skills in their learning both in social and academic domains. These students played an innovative role in the learning process through actively regulating their own learning. These digital skills appeared to be strong enablers for students to proactively manage their learning and confidently deal with the challenge of EMI (Vo et al., 2022).

CONCLUSION

The implementation of EMI in Vietnamese contexts still reflects a number of tensions in which teachers and students have to face myriad challenges (Vo et al., 2022). The study findings raise several implications for both institutional policy makers and EMI teachers in providing sufficient learning support for EMI students so that they can be well-prepared to participate in EMI courses. Given more knowledge about the students' deployment of SRL strategies and digital technologies, EMI teachers should create opportunities for students to engage in self-regulatory cycle with the assistance of different digital resources and devices. This also calls for further research on specific impacts of technology assisted SRL strategies on students' achievement of both content knowledge and language competent in EMI courses.

REFERENCES

Bernacki, M. L., Aguilar, A. C., & Byrnes, J. P. (2011). Self-regulated learning and technology- enhanced learning environments: An opportunity-propensity analysis. In G. Dettori & D. Persico (Eds.), *Fostering self-regulated learning through ICT* (pp. 1–26). IGI Global Publishers. doi:10.4018/978-1-61692-901-5.ch001

Broadbent, J., & Poon, W. L. (2015). Self-regulated learning strategies & academic achievement in online higher education learning environments: A systematic review. *The Internet and Higher Education*, 27, 1–13. doi:10.1016/j.iheduc.2015.04.007

ChanLin, L.-J.ChanLin. (2012). Learning strategies in web-supported collaborative project. *Innovations in Education and Teaching International*, 49(3), 319–331. doi:10.1080/14703297.2012.703016

Chapple, J. (2015). Teaching in English is not necessarily the teaching of English. *International Education Studies, 8*(3), 1–13. doi:10.5539/ies.v8n3p1

Chen, H., & Boore, J. (2010). Translation and back-translation in qualitative nursing research: Methodological review. *Journal of Clinical Nursing, 19*(1-2), 234–239. doi:10.1111/j.1365-2702.2009.02896.x PMID:19886874

Choi, H., Kim, J., Bang, K. S., Park, Y. H., Lee, N. J., & Kim, C. (2015). Applying the flipped learning model to an English-medium nursing course. *Journal of Korean Academy of Nursing, 45*(6), 939–948. doi:10.4040/jkan.2015.45.6.939 PMID:26805506

Chuang, Y.-T. (2017). MEMIS: A mobile-supported English-medium instruction system. *Telematics and Informatics, 34*(2), 640–656. doi:10.1016/j.tele.2016.10.007

Dafouz, E., & Smit, U. (2020). *ROAD-MAPPING English medium education in the internationalised university.* Springer. doi:10.1007/978-3-030-23463-8

Dang, T. H., & Moskovsky, C. (2019). English-medium instruction in Vietnamese higher education: A ROAD-MAPPING perspective. *Issues in Educational Research, 29*(4), 1319–1336.

Delgado-Márquez, B. L., Escudero-Torres, M. A., & Hurtado-Torres, N. E. (2013). Being highly internationalised strengthens your reputation: An empirical investigation of top higher education institutions. *Higher Education, 66*(5), 1–15. doi:10.100710734-013-9626-8

Hu, G., & Lei, J. (2014). English-medium instruction in Chinese higher education: A case study. *Higher Education, 67*(5), 551–567. doi:10.100710734-013-9661-5

Hua, T. (2020). Understanding the learning challenges of English-medium instruction learners and ways to facilitate their learning: A case study of Taiwan psychology students' perspectives. *Latin American Journal of Content & Language Integrated Learning, 12*(2), 321–340. doi:10.5294/laclil.2019.12.2.6

Hung, B. P., & Anh, D. P. T. (2022). Computer-mediated communication and second language education. In R. Sharma & D. Sharma (Eds.), *New trends and applications in Internet of things (IoT) and big data analytics* (pp. 109–122). Springer. doi:10.1007/978-3-030-99329-0_8

Hung, B. P., & Nguyen, L. T. (2022). Scaffolding language learning in the online classroom. In R. Sharma & D. Sharma (Eds.), *New trends and applications in Internet of things (IoT) and big data analytics* (pp. 45–60). Springer. doi:10.1007/978-3-030-99329-0_4

Jiang, A. L., & Zhang, L. J. (2019). Chinese students' perceptions of English learning affordances and their agency in an English-medium instruction classroom context. *Language and Education*, *33*(4), 322–339. doi:10.1080/09500782.2019.1578789

Karjanto, N., & Simon, L. (2018). English-medium instruction calculus: Is flipping helpful? arXiv.org. https://arxiv.org/pdf/1611.08377.pdf

Khoa, B. T., Hung, B. P., & Hejsalembrahmi, M. (2022). Qualitative research in social sciences: Data collection, data analysis, and report writing. *International Journal of Public Sector Performance Management*, *9*(4), 10038439. doi:10.1504/IJPSPM.2022.10038439

Kirkpatrick, A. (2011). English as an Asian lingua franca and the multilingual model of ELT. *Language Teaching*, *44*(2), 212–224. doi:10.1017/S0261444810000145

Kitsantas, A. (2013). Fostering college students' self-regulated learning with learning technologies. *Hellenic Journal of Psychology*, *10*, 235–252.

Kitsantas, A., & Dabbagh, N. (2011). The role of web 2.0 technologies in self-regulated learning. *New Directions for Teaching and Learning*, *2011*(126), 99–106. doi:10.1002/tl.448

Krefting, L. (1991). Rigor in qualitative research: The assessment of trustworthiness. *The American Journal of Occupational Therapy*, *45*(3), 214–222. doi:10.5014/ajot.45.3.214 PMID:2031523

Lai, C., & Gu, M. (2011). Self-regulated out-of-class language learning with technology. *Computer Assisted Language Learning*, *24*(4), 317–335. doi:10.1080/09588221.2011.568417

Larsen, M. A. (2016). Globalisation and internationalisation of teacher education: A comparative case study of Canada and Greater China. *Teaching Education*, *27*(4), 396–409. doi:10.1080/10476210.2016.1163331

Margaryan, A., Littlejohn, A., & Vojt, G. (2011). Are digital natives a myth or reality? University students' use of digital technologies. *Computers & Education*, *56*(2), 429–440. doi:10.1016/j.compedu.2010.09.004

Merriam, S. B. (2014). *Qualitative research a guide to design and implementation*. Jossey Bass Ltd.

Paliwoda-Pękosz, G., & Stal, J. (2015). ICT in supporting content and language integrated learning: Experience from Poland. *Information Technology for Development: ICT in Transition Economies*, *21*(3), 403–425. doi:10.1080/02681102.2014.1003521

Pelikan, E. R., Lüftenegger, M., Holzer, J., Korlat, S., Spiel, C., & Schober, B. (2021). Learning during COVID-19: The role of self-regulated learning, motivation, and procrastination for perceived competence. *Zeitschrift für Erziehungswissenschaft*, *24*(2), 393–418. doi:10.100711618-021-01002-x PMID:33686344

Pintrich, P. R. (2000). The role of goal orientation in self-regulated learning. In M. Boekaerts, P. R. Pintrich, & M. Zeidner (Eds.), *Handbook of self- regulation* (pp. 451–502). Academic Press. doi:10.1016/B978-012109890-2/50043-3

Puustinen, M., & Pulkkinen, L. (2001). Models of self-regulated learning: A review. *Scandinavian Journal of Educational Research*, *45*(3), 269–286. doi:10.1080/00313830120074206

Querol-Julián, M., & Camiciottoli, B. C. (2019). The impact of online technologies and English medium instruction on university lectures in international learning contexts: A systematic review. *ESP Today*, *7*(1), 2–23. doi:10.18485/esptoday.2019.7.1.1

Rose, H., & McKinley, J. (2018). Japan's English-medium instruction initiatives and the globalization of higher education. *Higher Education*, *75*(1), 111–129. doi:10.100710734-017-0125-1

Soruç, A., & Griffiths, C. (2018). English as a medium of instruction: Students' strategies. *ELT Journal*, *72*(1), 38–48. doi:10.1093/elt/ccx017

Tarnopolsky, O. B., & Goodman, B. A. (2014). The ecology of language in classrooms at a university in eastern Ukraine. *Language and Education*, *28*(4), 383–396. doi: 10.1080/09500782.2014.890215

Tran, L. T., & Nguyen, H. T. (2018). Internationalisation of higher education in Vietnam through English-medium instruction (EMI): Practices, tensions and implications for local language policies. In I. Liyanage (Ed.), *Multilingual Education Yearbook 2018* (pp. 91–106). Springer. doi:10.1007/978-3-319-77655-2_6

Valentín, A., Mateos, P. M., González Tablas, M. M., Pérez, L., López, E., & García, I. (2013). Motivation and learning strategies in the use of ICTs among university students. *Computers & Education*, *61*, 52–58. doi:10.1016/j.compedu.2012.09.008

Vo, T. (2021). *The use of digital technologies in an English-medium instruction context: A case study of vietnamese higher education teachers and students* [Doctoral dissertation, Open Access Te Herenga Waka-Victoria University of Wellington].

Vo, T. D., Gleeson, M., & Starkey, L. (2022). The glocalisation of English-medium instruction examined through of the ROAD-MAPPING framework: A case study of teachers and students in a Vietnamese university. *System, 108*, 102856. doi:10.1016/j. system.2022.102856

Vu, N. T., & Burns, A. (2014). English as a medium of instruction: Challenges for Vietnamese tertiary lecturers. *The journal of Asia TEFL, 11*(3), 1-31.

Winters, F. I., Greene, J. A., & Costich, C. M. (2008). Self-regulation of learning within computer-based learning environments: A critical analysis. *Educational Psychology Review, 20*(4), 429–444. doi:10.100710648-008-9080-9

Yeh, C.-C. (2014). Taiwanese students' experiences and attitudes towards English-medium courses in tertiary education. *RELC Journal, 45*(3), 305–319. doi:10.1177/0033688214555358

Zhou, S., & Rose, H. (2021). Self-regulated listening of students at transition from high school to an English medium instruction (EMI) transnational university in China. *System, 103*, 102644. doi:10.1016/j.system.2021.102644

Zimmerman, B. J. (2000). Attaining self-regulation: A social cognitive perspective. In M. Boekaerts, P. R. Pintrich, & M. Zeidner (Eds.), *Handbook of self-regulation* (pp. 13–40). Academic Press. doi:10.1016/B978-012109890-2/50031-7

ADDITIONAL READINGS

Ahmed, S. T., & Roche, T. (2021). Making the connection: Examining the relationship between undergraduate students' digital literacy and academic success in an English medium instruction (EMI) university. *Education and Information Technologies, 26*(4), 4601–4620. doi:10.100710639-021-10443-0

Airey, J. (2020). The content lecturer and English-medium instruction (EMI): Epilogue to the special issue on EMI in higher education. *International Journal of Bilingual Education and Bilingualism, 23*(3), 340–346. doi:10.1080/13670050.2020.1732290

Blau, I., Shamir-Inbal, T., & Avdiel, O. (2020). How does the pedagogical design of a technology-enhanced collaborative academic course promote digital literacies, self-regulation, and perceived learning of students? *The internet and higher education, 45*, 100722. doi:10.1016/j.iheduc.2019.100722

Ekoç, A. (2020). English Medium Instruction (EMI) from the perspectives of students at a technical university in Turkey. *Journal of Further and Higher Education*, *44*(2), 231–243. doi:10.1080/0309877X.2018.1527025

Farrell, T. S. (2020). Professional development through reflective practice for English-medium instruction (EMI) teachers. *International Journal of Bilingual Education and Bilingualism*, *23*(3), 277–286. doi:10.1080/13670050.2019.1612840

Galloway, N., & Ruegg, R. (2022). English Medium Instruction (EMI) lecturer support needs in Japan and China. *System*, *105*, 102728. doi:10.1016/j.system.2022.102728

Hooshyar, D., Pedaste, M., Saks, K., Leijen, Ä., Bardone, E., & Wang, M. (2020). Open learner models in supporting self-regulated learning in higher education: A systematic literature review. *Computers & Education*, *154*, 103878. doi:10.1016/j.compedu.2020.103878

Luu, T. Q. H., Sit, H. H. W., & Chen, S. (2023). Development of English Medium Instruction (EMI). *Cultural Interactions of English-Medium Instruction at Vietnamese Universities*, 11-38.

Rivers, D. J., Nakamura, M., & Vallance, M. (2022). Online self-regulated learning and achievement in the era of change. *Journal of Educational Computing Research*, *60*(1), 104–131. doi:10.1177/07356331211025108

Zhu, Y., Zhang, J. H., Au, W., & Yates, G. (2020). University students' online learning attitudes and continuous intention to undertake online courses: A self-regulated learning perspective. *Educational Technology Research and Development*, *68*(3), 1485–1519. doi:10.100711423-020-09753-w

KEY TERMS AND DEFINITIONS

English medium instruction (EMI): EIM is the educational system in which English is used as the main medium of instruction to teach and learn subject matter.

Self-Regulated Learning (SRL): the process in which learners deploy a variety of strategies to regulate their own learning.

Technology-Assisted: referring to the state in which a wide range of digital technologies are being integrated in teaching and learning practices.

Chapter 4

Boredom in Online Language Classrooms:
Vietnamese EFL Students' Perspectives

Sieu Khai Luong
University of Economics Ho Chi Minh City, Vietnam

Chau Thi Hoang Hoa
Tra Vinh University, Vietnam

ABSTRACT

This study investigates the causes of and suggested solutions to students' boredom in online EFL learning in Vietnam during Covid - 19. The study follows descriptive qualitative research design with the use of semi-structured interview as the sole instrument. Due to the social distance, online interviews with 38 student participants were conducted via Google Meet. Findings show that among the three factors (teacher-related, IT-related, and task-related), the teacher-related factors were the leading causes of students' boredom. Likewise, most of the students' suggestions to mitigate boredom are teacher-related: teachers' IT competency, interactive classrooms, real-life task types and authentic materials, games, bonus points, teachers' and students' relationships. On that basis, the study recommends teachers' training to improve online instruction, not only while but also post-Covid – 19 area because remote teaching using technology is an unavoidably rising trend in modern society.

INTRODUCTION

Recent research has shown the importance of emotions in second language (L2)

DOI: 10.4018/978-1-6684-7034-3.ch004

learning (Nakamura, 2018). Boredom, a negative feeling frequently experienced by students at school (Vogel-Walcutt et al., 2012), is a psychological phenomenon that students experience when facing unwanted moments at school and may result in detrimentally unexpected activities and learning outcomes (Pekrun et al., 2010; Tze et al., 2016). Likewise, in recent years, the high time of Covid – 19, language education has witnessed an emerging trend in studying students' mental health and psychological condition during online learning as a temporary solution due to social distancing (e.g., Kruk & Zawodniak, 2020; Pawlak et al., 2020). The unprecedented switch to online education or emergency remote teaching has challenged the efficacy of language education due to the lack of interaction and adverse effects on students' emotions (Nayman & Bavlı, 2022). In the Covid-19 and the emerging rise of online teaching thanks to technological development, it is necessary to raise the issue in the EFL local teaching context of Vietnam to explore the causes of and possible solutions to the boredom among Vietnamese EFL students. Specifically, the research issues are addressed in the two following research questions (RQ).

RQ1. What factors influence the trajectories of self-reported boredom in the online class experienced by Vietnamese EFL students?

RQ2. What solutions do Vietnamese EFL students suggest reducing experienced boredom in online classes?

BOREDOM AND ITS NEGATIVE EFFECTS

Boredom is commonly regarded as an unpleasant emotional state in which the individual experiences a general lack of interest and difficulty concentrating on the current activity. Nakamura et al. (2010) argue that "boredom is a multidimensional construct" (p. 2). It involves emotion, cognition, motivation, expression, and physiology. Boredom is also characterized in terms of symptoms, intensity, and manifestation during the learning process. Falman (2009) asserts that boredom manifests through various characteristics such as lethargy, displeasure, distraction, and a distorted perspective of time and space. In education, boredom is a state of weariness or ennui resulting from a lack of engagement with stimuli in the classroom (Kruk & Zawodniak, 2020; Westgate & Wilson, 2018). This definition is adopted in this study as the main framework to guide the discussion on causes of and solutions to boredom.

Boredom is generally considered one of the most undesirable psychological statuses and is often identified by individuals as the cause of feeling depressed (Tze et al., 2016). It breaks through instructive settings, which is problematic for personal academic work, contrarily influencing individual behavior, responsibility, perception, interest, inspiration, learning techniques, execution, and results (Daniels et al., 2015). As noted by Goetz et al., (2014), boredom makes students feel uncomfortable, resulting from lack of excitement and learning motivation. Thus, it leads to decreasing learning speed or dropout.

CAUSES FOR AND SUGGESTED SOLUTIONS TO THE STUDENTS' BOREDOM

The most common classification of boredom predictors was based on the physical boundary and psychological views of subjectivity – the learners: internal and external factors (see Chapman, 2013; Nakamura et al., 2021; Li & Han, 2022). However, this classification tends to serve the purpose of representation and simplification. In fact, in many L2 educational studies, causes of boredom are diverged and divided into internal and external factors. This categorization seems logical and clear-cut but they are too broad. That is why this study does not apply this classification of boredom factors but would instead explore and apply another framework based on related studies and educational context related to online learning of EFL teaching to young adult learners.

In the literature, boredom emerged from varied and interrelated sources (Tran & Bui, 2022). The three most common antecedents of boredom were related to teachers, lessons and the students themselves. For example, boredom can be liable to the teachers and their instructions (Chapman, 2013; Lewinski, 2005; Nakamura et al., 2020), the conduction and nature of lessons or tasks (Westgate & Wilson, 2018) and the students themselves (Derakhshan et al., 2021; Derakhshan, 2022). Levels and proneness of the three factors contributing to student boredom (teachers, lessons/ or tasks and the students) vary in the related literature. Teachers and tasks are interconnected, and the former is more a dominating and decisive factor than the latter (Lewinski, 2005; Nakamura et al., 2020; Tran & Bui, 2022). Interestingly, Zawodniak et al. (2021) pinpointed that teachers and students agreed that the primary cause of classroom boredom is related to lesson content that cannot adapt to current events.

Beside teachers' personalities and physical traits, their teaching methods decide the degree to which teachers can conduct the tasks that draw students' interest and engagement. Teachers' monotonous voices, authoritarian personalities, lack of creativity and sense of humor made students lose interest in their studies (Derakhshan et al., 2021; Tran & Bui, 2022). Teachers should be the ones to recognize and drive students out of boredom (Perkrun, 2010). Likewise, Lewinski (2015) argued that teachers could manage the classroom and their pedagogy and classroom environment to reduce boredom. Teachers can monitor learners and learning in the classroom to identify sources of boredom. According to Derakhshan et al. (2021), the tasks deployed lacked the enthusiastic participation of students because teachers did not ignite students' need to learn. Also, regarding the meaning and quality of the tasks, Perkrun's (2010) pointed out that monotonous and repetitive tasks could not evoke students' attention and interest to learn. Additionally, Pawlak et al. (2021) and Tran and Bui (2022) confirmed the repetitive nature of the tasks, the mismatch of the task requirement and respondents' L2 competency, and class arrangement led to the students' disengagement and resulted in boredom.

The other factors leading to students' ennui found in the literature related to logistical factors (Derakhshan et al., 2021). Logistical factor as a primary source of boredom has not been viewed as often as other factors (e.g. technical conditions and physical qualities), or other variables (e.g., class time, weather, excessive work). In fact, it was often the sub-findings of unlisted factors, known as "other factors" (see Kruk & Zawodniak, 2020; Tran & Bui, 2022). Of the logistical factors causing boredom among students in Vietnam, IT-related issues were the most emerging factor inducing this negative emotion. Considering the previous literature and the social and local context of Vietnam, this study follows the framework of teacher-related, task-related and IT-related for boredom causes and solutions.

METHODOLOGY

Setting and Participants

This study was conducted during the severe lockdown of Covid-19 in Vietnam in 2021. Students spent more than one year engaging in online education, a long temporary switch from offline teaching and learning mood. This situation caused many problems for students around Vietnam, millions of whom were unable to adapt to the situation and lost interest in online learning (Duc-Long et al., 2021).

The participants were 38 students, aged 10 to 13, selected from secondary schools in the locality. They studied English three periods a week, prescribed in the national curriculum. Though the students joined the interview at their own will, the researchers got their parents' permission prior to the virtual meeting as a part of research ethics. Also, the confidentiality of answers and the use of participants' pseudonyms were informed to parents and students.

Instrument

The instrument gathering data in this study was semi-structured interviews. The interview questions were divided into two parts. The first part, the opening, started with questions about ice-breaking and background information for demographic data. The second part consists of issues that focus on the two questions: causes of and suggested solutions to the students' boredom in online learning during Covid – 19. The interview scheme was adapted based on Kruk and Zawodniak (2020), Nakamura et al. (2021), Pawlak et al. (2021), and Tran and Bui (2022) with the three categories: task-related factors, teacher-related factors, IT-related factors, and other open for the different factors that the students might suggest.

The interviews were conducted in Vietnamese to make the participants feel confident and free to express themselves as much as they wished. Since the interview was conducted online, via Google Meet, which could be overwhelming to the students those days, the students and their parents decided the most convenient time. Each interview lasted from 20 to 25 minutes.

Data Analysis

The participants' responses were analyzed in this study based on two research questions that previously aimed to identify causes of and solutions to boredom in online English classes. The head of the research team reviewed all of the data analysis results to ensure the credibility of my research. The data were read through several times during the first step of the data analysis process to ensure consistency and to remove unnecessary or excessive data that could affect the research result. The researcher collected, excluded irrelevant data, and transcribed the collected data. The thematic content approach was taken to classify the collected data. Themes were based on the main framework of the semi-structured interviews focusing on the factors inducing boredom: task-related factors, teacher-related factors, IT-related factors, and other possible factors. Classifying, grouping, and calculating the coded data were double-checked by a research mate in the team before they were subjected to reports on the findings.

FINDINGS

The main findings in this section were derived from the interviewees' responses to questions about the causes and solutions to boredom in online English classes. The topics mentioned in the interviews were listed in order of frequency.

What Factors Influence the Trajectories of Self-Reported Boredom in the Online Class Experienced by Vietnamese EFL Students?

Based on thematic data analysis, three chosen aspects, which are teacher-related factors, IT - related factors, and task-related factors, are reported to address the first research question.

Teacher-Related Factors

Of the three clusters, teacher-related factors made the main sources of students' boredom. In fact, 26 students stated that they experienced this negative emotion due to teachers' instructions in terms of (1) inability to create meaningful interaction in the classroom, (2) lack of creativity to diversify class activities and (3) lack of necessary preparedness.

Thirteen of the participants agreed that the leading cause of their boredom was the teachers' failure to engage students' participation. They compared their language lessons to other theoretical base lessons, occupied with a high percentage of teachers' talking time. As they reported, there were no types of meaningful communication like dialogues and class discussions, but teachers' monologues and lecturing dominated even the sole activity in class. Noticeably, seven students reported that they rarely could talk to their teachers because teachers kept talking and disabled them from communicating with each other by muting them and locking their chat boxes to prevent distraction. During online studying, the students sat motionless, listened attentively, and looked at the screens continuously. Hence, it is understandable that the lack of chance to engage in the lessons and keep listening and watching for long hours made students tired and bored.

Monotonous ways of teaching were placed as second sources of teachers' instruction, as found in the report of nine students. They further explained that their teachers kept the same way of conducting class activities provided in the course books and workbooks from one lesson to another. Five students stated that their teachers only dealt with the coursebooks' content. In further discussion with the students, it was proven that teachers lose the balance of "teach and play"; meaning that they kept teaching and ignored "playing with" their students. They did not mean to compare their teachers with the former teachers who taught them offline but agreed that online classes were not filled with excitement as much as offline classes because of the lack of games.

The third aspect of teachers' instruction relating to students' boredom was that teachers were "not well-prepared enough" for online teaching (reported by seven students). Students acknowledged that their teachers sometimes enriched their teaching materials by searching for additional reading texts or just stories from the internet. However, they thought that teachers did not adapt and refine the collected materials and did not tailor the materials to be fitted with online lessons. As examples to illustrate, students experience moments of embarrassment when the materials included many new words far from their understanding ability. Teachers did not help them to clarify the added content but led students to drift away from the focus of the lesson. Furthermore, three students confessed that they were bored and failed to follow the lesson threads because their teachers' communication method was somewhat "cumbersome". Teachers' words were difficult to understand and they could not fully comprehend what they said. Hence they were unable to perform given tasks well, especially in grammar lessons. In sum, students' boredom resulted from not being able to follow teachers' instructions due to teachers' unpreparedness to give proper and appropriate elaboration and instruction in their online teaching.

IT- Related Factors

IT factors are the second source of students' boredom, which occurred in the answers of 15 students. In terms of technology, students reported two issues: teachers' IT capacity and internet bandwidth. According to eight students, their teachers were unable to properly and effectively use teaching technology for online teaching. Though applied platforms, Zoom, Microsoft Teams, and Google Meet platforms provided many instructional functions, their teachers used these applications for lecturing and transmitting their voice most often. Additionally, ten students mentioned when it came to boredom in an online English class related to the poor transmission signal which interfered and interrupted their learning. In fact, they could not hear the whole point of the teachers' lectures due to the lagging internet connection. Khanh (grade 6) further explained that almost every time he was given a chance to speak, he had problems with the connection. Sometimes, he was often dropped and thrown out of the online classroom and the teacher thought that he skipped the class. In conclusion, to the students, teachers' lack of skills to apply technology for online teaching and poor internet connection induced students' disengagement and boredom.

Task-Related Factors

Finally, the last factor the participants often mentioned was relevant to the tasks given (answers from 7 students). These issues raised were task type (4 students) and lack of flexible solutions to the tasks.

Students mentioned task types and agreed that lack of diversity discouraged students' engagement and interests. My (grade 6), one of the interviewees stated that "the teacher kept giving the same type of exercises. We had to apply the grammatical formula to get the keys''. She wondered why the teacher gave her a lot of written practice on grammar like filling in the blanks with the correct verb forms. Types of inapplicable exercises repeatedly occurring made her and her classmates feel too lazy to think and uninterested in doing given exercises.

Another factor leading to students' ennui was the requirement of task performance. Teachers tended to give strenuous exercises to challenge students and were not open to the students' responses or solutions (2 students). This type of challenge had a negative impact on students' emotions. In fact, they believed that the activities were not "necessary to be so complicated" and their teachers remained so strict with the answers. The students were discouraged and bored because they did not properly contact their teachers for interactive and responsive feedback. Ai (grade 6) stated that she did not understand why her answer was wrong because she was asked to give her answer briefly and orally (not visually) while doing exercises.

To conclude, students agreed that their ennui came from teachers' instruction, teachers' IT competency, poor internet connection, and English online tasks. Though the framework aims to classify the source of boredom from three aspects (teachers, tasks, and IT), from the student's reports, it was the teachers who were responsible for their boredom: inability to conduct meaningful communication and engage students' participants, incompetency to exploit technology in teaching to involve students and attract their attention, unpreparedness to tailor the extended parts of lessons to online class to fit students' language proficiency and preference and lack of devotion in designing, conducting and correcting the tasks given to students. The lagging internet connection was the only factor that brought boredom to students not coming from the teachers.

What Solutions do Vietnamese EFL Students Suggest Reducing Experienced Boredom in the Online Classes?

Findings show that teachers and their instruction were the leading causes of students' boredom among the three factors teacher-related, IT-related, and task-related. Likewise, most of the students' suggestions to mitigate boredom are teacher-related: teachers' IT competency, interactive classrooms, real-life task types and authentic materials, games, bonus points, teacher's and students' relationships.

Teachers' IT Competency

Though ranked third as the source of students' boredom, teachers' IT ability to handle online classes was put a great deal of attention (20 students). They explained that if the teachers had been technology-savvy, they could have made good use of a variety of educational functions of platforms and websites. For example, they could have put students into small group discussions to multiply the classroom interaction. In addition, they opined that if the teachers could use many of the features available on online applications, namely the Messenger and Zoom network platforms, they could attract students better and increase their ability to absorb knowledge and acquire better skills.

Interactive Classrooms

Conducting online lessons with a variety of patterns of meaningful communication was what the students expected to mitigate the boredom (14 students). The interviewees expressed that teachers should make the classroom more interactive and engaging. Simply, one participant suggested that the teacher should not unmute students but allow them to talk about learned topics, just for a change of voice for their ears or to wake them up.

Real-Life Task Types and Authentic Materials

As reported, the types of preferred lessons were skill lessons about common topics in their daily life, like giving directions, school life, family and friends. Nine students suggested that teachers could draw students' interests to the lessons by providing authentic materials with applicable content knowledge and real-life based tasks. According to Nam (grade 7), "It would be better and more interesting if teachers taught us something practical, applicable in everyday life. When I learn something new in English I am proud to tell my parents about it".

Games and the Like

Student participants agreed that clips and games could make the class more lively (14 students). The students pinpointed that fun music or videos could help them refresh their mind (8 students). Nine students suggested that games could help them to combat classroom boredom. If teachers integrated fun class games into their online lessons, the classrooms would be stimulating. Dang (grade 7), voiced his opinion "Class games helped me reduce stress and boredom and gave me time to relax my mind so that I could absorb the lesson more effectively".

Bonus Points

Bonus points were awarded to the following solution, which was provided by a few interview participants (2 students). Teachers could give the students additional points to boost engagement or to recognize good performance. This also encouraged students to engage in learning activities so that they would not be bored during online lessons.

Teachers and Students' Relationship

Fifteen students agreed that they wished for a friendly and open atmosphere in online classes where the teachers build rapport with their students attentively. They pointed out that teachers should put themselves in their students' situations to understand and empathize with them, which could improve student motivation and foster a good relationship between students and teachers (8 students). According to Lan (grade 6), "Teachers should understand students and be willing to make jokes in class or chat with them, so students did not experience boredom or even fear in the classroom." Also, seven students expected teachers should call for and appreciate the students' opinions. Being friendly, understanding, and sympathetic teachers could mitigate students' boredom and stress in a stressful pandemic situation.

The majority of students thought that the teachers should be responsible for their boredom, and their suggested solutions were relevant to their reported sources of tediousness. Noticeably, a good number of students expressed that they would be more interested in their lessons if the teachers were genuinely interested in establishing closer relationships with them.

DISCUSSION AND RECOMMENDATIONS

As stated in the results, the majority of the students believed that the boredom in the classroom was caused by the teacher, specifically the teacher's teaching method. The predictors to student boredom were teachers' lengthy and monotonous talk, lack of meaningful interaction and practice, lack of dedication to prepare appropriate tasks, incompetence to exploit technology to involve students in the meaningful tasks and interesting activities. Their suggested solutions to lighten this negative emotion were also teacher - related, which involved the improvement of teachers' IT competency, interactive classrooms, real-life task types and authentic materials, games, bonus points, teachers and students' relationship. Generally, these findings echoed previous research (e.g. Derakhshan et al., 2021; Tran & Bui, 2022; Kruk & Zawodniak, 2020; Nakamura et al., 2021; Pawlak et al., 2020; Zawodniak et al., 2021).

However, the three suggestions (improvement of teachers' IT competency, more games and bonus points) for lessening boredom could be considered new findings as suggestions to reduce students' boredom. Though the mentioned suggestions related to different areas of teachers' competence, assessment, and instruction, they shared one thing in common: they were exclusive to teaching language online to young adult learners.

Games and extra points were a kind of extrinsic motivation that teachers could consider as a part of their teaching and scoring system to instigate young adults' engagement, attention and interests. In most of the related research in the field of online teaching during Covid - 19, meaningful practice, social interaction, and class engagement are amongst the most common issues, but the solution to this issue relating to teachers' IT competence is less commonly found. The reason for this gap relies on the participant difference. Unlike the participants in most, if not all, previous studies, pre-adolescent learners tend to prefer a variety and diversity of work arrangements to expand the types and ways of interaction and enhance students' interests and attention.

Regarding teachers' and students' relationship as a therapy for students' boredom, this finding echoed the finding of Tran and Bui (2022). The relationship between teachers and students came from teachers' creativity and friendliness. The teacher-student connection in this study derived from allowance, acceptance, attention and appreciation of what students said, and the conduction of class games in which the students could "play with" their teachers. Adolescents are more likely to be more vulnerable and sensitive than adults, so they have a bigger demand for proper and personal care, attention, and understanding.

Also relating to teachers' personality to build better relationships, this study confirmed the findings of the previous research (Nakamura et al., 2021; Tran & Hung, 2022; Pekrun et al., 2010). Humor in lessons or conversations between teachers and students outside of homework can create sympathy and stimulate students' motivation. Hence, providing students with some recess and subtly incorporating humor into classroom activities could undeniably reduce the monotonous rhythm of classroom activities and provide students with time to breathe or for the brain to breathe in order to self-regulate attention and process newly acquired knowledge. Beside humor, the young adults expected care, sympathy, and understanding from their teachers.

Lack of class interaction and teacher-student communication as an inducing factor to boredom is commonly found in many studies (Nakamura et al., 2020; Nakamura et al., 2021; Tran & Hung, 2022; Pekrun et al., 2010). However, the causes found in this study were different. As reported, the shortage of classroom interaction in this study was due to the domination of teachers' talking time, over explanation of lengthy grammatical knowledge. Additionally, teachers' limited use of technology in language teaching could hinder simultaneous classroom interaction in pairs or groups in online learning.

In parallel to the findings and discussions, most of the recommendations are for teachers and related parts. As the causes for and suggestions to boredom reported from students, teachers' openness is recommended as the key solution. In fact, teachers should be open to a more flexible scoring system to encourage students' participation and performance and better practice more variety innovation of online teaching. Also, teachers should open their hearts to students: listening to them, recognizing students' aspirations and expectations, and adjusting their instruction suitably and effectively. There should be direct and warm communication between teachers and students to build mutual understanding and sympathy and teachers acknowledge their problems with teaching. Finally, the educational managers should provide training courses to facilitate teachers with instructional strategies focusing on enhancing students' interaction and engagement.

LIMITATIONS AND CONCLUSION

Aside from the key findings that this research article has contributed to, there are some limitations as follows. First, reliability and generality is limited due to the small size of participants and lack of triangulation of data collected. To the researcher' best effort, only 38 students were voluntarily to join this study as interviewees, and the interview is the sole research instrument. Second, that the interview was conducted online due to social distance may hinder communication between the researcher and the participants. Overwhelming online communication during Covid – 19 and limited social interaction could negatively affect the quality and size of data collected. Third, the framework for sources of students' boredom was limited to only the three factors; no other factors from students were elicited to enrich the reported data. Finally, the suggested strategies to improve classroom communication were not well-searched. Hence, this limitation could be further researched because building social interaction among the students in distant communication is a big issue relating to teacher's pedagogy.

In conclusion, online education has become more and more popular and the interaction in this teaching and learning mood is quite different from that of offline teaching. The lack of social interaction may lead to restriction of teaching and studying efficacy as well as the teachers' and students' well-being-ness. For that reason, the study of students' mental health and emotional state should not be encouraged. Online teaching as an emergency remote teaching has different features to prepared online education but the boredom of students may have rather similar causes and solutions. Hence the result could be a good reference to educators, teachers, and researchers and stakeholders who are interested in online education.

REFERENCES

Daniels, L. M., Tze, V. M. C., & Goetz, T. (2015). Examining boredom: Different causes for different coping profiles. *Learning and Individual Differences, 37*, 255–261. doi:10.1016/j.lindif.2014.11.004

Derakhshan, A., Fathi, J., Pawlak, M., & Kruk, M. (2022). Classroom social climate, growth language mindset, and student engagement: The mediating role of boredom in learning English as a foreign language. *Journal of Multilingual and Multicultural Development*, 1–19. doi:10.1080/01434632.2022.2099407

Derakhshan, A., Kruk, M., Mehdizadeh, M., & Pawlak, M. (2021). Boredom in online classes in the Iranian EFL context: Sources and solutions. *System, 101*(February), 102556. doi:10.1016/j.system.2021.102556

Duc-Long, L., Thien-Vu, G., & Dieu-Khuon, H. (2021). The impact of the COVID-19 pandemic on online learning in higher education: A Vietnamese case. *European Journal of Educational Research, 10*(4), 1683–1695. doi:10.12973/eu-jer.10.4.1683

Fahlman, S. A., Mercer, K. B., Gaskovski, P., Eastwood, A. E., & Eastwood, J. D. (2009). Does a lack of life meaning cause boredom? Results from psychometric, longitudinal, and experimental analyses. *Journal of Social and Clinical Psychology, 28*(3), 307–340. doi:10.1521/jscp.2009.28.3.307

Farmer, R., & Sundberg, N. D. (1986). Boredom Proneness-The Development and Correlates of a New Scale. *Journal of Personality Assessment, 50*(1), 4–17. doi:10.120715327752jpa5001_2 PMID:3723312

Kruk, M. (2016). Variations in motivation, anxiety and boredom in learning English in Second Life. *The EuroCALL Review, 24*(1), 25. doi:10.4995/eurocall.2016.5693

Kruk, M. (2022). Dynamicity of perceived willingness to communicate, motivation, boredom and anxiety in Second Life: The case of two advanced learners of English. *Computer Assisted Language Learning, 35*(1–2), 190–216. doi:10.1080/0958822 1.2019.1677722

Kruk, M., & Zawodniak, J. (2018). Boredom in practical English language classes: Insights from interview data. *Interdisciplinary Views on the English Language, Literature and Culture, January*.

Lewinski, P. (2015). Effects of classrooms' architecture on academic performance in view of telic versus paratelic motivation: A review. *Frontiers in Psychology, 6*, 746. doi:10.3389/fpsyg.2015.00746 PMID:26089812

Li, C., & Han, Y. (2022). *Learner-internal and learner-external factors for boredom amongst Chinese university EFL students.* Applied Linguistics Review. doi:10.1515/applirev-2021-0159

MacIntyre, P. D., Gregersen, T., & Mercer, S. (2020). Language teachers' coping strategies during the Covid-19 conversion to online teaching: Correlations with stress, wellbeing and negative emotions. *System, 94,* 102352. doi:10.1016/j.system.2020.102352

Nakamura, P. M., Pereira, G., Papini, C. B., Nakamura, F. Y., & Kokubun, E. (2010). Effects of preferred and nonpreferred music on continuous cycling exercise performance. *Perceptual and Motor Skills, 110*(1), 257–264. doi:10.2466/pms.110.1.257-264 PMID:20391890

Nakamura, S., Darasawang, P., and Reinders, H. (2021). The antecedents of boredom in L2 classroom learning. *System, 98,* 102469. doi:10.1016/j.system.2021.102469

Nayman, H., & Bavlı, B. (2022). Online Teaching of Productive Language Skills (PLS) during Emergency Remote Teaching (ERT) in EFL Classrooms: A Phenomenological Inquiry. *International Journal of Education and Literacy Studies, 10*(1), 179. doi:10.7575/aiac.ijels.v.10n.1p.179

Pawlak, M., Zawodniak, J., & Kruk, M. (2020). Boredom in the foreign language classroom: A micro-perspective. In Second Language Learning and Teaching. doi:10.1007/978-3-030-50769-5

Pawlak, M., Zawodniak, J., & Kruk, M. (2021). Individual trajectories of boredom in learning English as a foreign language at the university level: Insights from three students' self-reported experience. *Innovation in Language Learning and Teaching, 15*(3), 263–278. doi:10.1080/17501229.2020.1767108

Pekrun, R., Goetz, T., Daniels, L. M., Stupnisky, R. H., & Perry, R. P. (2010). Boredom in Achievement Settings: Exploring Control-Value Antecedents and Performance Outcomes of a Neglected Emotion. *Journal of Educational Psychology, 102*(3), 531–549. doi:10.1037/a0019243

Pekrun, R., Hall, N. C., Goetz, T., & Perry, R. P. (2014). Boredom and academic achievement: Testing a model of reciprocal causation. *Journal of Educational Psychology, 106*(3), 696–710. doi:10.1037/a0036006

Post, D., Carr, C., & Weigand, J. (1998). Teenagers: Mental health and psychological issues. *Primary care, 25*(1), 181–192. doi:10.1016/S0095-4543(05)70331-6 PMID:9469922

Resnik, P., & Dewaele, J. M. (2021). *Learner emotions, autonomy and trait emotional intelligence in "in-person" versus emergency remote English foreign language teaching in Europe.* Applied Linguistics Review., doi:10.1515/applirev-2020-0096

Tran, N. H., & Bui, H. P.(2022) *Causes of and Coping Strategies for Boredom in Language Classrooms: A Case in Vietnam,* 0-196.

Tze, V., Daniels, L. M., & Klassen, R. M. (2016). Evaluating the relationship between boredom and academic outcomes: A meta-analysis. *Educational Psychology Review, 28*(1), 119–144. doi:10.100710648-015-9301-y PMID:28458499

Vogel-Walcutt, J. J., Fiorella, L., Carper, T., & Schatz, S. (2012). The definition, assessment, and mitigation of state boredom within educational settings: A comprehensive review. *Educational Psychology Review, 24*(1), 89–111. doi:10.100710648-011-9182-7

Westgate, E. C., & Wilson, T. D. (2018). Boring thoughts and bored minds: The MAC model of boredom and cognitive engagement. *Psychological Review, 125*(5), 689–713. doi:10.1037/rev0000097 PMID:29963873

Zawodniak, J., Kruk, M., & Pawlak, M. (2021). Boredom as an Aversive Emotion Experienced by English Majors. *RELC Journal.* doi:10.1177/0033688220973732

ADDITIONAL READINGS

de Zordo, L., Hagenauer, G., & Hascher, T. (2019). Student teachers' emotions in anticipation of their first team practicum. *Studies in Higher Education, 44*(10), 1758–1767. doi:10.1080/03075079.2019.1665321

Nayman, H., & Bavlı, B. (2022). Online Teaching of Productive Language Skills (PLS) during Emergency Remote Teaching (ERT) in EFL Classrooms: A Phenomenological Inquiry. *International Journal of Education and Literacy Studies.*

Nett, U. E., Goetz, T., & Daniels, L. M. (2010). What to do when feeling bored? Students' strategies for coping with boredom. *Learning and Individual Differences, 20*(6), 626–638. https://doi.org/10.1016/j.lindif.2010.09.004

Nett, U. E., Goetz, T., & Hall, N. C. (2011). Coping with boredom in school: An experience sampling perspective. *Contemporary Educational Psychology, 36*(1), 49–59. https://doi.org/10.1016/j.cedpsych.2010.10.003

Pawlak, M., Zawodniak, J., & Kruk, M. (2020). The neglected emotion of boredom in teaching English to advanced learners. *International Journal of Applied Linguistics (United Kingdom)*, *30*(3). https://doi.org/10.1111/ijal.12302

Pekrun, R. (1992). The Impact of Emotions on Learning and Achievement: Towards a Theory of Cognitive/Motivational Mediators. *Applied Psychology*, *41*(4). https://doi.org/10.1111/j.1464-0597.1992.tb00712.x

Pekrun, R., Goetz, T., Titz, W., & Perry, R. P. (2002). Academic emotions in students' self-regulated learning and achievement: A program of qualitative and quantitative research. *Educational Psychologist*, *37*(2), 91–105. https://doi.org/10.1207/S15326985EP3702_4

Robinson, W. P. (1975). Boredom at school. *The British Journal of Educational Psychology*, *45*(2), 141–152. https://doi.org/10.1111/j.2044-8279.1975.tb03239.x

Tze, V. M. C., Daniels, L. M., & Klassen, R. M. (2016). Evaluating the relationship between boredom and academic outcomes: A meta-analysis. *Educational Psychology Review*, *28*(1), 119–144. https://doi.org/10.1007/s10648-015-9301-y

Villavicencio, F. T., & Bernardo, A. B. I. (2012). Positive academic emotions moderate the relationship between self-regulation and academic achievement. *The British Journal of Educational Psychology*, *83*(2), 329–340. https://doi.org/10.1111/j.2044-8279.2012.02064.x

Zembylas, M. (2008). Adult learners' emotions in online learning. *Distance Education*, *29*(1), 71–87. https://doi.org/10.1080/01587910802004852

KEY TERMS AND DEFINITIONS

Bonus points: These are extra scores given to students to recognize their positive performance or contribution during their studying, not in testing.

Boredom: This is a state of weariness or ennui resulting from a lack of engagement with stimuli in the classroom.

Games: This denote types of language class activities explicitly and purposefully designed to balance study and play.

Chapter 5
The Impact of 4English Mobile App of EFL Students' Reading Performance in a Secondary Education Context

Nguyen Ngoc Vu
Ho Chi Minh City University of Foreign Languages and Information Technology, Vietnam

Dang Thanh Tam
Chu Van An High School, Vietnam

Le Nguyen Nhu Anh
Ho Chi Minh City University of Education, Vietnam

Nguyen Thi Hong Lien
Hoa Sen University, Vietnam

ABSTRACT

The main goal of this study was to examine how the 4English mobile app affected 10th-grade students' reading abilities and how they felt about using it to learn to read. At a high school in the province of An Giang, Vietnam, 90 10th graders are chosen and split equally into two groups. The curriculum, materials, school resources, and classroom instruction are the same for both groups. The experimental group is instructed to use the 4English mobile app to improve their reading skills, while the control group is given traditional reading assignments from the teacher. Four research tools employed in the study were a questionnaire, an interview, a pretest, and a posttest. The results demonstrate that the experimental participants performed better in the posttest and had favorable sentiments toward using the 4English mobile app to teach reading.

DOI: 10.4018/978-1-6684-7034-3.ch005

INTRODUCTION

All English learners must be proficient readers if they wish to learn the language since reading is seen as a way for learners to acquire language inputs like vocabulary, grammar, pronunciation, and other language components to support the other skills (Aryadoust, 2019). Reading is therefore given particular attention among English language learners. However, actual classroom activities demonstrate that the traditional teacher-centered approaches continue to predominate, which limits the outcomes of English reading comprehension (Kumar et al., 2018; Wixson, 2017). Numerous scholars have worked hard to identify more efficient language learning techniques in an effort to address this issue. To enhance reading instruction, more effective methods should be investigated and incorporated into the situations in which EFL students are currently learning to read. According to Pardo (2004), prior knowledge, experience, and approaching a book through a variety of cognitive processes are three variables that may impact the success of understanding a text. Other studies confirm that one of the most crucial subskills that all language learners must master in order to learn a language is reading comprehension (Daniels et al., 2019; Miller & Pennycuff, 2008). Without the ability to comprehend what they are reading in their textbooks, students will not be able to achieve in the classroom. Reading, therefore, is a fundamental ability for success in school and the workplace. The 4English mobile app is anticipated to satisfy students' needs and complement their learning styles because it is based on a social constructivism approach, promoting linguistic knowledge acquisition through independent learning.

Students have access to a wide selection of helpful materials that are available to them in both English and Vietnamese thanks to the 4English mobile app. It has numerous sections, including dictionaries, games, webpages, blogs, and YouTube videos, to mention a few. One of the biggest benefits of this mobile app is that students can read English words and instantly seek the Vietnamese meaning by clicking on any words, phrases, sentences, or even paragraphs they do not understand. Additionally, this app has a blog where the reader may get a wealth of useful lessons on pronunciation, grammar, and vocabulary. Since this software is mobile, students can access the content whenever they want and wherever they want. They can also keep freshly learnt English terms in a designated vocabulary corner for subsequent reference. As a result, because of the benefits described above, the 4English mobile app can be a useful and appropriate tool for English learners to enhance their language learning, particularly reading skills. With the use of the two research questions below, the purpose of this study is to investigate the potential of the 4English mobile app to assist students in improving their reading comprehension:

- To what extent does the 4English mobile app affect students' reading performance?
- How do students perceive 4English mobile app in learning reading?

LITERATURE REVIEW

Mobile Language Learning

Mobile devices have long been used in language instruction, and educational technology is developing quickly. In research by Vu (2016), when asked if participants use smartphones for language learning, 58 out of the 87 participants (or 67%) responded in the affirmative. Another study also finds that nearly 90% of K12 students have access to smart devices (Valk et al., 2010). There is no doubting that with the widespread use and quick development of mobile devices, educational experts have grown increasingly interested in mobile learning, particularly in the teaching of languages (Cochrane, 2010; Pegrum, 2020). Mobile devices can be broadly and functionally categorized into the following categories in terms of language instruction: cellphones, smartphones, tablets or e-book readers, netbooks or laptops, desktops, personal media players, and PDAs (Personal Digital Assistants). These devices could be used effectively for mobile language instruction (Kukulska-Hulme & Traxler, 2005).

M-learning is described as a learning form employing mobile devices and wireless transmission, as opposed to e-learning, which is the learning assisted by digital media (Peng et al., 2009). The delivery and facilitation of learning through mobile devices is referred to as "mobile learning" (Ally, 2014; Herrington & Herrington, 2009). The fact that mobile-based EFL learning is frequently addressed in terms of technology, particularly the adoption of numerous mobile device applications for English learning, is evidence of its benefits. For educational use, Kukulska-Hulme and Traxler (2005) described some popular mobile device types, including laptops, iPads, tablets, phablets, and smartphones. The concept of M-learning has frequently been described as learning using media-mobile devices (Kearney et al., 2012). Vu (2016) argued that M-learning should place more of an emphasis on the learners' mobility. He added that learners might learn whenever, whenever, and according to their own convenience thanks to mobile gadgets. Additionally, mobile learning, according to Cook (2010), has a favorable impact on EFL teaching and learning. In point of fact, mobile learning not only fills the void that is left by conventional education but also establishes a new benchmark for educational practice on account of the mobility, convenience, and accessibility benefits it offers (Udell & Woodill, 2014; Uther, 2019).

Mobile Learning and Constructivism

Constructivism as a learning theory, according to Qiu (2019), has a close relationship with the student-centered approach, in which students generate their own knowledge based on prior personal experience. Additionally, the constructivism hypothesis proposes that humans making sense of meaning in cultural and social communities of discourse construct knowledge through emergent and developing, non-objective, and workable constructed explanations. Peng et al. (2009) also confirms that meaning-making is at the center of constructivism. Besides, Qiu (2019) states that to construct knowledge, all learners have to:

- investigate information to build up its meaning
- gather necessary valuable sources and materials
- critically analyze, evaluate and invent new knowledge

It is clear that constructivism encourages students to connect, interact, process, communicate meanings, and personalize in order to build knowledge from their own experiences. Additionally, social constructivism-based learning can take place in the contexts of students' participation in social activities, where meaning is developed both individually and socially in the processing of knowledge (Jie et al., 2020; Reed et al., 2008).

Mobile language education is compatible with constructivist learning approaches because of the capabilities of mobile devices that have been discussed above. Numerous researches have shown that m-learning adheres to connection, interaction, learner autonomy, and learner-centered trends that help create individual learning environments, promote reciprocal interactions, and create linkages with the natural world (Klopfer et al., 2002; Peng et al., 2009; Baharom, 2013). Additionally, learning via mobile devices tends to fit different learning preferences. Therefore, it is crucial for teachers to design a decent learning environment filled with a range of activities and facilitate students in building their own knowledge (Schwandt, 1994).

The fact that children may need learning environments outside of the classroom is another obvious problem. According to earlier research, mobile learning significantly enhances the learning outcome by filling the void between formal and informal learning (Kukulska-Hulme & Traxler, 2005). This means that m-learning can provide students with more opportunities to integrate the physical world into their fundamental background knowledge. The creation of a supportive learning environment where connection, interaction, self-reflection, and learner autonomy is well aligned with the principles of M-learning on constructivism just described above.

English Mobile App

The 4English mobile app is a new application that supports mobile-assisted language study. It offers an easy-to-use navigational design and gives students quick access to a variety of learning resources. Different kinds of reading materials in both English and Vietnamese are available. There are numerous sections including dictionaries, games, webpages, blogs, and YouTube videos, to mention a few. One of this app's benefits is that students can read the materials created in English and instantly seek up the Vietnamese meaning by clicking on any words, phrases, sentences, or even paragraphs that they don't understand. Additionally, this app has a blog where the reader may get many useful lessons on vocabulary, grammar, etc. Since this software is a portable tool, students can access the content whenever and wherever they want. In addition, students can save recently learned English terms in a designated vocabulary corner called "Tu Vung Cua Ban" for subsequent review.

Figure 1. A screenshot of 4English mobile app

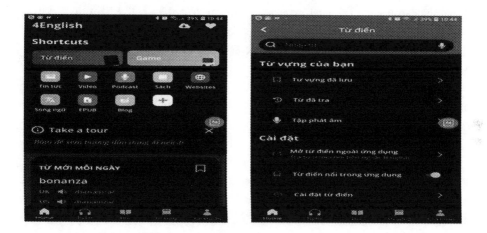

The 4English mobile app shows promise as a tool for teaching English reading since it gives students the chance to master the skill in a fun and varied manner. It is well accepted that if students read something for enjoyment, they retain the content better. Also, thanks to the portability, and flexibility of their mobile phones, students may read and enjoy these reading materials whenever and wherever they like. There is no denying that the rapid technological advancements of recent years have made mobile apps, such as the 4English smartphone app, beneficial for English language learners. Additionally, despite the lowering cost, mobile device quality continues to improve. These aspects have an impact on learners' ways of learning in settings where m-learning is deployed, along with a pervasive broadband internet connection. Further, Vu (2016) asserts that students utilized their mobile devices for a variety of purposes, with the majority of them also utilizing them pretty regularly for educational purposes. These educational purposes included things like checking emails and using learning programs. Furthermore, due to its portability and flexibility, a mobile phone has a lot of advantages over a PC when it comes to supporting different types of language acquisition (Herrington & Herrington, 2009; Stockwell, 2007).

METHODOLOGY

Research Site and Participants

The study was conducted in a classroom environment for 10th-grade English students at a K12 school in the province of An Giang, Vietnam. With the exception of a small group of high performers who desire to enhance their English skills for future usage, most students are not motivated to learn English. In this context, most students, notably the school's low performers, emphasize GCSE (General Certificate of Secondary Education) and university entrance tests because of their high stakes.

The current Pearson-designated textbook spans ten units and includes a variety of topics from daily life. It is meant to assist students in achieving level B1 (CEFR scale) after completing secondary education. The majority of these participants are at English levels A2 to B1 (CEFR scale). The book is divided into units focusing on a specific language aspect, such as grammar, pronunciation, vocabulary, reading, speaking, listening, and writing. Each unit also includes two additional sections: (1) communication and culture, which offers additional reading to help students learn the cultures of the target language, and (2) looking back and project, which allows students to review previous lessons and practice working collaboratively on projects.

The participants used the new edition of the ten-year English program textbook during this study. The research sample consisted of 90 students from the tenth grade, and they were split evenly between an experimental group (EG) and a control group (CG). Most of the participants were between the ages of 15 and 16. Both groups got the same curriculum, instruction, resources, and textbooks as previously described; however, the experimental group was required to complete their additional reading assignments via the 4English mobile app.

Research Instruments

The reading pretest and reading posttest were used to collect quantitative data. In a subsequent stage of the research, the outcomes of these tests were compared to determine the treatment's impact. The second data collection instrument was a poll conducted online in an effort to learn more about how participants felt about the mobile activities of the 4English app. In the final step of the research, a semi-structured interview was conducted with the participants to investigate their thoughts about the intervention.

The Tests

Pretests and posttests were used in the study to collect information on the students' reading scores and compare the control and experimental groups to determine if the treatment was effective. Each test contains two reading activities with an equal level of difficulty.

The two reading tests were taken from standardized reading test banks to ensure the tests' validity and reliability.

The Questionnaire

Adapted from Hsieh and Ji (2013), the questionnaire is designed to look into how participants feel about utilizing the 4English mobile app to learn to read. To gather participant's evaluation of the 4English mobile app, the questionnaire employed a five-point Likert scale. It had ten questions, which were divided into two sections: questions about how the 4English mobile app affected students' reading ability and questions about how they saw the app's potential role in reading support. The questionnaire is in Vietnamese to ensure participants understood the questions quickly.

Semi-Structured Interview

The interview in this study was given to gain more insights about the effectiveness of 4English app-based reading activities and get a comprehensive understanding of the current context in reading learning with the help of the app. During the interview, which consisted of five questions, an attempt was made to investigate how the app assisted students in their learning in more depth. During the interview, students were prompted to discuss their perspectives on a variety of reading-related topics associated with the 4English mobile app.

RESULTS

Tests

Pretest Results

A T-test was carried out in order to investigate whether or not there are differences that are statistically significant between the mean score that EG received on the reading pretest and the score that CG received. According to the information presented in Table 1, the average score on the EG pretest is 6.0667, whereas the average score on the CG pretest is 6.1889. Therefore, the mean value of CG is just a little higher than that of EG.

Table 1. Descriptive statistics of pretests scores

Group Statistics					
	CLASS	N	Mean	Std. Deviation	Std. Error Mean
Pre_test results	Experimental class	45	6.066 7	1.68077	.25055
	Control class	45	6.188 9	1.43531	.21396

The pretest scores of the EG and CG were evaluated using a normal Q-Q plot in order to check the assumptions that were made before to doing the T-test analysis. According to the data presented in Figure 2, both groups' scores were distributed along a somewhat straight line. This demonstrates that the scores were satisfactory for carrying out a T-test analysis.

From Table 2, the Sig. value of Levene's Test is .223 (Sig. >.05). As a result, one might draw the conclusion that the requirements for an equal variance were successfully met. Consequently, based on the findings of the T-test to determine whether or not the Means are equal, the Sig. (2-tailed) value is.712 (more than.05). This indicates that the variations between the means of EG and CG do not constitute a statistically significant difference. That is to say, prior to receiving the therapy, CG and EG exhibited reading performance that was equivalent to one another.

Figure 2. Normal Q-Q plots for the reading pretest results

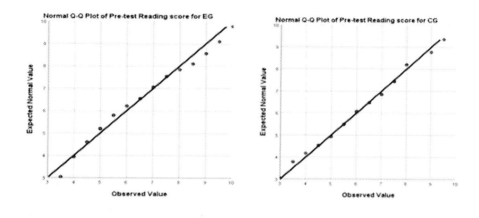

Table 2. Independent Samples T-Test of the results of the pretests

		Levene's Test for Equality of Variances		t-test for Equality of Means					95% Confidence Interval of the Difference	
		F	Sig.	t	df	Sig. (2-tailed)	Mean Difference	Std. Error Difference	Lower	Upper
Pre-test Results	Equal variances assumed	1.508	.223	-.371	88	.712	-.12222	.32948	-.77700	.53255
	Equal variances not assumed			-.371	85.894	.712	-.12222	.32948	-.77722	.53278

Posttest Results

After the findings of the Pearson correlational test had been presented, general descriptive statistics on the posttest scores of CG and EG were addressed. These data were compared and contrasted. According to Table 3, the posttest mean score of EG is considerably higher than that of CG (M=7.2667, SD=1.37593), which can be proven to be only 6.4000, SD=1.24133. This is a statistically significant difference. Following that, a test would be run to evaluate whether or not there was a statistically significant discrepancy between the mean scores of CG and EG. This would be done by comparing the two sets of results.

The normality test was utilized, in a manner analogous to that of the preceding section, in order to investigate the distribution of the EG and CG posttest results. The Q-Q plot showed that the data from each group clustered around a straight line,

Table 3. Descriptive statistics of posttests scores

Group Statistics					
	CLASS	N	Mean	Std. Deviation	Std. Error Mean
Post_test_results	Experimental class	45	7.2667	1.37593	.20511
	Control class	45	6.4000	1.24133	.18505

which was consistent with the findings of the plot. Because of this, it was possible to draw the conclusion that the posttest scores of both groups followed a normal distribution and that independent samples T-tests could be utilized.

Figure 3. Normal Q-Q plots for the reading posttest results

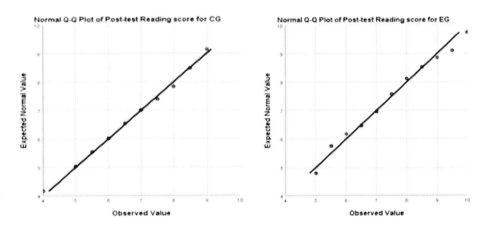

The first thing that was looked at was the outcome of Levene's test to determine whether or not the variances were equal. The significance level (.597) is more than (.05). Therefore, it is reasonable to infer that the variances are equal. In addition, the significance level for the two tails of the test, which is.002, is lower than (.05). This demonstrates that there was a difference that might be considered statistically significant between the posttest mean scores of EG and CG. Following the administration of the drug, there was a discernible difference in the reading performance of EG as compared to that of CG, as can be shown in Figure 4 below.

Table 4. Independent Sample T-Test of posttests results

		Levene's Test for Equality of Variances		t-test for Equality of Means					95% Confidence Interval of the Difference	
		F	Sig.	t	df	Sig. (2-tailed)	Mean Difference	Std. Error Difference	Lower	Upper
Post-test	Equal variances assumed	.282	.597	3.137	88	.002	.86667	.27625	.31768	1.41565
	Equal variances not assumed			3.137	87.084	.002	.86667	.27625	.31760	1.41573

Figure 4. Comparison of means of the pretest and posttest scores

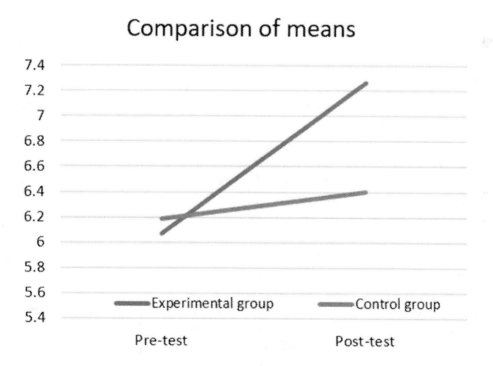

Questionnaire

The 4English Mobile App Influence On Reading Performance

The purpose of this section was to investigate the students' perspectives on the impact that using the 4English mobile app has had on their reading performance

Figure 5. 4English mobile app influence on students' reading performance

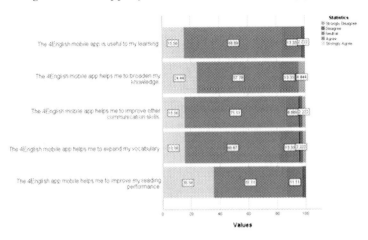

using five different items. The results of the data analysis carried out on items 1-5 are summed up and explained in Figure 5 which may be seen below.

The responses of the students to the five questions (1–5) that were asked on the impact that the 4English mobile app had on their reading abilities are displayed in Figure 5. The fact that every item's mean score was higher than 3.0 as a whole suggests that the vast majority of students believed that practicing their reading skills with the 4English mobile app exercises enhanced their abilities overall. Students reported an improvement in their reading abilities after using the 4English mobile app to master the skill. This is seen in the image below. This statement was accepted by 51.1% of the student body, with 35.6% of those students giving it their full support. However, there was just one kid who refuted the information. Only 11% of the pupils were still uncertain as to whether or not they had improved their reading abilities.

Similarly, it can be seen from the replies to item 2 that many respondents believed that learning new words would be beneficial. It's interesting to note that 15,6% of students indicated they would be open to learning vocabulary using 4English mobile app activities, and 66,7% expressed a strong interest in doing so. Additionally, 2,2% of respondents strongly disagreed and disagreed, demonstrating that few students disputed the value of the 4English mobile app activities in helping them increase their vocabulary. Similar to item 2, item 3 had a similar proportion of students who expressed disagreement and strong disagreement with only 2,2%. This indicated that the majority of students believed their communication abilities had considerably improved as a result of the 4English mobile app exercises. In the next item, the majority of students expressed satisfaction with the 4English mobile app activities' ability to help them expand their knowledge, garnering 57,8% agreement and 24,4%

Figure 6. Perceptions of the 4English mobile app in reading learning

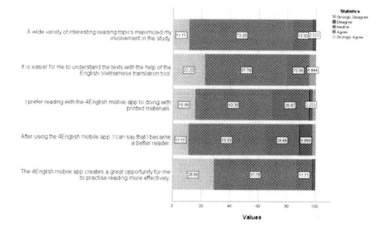

high agreement. Few students, as evidenced by the last item, had negative impressions of the reading-related exercises in the 4English mobile app.

Students' Perceptions of The English Mobile

The students' perceptions of using the 4English mobile app were shown in figure 6. To investigate their perceptions of 4English mobile app activities in reading learning, five items (6-10) were used.

For the first question, the majority of students said that regular use of the 4English mobile app had significantly improved their reading abilities. Studying using the app to traditional classroom instruction gave students additional opportunities for reading practice, which was supported by 57.8% of agreed respondents and 28.9% of strongly agreed respondents. The third item also mentioned that using the 4English mobile app allowed them to develop their reading skills. 53.3% agreed with this statement, and 11.1% strongly agreed. The data revealed that up to 53.3% of respondents agreed that students enjoy reading texts in apps more than they do with printed materials. A further 15.6% indicated their strong agreement. Only 2.2% of respondents indicated strong disagreement and a further 2.2% indicated disagreement with the use of apps. Additionally, it is mentioned that 57.8% and 22.2% of students, respectively, agreed and highly agreed with the 4English app's positive aspects in helping them understand the texts quickly due to its bilingual translation feature. Finally, the 4English mobile app offered students a variety of reading themes that piqued their attention, which increased their engagement in the course's reading exercise. Evidently, 73.3% of respondents chose agreement, while 11.1% chose a strong agreement.

Interview

Students' Evaluation of Reading Learning with the 4English Mobile App

When asked what they thought of using the 4English mobile app for their reading practice, a substantial majority of students voiced their good attitudes toward it when questioned about what they thought of using the program. Reading comprehension was significantly improved by the use of the 4English mobile app, which was lauded positively by almost all of the respondents. All of the responders, in general, have the same point of view, which is that their reading ability has significantly increased as a result of using this innovative approach to learning using the 4English mobile app. In addition, some of them also showed special interest in using the app:

I think it's very good and suitable for me. After using this app, I have made much progress in my study because I have had a valuable opportunity to get access to useful and updated sources of reading materials. Moreover, by this way of learning, I can enrich my vocabulary thanks to the unique translation tool and review new words stored in my vocabulary corner in the app. The thing worth mentioning here is that Reading on this app helps me relax and develop my reading skills. (S4)

However, one student (S1) mentioned that while using the program for educational purposes, they discovered that it was a little bit challenging for them:

This learning method is quite new to me, so I feel nervous when using a mobile app as a means for my language learning. It is a bit hard for me to decide what to read since there is a wide range of reading texts on the app. I sometimes get troubles with the internet connection.

It is conceivable to reach the conclusion that, on the whole, the students who were allotted to the experimental group had a favorable attitude toward the use of the 4English mobile app activities. This is a conclusion that can be drawn since it is possible to draw this conclusion. They all came to the conclusion that the reading activities that were provided on the app were not only useful for enhancing their ability to read, but that they were also enjoyable and useful for boosting their overall reading performance.

Students' Favorite Activities in the 4English Mobile App

When asked what they thought of using the 4English mobile app for their reading practice, a substantial majority of students voiced their good attitudes toward it when questioned about what they thought of using the program. The vast majority of respondents said that they were impressed by the way in which the 4English mobile app might be utilized for reading activities. All of the responders, in general, have the same point of view, which is that their reading ability has significantly increased as a result of using this innovative approach to learning using the 4English mobile app. In addition to that, several of them demonstrated an especially strong enthusiasm in making use of the application.

I think it's very good and suitable for me. After using this app, I have made much progress in my study because I have had a valuable opportunity to get access to useful and updated sources of reading materials. Moreover, by this way of learning, I can enrich my vocabulary thanks to the unique translation tool and review new words stored in my vocabulary corner in the app. The thing worth mentioning here is that Reading on this app helps me relax and develop my reading skills. (S3)

However, one student (S2) stated that while using the program for educational purposes, they discovered that it was a little bit challenging.

This learning method is quite new to me, so I feel nervous when using a mobile app as a means for my language learning. It is a bit hard for me to decide what to read since there is a wide range of reading texts on the app. I sometimes get troubles with the internet connection.

It is feasible to get at the conclusion that, on the whole, the students who were given roles in the experimental group had a favorable attitude about the participation in the activities provided by the 4English mobile app. They all came to the conclusion that the reading activities that were provided in the app were not only helpful in enhancing their ability to read, but that they were also enjoyable and useful in terms of enhancing their reading performance.

Students' Favorite Activities in the 4English Mobile App

Among five selected students, three of them (S1, S2, S5) had the same idea that reading world news was their favorite activity since it provided them with updated information about what was happening in the world. "*Reading news enables me to know more about the world. In this way, it is helpful for me to practice my reading skills"* (S5). Another student revealed that the app helped her to expand her vocabulary:

Reading helps me expand not just my vocabulary but also my understanding of various sentence forms. In point of fact, I am able to pick up quite a bit from reading the subtitles that are provided on videos. This is due to the fact that I am able to comprehend the meanings of specific words, sentences, and even passages with relative ease, all thanks to the rapid translation provided (S2).

Meanwhile, S3 was in favor of reading news in bilingual newspapers because it allowed him to understand the situation of the country, and the writing style the authors used was quite familiar with and easily understandable for him. Furthermore, the images in each newspaper make him more interested in reading.

One of the activities I am most fond of is reading news in bilingual newspapers because with colorful demonstration pictures, the reading texts seem to be more comprehensible, and more importantly, I can read the news in both languages at the same time. As a result, my reading becomes better and better. (S3).

DISCUSSION, CONCLUSION, AND RECOMMENDATIONS

After receiving the treatment, it was discovered using t-tests that the individuals' reading abilities had greatly improved. Despite the fact that the mean score of CG (M=6.1889) is only slightly higher than the mean score of EG (M=6.0667) in the findings of the pretest, the significance level for this comparison is.712, which is greater than.05. Before receiving therapy, there was not a discernible gap in terms of reading performance between CG and EG. This is something that can be asserted with certainty. The outcomes of the posttest, on the other hand, indicate that EG's mean score (M=7.2667) is significantly higher than CG's (M=6.4000), and the Sig. (2-tailed) value is.002, which is less than.05. It was demonstrated that as a result of the therapy, there was a difference that could be considered statistically significant between the mean scores obtained by EG and CG. Therefore, it is reasonable to draw the conclusion that students' reading abilities increased as a result of utilizing the 4English mobile app. Reading comprehension can be improved for students when they use the app in conjunction with standard classroom training. This is in comparison to when students solely receive traditional language instruction.

Almost all of the students who participated in the survey provided feedback indicating that they had improved their reading abilities as a direct result of utilizing the 4English mobile app. They actually demonstrated their favorable attitudes for the app, and they were confident that the app had significantly improved their reading abilities. Additionally, they discovered that the mobile app was effective enough to support them in overcoming their reading challenges. Second, these students emphasized how helpful this software was for them in many facets of language acquisition. For instance, it widened their understanding and increased their vocabulary. They largely concurred, in the end, that the 4English mobile app could improve language acquisition in general and the growth of communication skills, including reading in particular. With the ease of a multilingual translation tool, this software truly provided students with a good learning environment and aided their understanding of the reading texts.

In addition to the two instruments stated above, informal interviews were used to elicit more information and provide justifications for the responses provided by the students in the questionnaire. The majority of respondents agreed that the reading exercises are beneficial to their reading acquisition. The internet blackout slightly irritated a few interview subjects. It would have been better if the network had been improved. Besides, when asked if they preferred studying reading with the 4English mobile app over the traditional method, 68,9% of the students said they did. Since they could use the 4English mobile app to enhance their reading skills, many students had positive perceptions of the app even though there were a few issues, including internet connectivity, a lack of writing and public speaking skills. So, the interview results confirmed the beliefs expressed by the students in the questionnaires.

From these findings, we recommend that English teachers adopt a 4English mobile app to engage students in reading activities and, if possible, to aid in developing students' vocabulary and other communication. Additionally, this mobile learning approach improved students' reading ability. The findings have some implications for students as well. First, learning to read with the 4English mobile app is a novel approach built on constructivism that facilitates student language acquisition. Regular practice, diligent work on assignments, reasonable efforts, and optimistic mindsets were prerequisites if they were to improve their reading proficiency. Second, based on the survey and interview results, students should also improve their writing and speaking abilities and get adequate language practice so they can complete their assigned reading tasks more successfully.

REFERENCES

Aryadoust, V. (2019). An Integrated Cognitive Theory of Comprehension. *International Journal of Listening*, *33*(2), 71–100. doi:10.1080/10904018.2017.1 397519

Baharom, S. S. (2013). Designing Mobile Learning Activities. In *The Malaysian He Context: A Social Constructivist Approach* (p. 395). Salford Business School University of Salford.

Cochrane, T. D. (2010). Exploring mobile learning success factors. *ALT-J: Research in Learning Technology, 18*(2), 133–148. doi:10.1080/09687769.2010.494718

Cook, G. (2010). Sweet talking: Food, language, and democracy. *Language Teaching, 43*(2), 168–181. doi:10.1017/S0261444809990140

Daniels, K., Elliott, C., Finley, S., & Chapman, C. (2019). Learning and Teaching in Higher Education. In *Learning and Teaching in Higher Education*. Edward Elgar Publishing. doi:10.4337/9781788975087

Herrington, J., & Herrington, J. (2009). New technologies, new pedagogies : Mobile learning in higher education. In *World, 0.* https://ro.uow.edu.au/newtech/

Jie, Z., Puteh, M., & Hasan Sazalli, N. A. (2020). A social constructivism framing of mobile pedagogy in english language teaching in the digital era. *Indonesian Journal of Electrical Engineering and Computer Science, 20*(2), 830–836. doi:10.11591/ijeecs.v20.i2.pp830-836

Kearney, M., Schuck, S., Burden, K., & Aubusson, P. (2012). Viewing mobile learning from a pedagogical perspective. *Research in Learning Technology, 20*(1), 1–17. doi:10.3402/rlt.v20i0.14406

Klopfer, E., Squire, K., & Jenkins, H. (2002). Environmental Detectives: PDAs as a window into a virtual simulated world. *Proceedings - IEEE International Workshop on Wireless and Mobile Technologies in Education, WMTE 2002,* (pp. 95–98). IEEE. 10.1109/WMTE.2002.1039227

Kukulska-Hulme, A., & Traxler, J. (2005). *Mobile Learning: A Handbook for Educators and Trainers*. The Open and Flexible Learning Series.

Kumar, V., Boorla, K., Meena, Y., Ramakrishnan, G., & Li, Y. F. (2018). Automating reading comprehension by generating question and answer pairs. Lecture Notes in Computer Science (Including Subseries Lecture Notes in Artificial Intelligence and Lecture Notes in Bioinformatics), 10939 LNAI, 335–348. doi:10.1007/978-3-319-93040-4_27

Miller, S., & Pennycuff, L. (2008). The Power of Story : Using Storytelling to Improve Literacy Learning. *Journal of Cross-Disciplinary Perspectives in Education, 1*(1), 36–43.

Pardo, L. S. (2004). What Every Teacher Needs to Know About Comprehension. *The Reading Teacher, 58*(3), 272–280. doi:10.1598/RT.58.3.5

Pegrum, M. (2020). Mobile lenses on learning: Languages and literacies on the move. In *Mobile Lenses on Learning*. Languages and Literacies on the Move. doi:10.1007/978-981-15-1240-7

Peng, H., Su, Y. J., Chou, C., & Tsai, C. C. (2009). Ubiquitous knowledge construction: Mobile learning re-defined and a conceptual framework. *Innovations in Education and Teaching International, 46*(2), 171–183. doi:10.1080/14703290902843828

Qiu, J. (2019). A preliminary study of english mobile learning model based on constructivism. *Theory and Practice in Language Studies, 9*(9), 1167–1172. doi:10.17507/tpls.0909.13

Reed, P., Smith, B., & Sherratt, C. (2008). A New Age of Constructivism: 'Mode Neutral.'. *E-Learning and Digital Media, 5*(3), 310–322. doi:10.2304/elea.2008.5.3.310

Ririn, D. (2020). European Journal of Education Studies. *European Journal of Education Studies, 7*(1), 326–337. doi:10.5281/zenodo.582328

Schwandt, T. A. (1994). Constructivist, interpretivist approaches to human inquiry. *Handbook of Qualitative Research, January 1994*, 118–137. http://psycnet.apa.org/psycinfo/1994-98625-006

Stockwell, G. (2007). Vocabulary on the move: Investigating an intelligent mobile phone-based vocabulary tutor. *Computer Assisted Language Learning, 20*(4), 365–383. doi:10.1080/09588220701745817

Udell, C., & Woodill, G. (2014). Mastering Mobile Learning. In C. Udell & G. Woodill (Eds.), *Mastering Mobile Learning*. John Wiley & Sons, Inc., doi:10.1002/9781119036883

Uther, M. (2019). Mobile learning—Trends and practices. *Education Sciences, 9*(1), 33. doi:10.3390/educsci9010033

Valk, J. H., Rashid, A. T., & Elder, L. (2010). Using mobile phones to improve educational outcomes: An analysis of evidence from Asia. *International Review of Research in Open and Distance Learning, 11*(1), 117–140. doi:10.19173/irrodl.v11i1.794

Vu, N. N. (2016). Mobile Learning in Language Teaching Context of Vietnam: an Evaluation of Students' Readiness. *Journal of Science, HCMC University of Education, 7*(85), 16–27. https://www.vjol.info/index.php/sphcm/article/viewFile/24861/21273

Wixson, K. K. (2017). An interactive view of reading comprehension: Implications for assessment. *Language, Speech, and Hearing Services in Schools, 48*(2), 77–83. doi:10.1044/2017_LSHSS-16-0030 PMID:28395296

Zaki, A. A., & Md Yunus, M. (2015). Potential of mobile learning in teaching of ESL academic writing. *English Language Teaching, 8*(6), 11–19. doi:10.5539/elt.v8n6p11

ADDITIONAL READINGS

Al-Rahmi, A. M., Al-Rahmi, W. M., Alturki, U., Aldraiweesh, A., Almutairy, S., & Al-Adwan, A. S. (2022). Acceptance of mobile technologies and M-learning by university students: An empirical investigation in higher education. *Education and Information Technologies, 27*(6), 7805–7826. doi:10.100710639-022-10934-8

Amineh, R. J., & Asl, H. D. (2015). Review of constructivism and social constructivism. *Journal of Social Sciences, Literature and Languages, 1*(1), 9–16. http://blue-ap.org

Cross, M. K. D. (2021). Social constructivism. In Palgrave Studies in European Union Politics (pp. 195–211). doi:10.1007/978-3-030-51791-5_10

Ganesan, M., Singh, V. K., & Biswas, S. (2021). Mobile learning as the future of e-learning. In E-learning Methodologies: Fundamentals, technologies and applications (pp. 133–146). doi:10.1049/PBPC040E_ch6

Goundar, M. S., & Kumar, B. A. (2022). The use of mobile learning applications in higher education institutes. *Education and Information Technologies, 27*(1), 1213–1236. doi:10.100710639-021-10611-2

Goyal, M., Krishnamurthi, R., & Yadav, D. (2021). E-learning methodologies: Fundamentals, technologies and applications. In E-learning Methodologies: Fundamentals, technologies and applications. doi:10.1049/PBPC040E

Ninghardjanti, P., & Dirgatama, C. H. A. (2021). Building Critical Thinking Skills Through a New Design Mobile-Based Interactive Learning Media Knowledge Framework. *International Journal of Interactive Mobile Technologies, 15*(17), 49–68. doi:10.3991/ijim.v15i17.23801

Păcurar, E. (2018). *Steps towards flipping classes in Higher Education (ESP). 3,* 47–40. doi:10.29007/2m9h

Pegrum, M. (2021). Mobile learning: what is it and what are its possibilities? In Teaching and Digital Technologies (pp. 142–154). doi:10.1017/CBO9781316091968.015

Perguna, L. A., Idris, I., & Widianto, A. A. (2021). From Paper to Screen: Encouraging Theory of Sociology through Sosiopedia by Heutagogy Approach. *International Journal of Interactive Mobile Technologies, 15*(1), 155–167. doi:10.3991/ijim. v15i01.14357

Vinet, L., & Zhedanov, A. (2011). A "missing" family of classical orthogonal polynomials. *Journal of Physics. A, Mathematical and Theoretical, 44*(8), 085201. doi:10.1088/1751-8113/44/8/085201

KEY TERMS AND DEFINITIONS

Blended Learning: This type of learning combines online learning with face-to-face training. Learners who use a blended method receive thorough, multimodal teaching.

Collaborative Learning: Collaborative learning is a community-based educational strategy in which each organization member shares their knowledge and expertise with others. Social learning and peer learning are closely related to collaborative learning.

E-learning: E-learning (short for electronic learning) refers to digital instruction. E-learning is largely synonymous with distance learning.

Gamification: Gamification is the use of game principles from video games or mobile applications to design an entertaining and difficult user experience. Modern education is largely founded on gamification.

Mobile Learning: Learning that is mobile in nature created with smartphones and tablets in mind. Although it is mobile-first in nature, it can also be accessible via laptop computers. This includes learning on small screens, requiring less concentration, using more gamified activities to onboard learners, etc.

Synchronous: In contrast to "asynchronous," the term "synchronous" describes learning in which teachers and students converse simultaneously. Synchronous training includes activities like conventional in-person classes.

Chapter 6

EFL Students' Perceptions and Practices Regarding Online Language Learning:
A Case in Vietnam

Thanh Nguyet Anh Le
Dong Thap University, Vietnam

ABSTRACT

Virtual teaching and learning have become a hot topic in the Covid-19 pandemic. This chapter will present an investigation of learners' perceptions and practices of learning online. The current research was conducted with 161 EFL students at Dong Thap University, Vietnam. Questionnaires and observations were set to collect data. The findings showed that most students indicated benefits of online learning, such as saving time and money, protecting their health from Coronavirus, and being a suitable way of learning during lockdown time. However, in practice, they felt stress in long online classes. They could not fix technological problems as well as understand rules, technology culture, and attitudes when learning online. Furthermore, freshmen met several difficulties in learning English major online rather than seniors. Participants also showed their wishes and suggestions to improve virtual language teaching and learning platforms in the distant future.

INTRODUCTION

Online teaching and learning have become popular in modern life, especially during the Covid-19 epidemic. For recent years, Coronavirus has covered broad regions all

DOI: 10.4018/978-1-6684-7034-3.ch006

over the globe, which has made several institutions in many countries face an extremely difficult setting of the advancement of virtual teaching and learning environments. Universities in Vietnam have met the same difficulties as others in various areas of the world. Hence, accompanying an official letter No.1061/BGDDT-GDTrH, 25 March 2020 issued by the Vietnamese Ministry of Education and Training (Vietnamese MOET, 2020), schools as well as universities in Vietnam were advised to apply the different varieties of E-teaching and E-learning platforms, namely Google Meet, Zoom, Learning Management System (LMS), Learning Content Management System, Microsoft Team, and so on, to teach students and give assignments and tests. In addition, the official letter No.606/ BGDDT-GDTrH, 18 February 2021, issued by the Vietnamese Ministry of Education and Training (Vietnamese MOET, 2021) focused on the importance of online teaching and learning, and the balancing the content of all subjects and students' background.

However, for some reasons, the wireless Internet access serving online teaching and learning in Vietnam, especially in colleges or institutions in rural or remote regions, might be a very bad connection and still has several difficulties. Similarly, students at Dong Thap University (DTU), a rural university in the Mekong Delta, South Vietnam, have had many problems when learning online. Up to now, although many studies about the merits and drawbacks of virtual teaching and learning have been researched in the world in general and in Asian settings in particular (i.e., Gao & Zhang, 2020; Le, 2021; Mukhtar et al., 2020; Nartiningrum & Nugroho, 2020; Pham et al., 2022; Tran & Nguyen, 2022), there has still been little research of this field in Mekong Delta area of Vietnam. Some recent research on online teaching and learning was conducted in some big cities in Vietnam, such as Ngo (2021)'s, Pham et al. (2022)'s, and Tran and Nguyen (2022)'s studies happening in Ho Chi Minh City, one of the big cities in Vietnam. Additionally, universities have evaluated virtual teaching and learning process, especially learners' attitudes and performance, and rarely focused on the rules and cultures of eLearning. Hence, exploring EFL students' understanding of and taking part in online classes in a setting of a remote university in Vietnam will supply more insights into literature in this field, especially factors of technological rules and cultures.

LITERATURE REVIEW

Online Learning In Tertiary Education

These days online learning has been a hot trend among universities as an alternative way of teaching and learning processes, especially during the outbreak of the Covid-19 pandemic. The definition of online learning has still been a controversial issue among researchers around the globe. Singh and Thurman (2019) indicate that "online learning as a concept and as a word has consistently been a focus of education for over two decades" (p. 289). Additionally, these scholars conducted a study to compile and interpret 46 definitions of online learning in 37 previous studies from 1988 to 2018 to enhance people's perceptions of this concept. They found 19 terms utilized to define the term "online learning", namely "online learning, e-learning, blended learning, online education, online course, distance education, distance learning, web-based learning, computer-assisted instruction, web-based training, web-based education, web-based instruction, computer-based training, web-enhanced learning, resource-based learning, e-tutoring, computer-based learning, distributed learning, and computer-assisted learning" (Singh & Thurman, 2019, p. 294). After analyzing the content of those definitions as well as their accepted aspects used to define the term "online learning," Singh and Thurman (2019) suggested three choices for this field. The researchers assume that "online education is defined as education being delivered in an online environment through the use of the Internet for teaching and learning. This includes online learning on the part of the students that are not dependent on their physical or virtual co-location. The teaching content is delivered online, and the instructors develop teaching modules that enhance learning and interactivity in the synchronous or asynchronous environment" (Singh & Thurman, 2019, p. 302).

Besides, according to Famularsih (2020), the conception of online learning at institutions has pertained to 21st-century education and 21st-century chops. It means that technology as well as skills in using technology, should be integrated into content and tutoring methodology. In online teaching and learning, lecturers often use LMS as a virtual platform for activities. In addition, students should develop their skills to explore technological tools (Hwee et al., 2016).

Furthermore, Ngo (2021) states that online learning in higher education relates to some core, complex, potential factors, namely technology experience, interactions, motivation, and engagement. Of the rudiments, learners' engagement in virtual learning plays an important part in leading them to enjoy their learning as well as their learning outcomes.

Advantages and Disadvantages of Online Learning

Regarding online learning assets, Bertea (2009) shows some positive impacts such as the flexible timetable for students to attend virtual studying every time and everywhere by connecting to the Internet, decreasing expenditure, and saving time. Furthermore, Chakraborty and Muyia (2014) acknowledge the number of components that can construct alluring studying practices for virtual students, such as designing and continuing a beneficial learning setting, creating studying associations, supplying timely persistent comments, and utilizing the appropriate technological tools to distribute the suitable material. Moreover, Mustafa (2015) indicates that via utilizing one of the modes of LMS, learners do not sense boredom when participating in activities in a virtual learning platform. In addition, Haron et al. (2015) conclude that a virtual studying environment is a right way to deal with natural space and help students get closer to each other by conducting online learning activities. In the same vein, Cakrawati (2017) states that in teaching and learning English, learners experience an e-learning environment to train their language skills and obtain new English words.

On the other hand, Gillett-Swan (2017) collects from previous studies and shows a vast number of students' problems in online learning as isolated learners, namely "anxiety associated with using technology; being out of one's comfort zone; (perception of) inequity in assessment, particularly in "group" assignments; and, the (perceived) inability or difficulty in peer interaction, particularly in presentations" (p. 21). Likewise, Kebritchi et al. (2017) synthesize the results of 104 articles about online teaching and learning and state that "learners' expectations, readiness, identity, and participation in online courses" (p. 7) are challenges. Furthermore, Famularsih (2020) found that students complained about a great deal of homework given and a lack of learning conditions, making them feel difficulty carrying out online learning. They revealed that their reading and writing ability was better than their listening and speaking skills due to the disadvantages of online activities.

Students' Perceptions of Online Learning

Bertea (2009) reported the results from the research about outstanding e-students' characters carried out at Bloomsburg University of Pennsylvania, such as "self-motivation, patience, self-discipline, easiness in using software, good technical skills abilities regarding time management, communication, organizing" (p. 2). She states that these elements affect e-learners' attitudes immediately. Hence, students can feel effective and happy if the new learning mode is suitable for their needs as well as attributes, and vice versa if they cannot suit the new learning environment since they do not possess those factors. She also mentions that exploring learners' perceptions via attitude regarding virtual learning is extremely crucial by virtue of its impact on their behaviour. She emphasizes that "attitude indicates in a certain degree the possibility of adopting certain behavior", and in online learning, "a favorable attitude of students shows a greater probability that they will accept the new learning system" (p. 2). Additionally, she introduced two modes of gauging students' attitudes which were grown by Rosenberg (through "the perceived utility of the object and the value of importance") and Fishbein (through "beliefs and evaluations") (Bertea, 2009, p. 2).

Boca (2021) made a collection of previous studies to have a statistic on some elements affecting students' perceptions and their behaviour in using technology, including perceived usefulness and perceived ease of use (Davis, 1993; Al Kurdi et al., 2020; Mailizar et al., 2020), perceived interaction (Liu et al., 2010), enjoyment, system interactivity, computer anxiety, technical support (Al Kurdi et al., 2020). According to Boca's research (2021), those studies' findings served as a deeper understanding of students' intention and acceptance of using e-learning.

Students' Practices of Online Learning

Croft et al. (2010) emphasize that if students do not interact and discuss in e-learning classes, their learning experience is not rich; moreover, interaction and discussion are two of the most important factors of the constructivist method in the teaching and learning process. in the same vein, okita (2012) states that interaction between student and student or teacher and students has been proven to be extremely useful in supporting students to control their thinking, show their comprehension, and look for their weaknesses reasonably. besides, butler (2012) indicates that giving learning materials on a virtual platform is inadequate, leading learners to be disinterested and miss obstacles in teaching and learning procedures.

in addition, boca (2021) listed some e-learning components impacting learners' behaviour and attitude, namely online support service quality, online learning acceptance, and student satisfaction (hung & nguyen, 2022; lee, 2010), attitudes, curriculum, motivation, and technology training (zia, 2020), service quality, information quality and self-efficacy, satisfaction (alzahrani & seth, 2021), the effort expectation, the performance expectation, social influence and facilitating conditions (md yunus et al., 2021).

reviewing components of virtual teaching and learning is an attempt. yet, it seems difficult to determine the most significant elements of teaching and learning language online. hence, based on the aforementioned, the current study explored language learners' perceptions and practices regarding online learning and specially focused on technological rules and cultures.

METHODOLOGY

Research Questions

1. What are English-major students' perceptions of online learning at DTU?
2. What are English-major students' practices of online learning at DTU?

3.   What are the differences among freshmen's, sophomores', juniors', and seniors' practice of online learning at DTU?

The Setting

DTU first pioneered LMS in 2019. Some leading lecturers at DTU were trained in using LMS for three days. At the beginning of 2020, the Covid-19 pandemic first spread to Vietnam, which made DTU decide to change from traditional courses to online ones. All lecturers were instructed to use the functions of LMS for two days as well as asked to use Zoom or Google Meet in case they could not access LMS. Additionally, teachers utilized Zalo (a popular social network in Vietnam) or Facebook to connect with their students more efficiently. However, because of some reasons, both teachers and students had challenges when teaching and learning on virtual platforms. Especially, most teachers had no experience with virtual teaching and learning before Covid-19 time. In Le's study (2021), teachers and students were not trained in attitudes, online teaching and learning behaviour, and technical competence. As a result, although online courses were introduced to educate learners at DTU in 2019, many challenges emerged with negative impacts on their academic achievements.

Participants

The participants were 161 EFL students from the Faculty of Foreign Language Education at DTU in Vietnam, where the researcher of this study has worked as a lecturer of English for over ten years. The students (55 freshmen, 55 sophomores, 31 juniors, and 20 seniors) were 56 males and 105 females from 18 to 22 years old (2021-2022 academic year). They were from many provinces from the end of the Southern to the Southeast of Vietnam (including Ca Mau, Bac Lieu, Soc Trang, Tra Vinh, Can Tho, Vinh Long, Hau Giang, An Giang, Kien Giang, Tien Giang, Long An, Tay Ninh, Ho Chi Minh City, and Dak Lak). All of them voluntarily participated in this study and answered items in the survey related to their perceptions of online learning and what they had done in their virtual learning. Besides, none of them had ever attended any training programs on online learning. They were just instructed on how to enroll in online courses on LMS in 2021.

Data Collection and Analysis

The mixed research method was used to conduct this study. Two tools were used to gather students' information on online learning. The first one was a set of questionnaires with 26 items designed according to the framework in the literature review. It included 10 items in the perception session and 17 items in practice one. The survey was conducted through Google form because of the lockdown at that time. The second one was virtual classroom observation. In the first semester of the 2021-2022 academic year, the researcher taught eight classes with three subjects: one assessment and testing class, 3 writing No.3 classes, and four teaching practice No.2 classes. Data for the study was collected through a Google Form survey instrument and analyzed using a statistical package for the social sciences to code and give statistics. This analysis indicated the means, standard deviations, tests, and ANOVA. Furthermore, data analysis was based on the results from the researcher's observation in online classes. Besides, the reliability of these statistics was paid attention to because they "lead to meaningful interpretations of data" (Creswell, 2014, p. 200). The results of this data analysis answered research questions 1, 2, and 3.

FINDINGS

Data From the Questionnaire Survey

Students' Perceptions

Table 1 shows that ten items of English-major students' perceptions of virtual learning obtain a high Cronbach's alpha coefficient at 0.801 > 0.6, so 10 items in the questionnaire have an acceptable value and show good internal consistency among them. Additionally, they indicate that all items measure the same thing and have high correlations. Besides, the mean of 8 items is from 3.33 to 4.30; only item 4 has mean score of 2.76, and item 9 has mean score of 2.84. This proves that most students had positive thinking about the benefits of online learning. However, the responses in questions 4 and 9 show that learners did not believe in the quality of virtual learning and assessment and testing. They still prefer to study face-to-face rather than online.

Students' Practices

It can be seen in table 2 that Cronbach's alpha coefficient of 17 items at 0.756 > 0.6 is reliable. In addition, 13 items (Q 12, 13, 15, 16, 17, 18, 19, 21, 22, 23, 24, 26, 27) have mean scores from 3.08 to 4.27, and 4 of them (Q 11, 14, 20, 25) have mean score 2.94, 2.61, 2.90, and 2.98 respectively. This shows that learners did not acquire online lessons, and deal with electronic devices' accidents. They also met low, unstable Internet connections, and disliked studying virtually.

Table 1. Students' perceptions of online learning

No.	Items	Mean	SD	Cronbach's alpha if item deleted
1	I think that online learning is a suitable learning way for the outbreak of Covid-19.	4.30	0.716	,777
2	I think I can save time commuting when learning online.	4.27	0.835	.774
3	I think I can save money for daily life when learning online.	3.91	0.977	.793
4	I think I study well on the online platform rather than face-to-face learning.	2.76	1.053	.772
5	I think I know the rules of online learning.	4.01	0.711	.781
6	I think I understand the culture of online learning.	3.95	0.696	.786
7	I know I need to have learner autonomy ability when learning online.	4.09	0.789	.786
8	I think I can interact well with my teachers and my friends in online classes.	3.33	0.960	.790
9	I think that types of online tests and assessments can exactly evaluate students' abilities.	2.84	0.912	.796
10	I think online learning helps me not have Covid-19 because I do not have to go to university or contact with other people.	4.20	0.982	.780

Differences Among Freshmen's, Sophomores', Juniors', and Seniors' Practice of Online Learning

Results from the Test of Homogeneity of Variances, ANOVA, and Robust Tests of Equality of Means in Table 3 show that there is no difference among mean scores of 4 groups of participants (the first-year students, the second-year ones, the third-year ones, and the fourth-year ones) in responses of 13 items from 14,15,17 to 27. Meanwhile, mean scores in the rest of items 11, 12, 13, and 16 of those four groups of students are different, and this is clearer via LSD - Multiple Comparisons.

Item Q11: I Can Understand Online Lectures Completely.

The sig value (Sig. = 0.001 < 0.05) proves that there is a difference in statistical meaning between freshmen and sophomores, while mean score of freshmen is lower than that of sophomores with a mean difference of 0.538. Moreover, between freshmen and seniors have a difference when they answered Q11 (Sig. = 0.006 < 0.05). According to table 3, the mean difference between the first-year students and the fourth-year ones is 0.629. It means that the mean score of the fourth group is higher than the first one.

Table 2. Students' practices of online learning

No.	Items	Mean	SD	Cronbach's alpha if item deleted
11	I can understand online lectures completely.	2.94	0.933	.727
12	I can take notes online lectures carefully.	3.39	0.889	.724
13	I have enough electronic devices to learn online.	3.47	0.975	.738
14	I can access the Internet or Wifi at my house easily to serve online learning.	2.61	1.013	.744
15	I cannot understand lessons completely because of the low quality of my Internet or power cut.	4.12	0.883	.769
16	I can interact well with my teachers as well as my friends in online classes.	3.38	0.894	.732
17	My teachers often hold effective discussion activities in breakout rooms or in the chat box.	4.27	0.714	.743
18	Those discussion activities are effective.	3.42	0.933	.740
19	My learning results through online tests, quizzes, or assessments are reliable and exact.	3.08	0.908	.730
20	I feel comfortable and like to learn online for a long time.	2.90	1.147	.737
21	I feel stressed and tired when learning online for a long time.	3.75	1.037	.788
22	I read materials before attending online classes.	4.15	0.700	.743
23	I submit assignments on time.	3.71	0.764	.737
24	I learn other aspects related to my subjects autonomously.	3.62	0.798	.738
25	I can fix some problems with my computer, electronic devices, or the Internet when learning online.	2.98	1.046	.744
26	I hope to learn my courses online next time.	3.19	1.238	.758
27	I hope that I will be trained in technological culture as well as some ways to fix technological problems when learning online.	3.68	0.802	.751
	Other ideas:			

Item Q12: I Can Take Notes Online Lectures Carefully.

When freshmen and seniors responded to Q12, the statistic meaning showed the sig value (Sig. $= 0.010 < 0.05$) and it means the difference among them. Besides, the mean difference between these two groups is 0.574, which showed that the mean score of freshmen is lower than that of seniors.

Table 3. Tests of comparison among means of 4 groups of students

	Test of Homogeneity of Variances				ANOVA (Sig.)	Robust Tests of Equality of Means (Sig.)	LSD - Multiple Comparisons			
	Levene Statistic	df1	df2	Test of Homogeneity of Variances (Sig.)			(I) – students in the year	(J) – students in the year	Sig.	Mean Difference (I-J)
Q11	1.162	3	157	0.326	0.003		1	2	0.001	- 0.538
							1	4	0.006	- 0.629
Q12	1.030	3	157	0.381	0.036		1	4	0.010	- 0.574
Q13	7.421	3	157	0.000		0.002	1	4	0.002	- 0.751
Q14	0.062	3	157	0.980	0.954					
Q15	0.335	3	157	0.800	0.983					
Q16	2.151	3	157	0.096	0.010		1	4	0.001	- 0.749
							2	4	0.008	- 0.585
Q17	0.267	3	157	0.849	0.660					
Q18	1.568	3	157	0.199	0.085					
Q19	0.471	3	157	0.703	0.106					
Q20	0.399	3	157	0.754	0.161					
Q21	2.163	3	157	0.095	0.957					
Q22	0.267	3	157	0.849	0.133					
Q23	5.732	3	157	0.001		......				
Q24	5.071	3	157	0.002		0.056				
Q25	0.482	3	157	0.695	0.235					
Q26	0.177	3	157	0.912	0.804					
Q27	1.308	3	157	0.274	0.085					

Item Q13: I Have Enough Electronic Devices to Learn Online.

Like the two above items, there is a difference between the first group's and the fourth one's statistic meaning (Sig. $= 0.002 < 0.05$). In addition, the mean difference between these two groups is 0.751, and this means that the mean score of freshmen is lower than that of seniors.

Item Q16: I Can Interact Well With My Teachers as Well as My Friends in Online Classes.

In this item, there is a difference between freshmen's and seniors' answers through the sig value (Sig. $= 0.001 < 0.05$). Furthermore, with a mean difference of 0.749, the mean score of freshmen is lower than that of seniors. Additionally, the sig value between sophomores and seniors (Sig. $= 0.008 < 0.05$) proves the difference in their responses to this question. As a result, the mean score of the second-year students is lower than that of seniors based on the mean difference of 0.585.

Other Ideas

When asked for other ideas, on the one hand, 37 students listed the benefits of online learning which were the same as the author mentioned in the above survey.

On the other hand, in terms of knowledge acquisition, the participants responded that they could not follow teachers' lectures (88 students), and 51 of them complained that they were hard to manage their online learning. 75 of them said that the virtual learning platform was ineffective for those who did not have learner autonomy ability. Even 87 of them showed their anxiety that they would not apply the knowledge they studied online when they took the final exam or visited somewhere for practice or for an internship because they hardly understood online lectures. Most of them (145 students) agreed that learning online was unhappy and boring.

The next aspect students complained about is that many of them (91 students) had problems with electronic devices or Internet quality; for instance, they could not turn on the micro or webcam whenever they needed regardless of having checked it before. Some students did not have computers or laptops. Instead, they had to use their cell phone to learn online, and when they attended five classes in the morning, their cell phone became very hot. As a result, they could not continue to join online classrooms via their cell phone in the afternoon. Sometimes, some students (50 of 161) could not participate in classes at the beginning because of low internet connection, and after that when their Internet at home is strong enough, their teachers did not permit them to join. As a result, they missed those classes. In some cases, while learning, their computer, laptop, or cell phone had accidents immediately, so they could not carry on with their classes or submit their assignments, testing, and quizzes on time. Furthermore, many of them (112) did not master informative technology as well as technological techniques, so they met difficulties in using computers or laptops.

One more important factor which is common in online learning is health. Most of them (151 students) revealed that they met both physical and mental health, namely myalgia, bleary eyes, eyestrain, headache, and stress when they prolongedly stared at the screen. One more difficulty is that they could hardly concentrate on learning because of factors around them such as their family members' conversation or observation, noise, sudden calls, the sound of rain, the sound of thunder, and so on.

In addition, learners met obstacles in interaction or group work. More specifically, they hardly interacted with their teachers or their classmates and discussed in groups when utilizing a cell phone. In groups, some students discussed and did tasks given by teachers, but others did not. 10 students claimed that they forgot registering before attending the online classes, so their teacher did not admit their attendance though they studied those. More than half of them (102 students) admitted that they did not study in a serious and strict manner. It meant that they did other things such as surfing Facebook, chatting with others, or sending messages, etc. while they were in classes.

Finally, 98 of the informants stated that it was hard to evaluate the quality of virtual assessment and testing because some of them used other devices to search the answers on the Internet or they hardly took notes on the text on file to analyze and find out the keys as they did on paper materials. In other cases, some good students did the tests or quizzes, and then sent answers to their friends through the social network; therefore, weaker ones who did nothing, and just waited for the keys, still finished, and submitted their tests or quizzes. Or, some students who had a weak background or were lazy could have other people do their tests and quizzes as no one observed them when they took online exams.

Additionally, the majority of participants (159 students) showed that they desired to be able to go to school as soon as possible because they still preferred to study in a real classroom directly than in a virtual one. 5 students suggested that the Department of testing and quality assurance should adjust the mode of online testing because many students did not have a computer or laptop to take the semester exam. 100 of them thought that they would be trained in techniques to utilize electronic devices in order to fix technological problems immediately and in time, rules as well as a culture when learning online the next time. Besides, most of them (140 students) said that Dong Thap University should upgrade the LMS system to serve learners better. Finally, they wish they had been instructed on learner autonomy ability.

The Results From the Author's Observing Virtual Classrooms

Students' Practices in Online Classes

As mentioned above, the author taught 8 classes with 3 subjects on the LMS platform. After the first semester of the 2021-2022 academic year, the researcher found some advantages and disadvantages of online learning from observing students' studying. The findings of the benefits of virtual platforms echoed ones in the survey. However, there were still many drawbacks for students like the following situations:

Situation 1

Some students logged in to the class on LMS, but when the researcher (she) called him/her, they did not respond. After the class finished, they sent a message to her and said that they had some business to solve, so they went out.

Situation 2

Some students' electronic devices often disconnected from the Internet due to the low quality of Internet service. They logged in and out of the classes many times during a period.

Situation 3

Once, in the evening, a student in Teaching Practice No.3 class sent her a message to ask about the requirements of an assignment she gave in an online class in the afternoon. He said that because the Internet service at his house lagged at that time, he did not listen to what she talked to. She replied that at the end of the lecture, she always asked students if they had any questions about the lesson, please let her know, and she would answer them. Moreover, she showed PowerPoint of the lesson on the screen so that students could take notes or photos easily. In addition, at the beginning of the course, she always told students if they met any problems with their electronic devices or the Internet, they had to tell her or their classmates, and they as well as she would support them. Yet, in that class, he said nothing. After that, he became angry with her, posted some impolite words on Zalo – a famous social network in Vietnam, and often sent bad words in the chat box class.

Situation 4

In the writing No.3 course, in each class, students had to submit an essay at the end of the lesson on the LMS system; however, some of them did not do that. After that, a student said that since she had to carry her family member to the hospital, she forgot pressing the submissive button. Another responded that he wrote an unfinished essay on LMS and went out without submitting it because he thought that the LMS system would automatically save it. Others met the power which went out while they were working online.

Situation 5

The researcher often divided the class into groups to discuss some tasks. When she joined each group to observe what they were doing and found that some of the students did nothing and kept silent. Or, when they saw her join their group, they started discussing questions.

Situation 6

Many students did not know how to use the functions on the LMS system when they enrolled in the courses, presented their presentations, showed their work on the screen, did quizzes, or submitted their assignments although, at the beginning of the semester, DTU organized to train them technological techniques in 100 minutes, and posted a video to instruct them to utilize the LMS system. However, when asked whether they attended that training class, they said that they did not participate in it, and just watched an instruction video.

Situation 7

Some students logged in and out of the classes because of Internet services or their own personal reasons; however, they did not ask the teacher's permission.

Situation 8

Sometimes LMS system had problems. For instance, when there were many online classes at the same time, the system had a low-speed connection, and the server was low. On some days, both teachers and students could not log in to online classrooms, or teachers or students could not sometimes present the content of the lectures on the screen because of an unstable internet connection. This made both teachers and students uncomfortable and unhappy.

DISCUSSION

Students' Perceptions of Online Learning

The findings from the questionnaire and observation show that most of the participants have positive perceptions of online learning, especially in Covid19 pandemic lockdown period. They claimed that this is the only, suitable, convenient, and safest way to protect citizens, especially students at that time. The advantages, as well as disadvantages of online learning, echoed previous studies as mentioned in the literature review (Bui, 2022; Bertea, 2009; Haron et al., 2015; Gao & Zhang, 2020; Le, 2021; Mukhtar et al., 2020; Nartiningrum & Nugroho, 2020; Pham et al., 2022; Tran & Nguyen, 2022).

Students' Practices of Online Learning

In reality, students met other drawbacks, namely mental health, limited ability in using technology, impolite behaviors in online classes, Internet connection, lack of learner autonomous ability, and so on. Especially, technological culture is the new view in this study compared with previous ones (Bertea, 2009; Haron et al., 2015; Gao & Zhang, 2020; Le, 2021; Mukhtar et al, 2020; Nartiningrum & Nugroho, 2020; Pham et al, 2022; Tran & Nguyen, 2022). As mentioned above, because those informants were from other regions in Vietnam, they brought a variety of characteristics, customs, cultures, as well as English backgrounds to classes at DTU's LMS system. Yet, they had a common thing which was that they possessed a low level of the use of technology.

Additionally, via observing contexts in virtual classrooms, it can be denied that some students did not know both rules as well as cultures when attending those ones. It is easy to understand because they have not been instructed or trained in these things before.

Differences in Practices of Online Learning Among Students From the First Year to the Fourth Year

The results from participants' other ideas and tests to compare as well as find out whether there are some differences among students in different years indicate that there are some new points compared with prior research in this field (Gillett-Swan, 2017; Famularsih, 2020; Gao & Zhang, 2020; Le, 2021; Mukhtar et al., 2020; Nartiningrum & Nugroho, 2020; Pham et al., 2022; Tran & Nguyen, 2022). More specifically, the rate of the first- or second-year students understanding lessons, taking notes carefully, having enough learning conditions (e.g., electronic devices, Wifi, Internet service, etc.), and interacting with teachers and classmates in the virtual platform is lower than the four-year ones (see Table .3). It means that freshmen have a larger number of difficulties in learning online than others.

From the results of the current study, the author believes that DTU should organize some technology training courses or workshops, including usages, rules, and culture of virtual platforms in the next time for students, especially freshmen at the beginning of the school year. In addition, they should be instructed on learner autonomy skills so that they can study autonomously inside or outside the classrooms as well as in online or offline classes. This is very essential and useful for DTU EFL students if they would like to achieve in their studies at university as well as their future job because Bertea (2009) found some qualities of successful e-learners, namely self-motivation, patience, self-discipline, easiness in using software, good technical skills abilities regarding time management, communication, organizing. Moreover, she emphasizes that students will accept the new learning environment if it is appropriate for their needs, their attributes, and makes them happy, comfortable, and effective, and vice versa.

One more crucial component not only DTU but also other institutions should consider is that there should be mental health care centers at schools or at universities to take care of students' spirit aspect in a busy, modern, pressured life these days. Many students had mental health problems but did not find whom to share their feelings with. This may lead to their study results being reduced.

In addition, types of assessment and testing should be designed reasonably and appropriately for real settings. It is important that how to make the results from those forms of assessment and testing still reliable notwithstanding students take them online.

In conclusion, EFL students at the Faculty of Foreign Languages at DTU had positive thoughts about virtual learning, but they met a range of problems when attending online classes. Hence, DTU administrators as well as lecturers should have effective measures to help them overcome those, especially the first-year students. When freshmen came to a new learning environment, they had some difficulties, namely accommodations, learning methods, learning conditions, necessary skills for the youth in the 21st century, learner autonomy ability, technological skills, homesickness, relationships, interaction, physical and mental health, and so on. It is believed that if freshmen are instructed, equipped, and trained in essential skills at the beginning of the first year, they will be able to overcome the drawbacks during university time. One more thing is that DTU should upgrade and improve the LMS system to serve to teach and learn online these days and in the future better.

CONCLUSION

Exploring EFL students' perceptions and practices of virtual learning plays a key role in the teaching and learning process, especially in difficult situations such as Covid 19 time. The findings in this study provide the local context in the literature review in this field. One new view in this research is that institutions as well as educators should pay attention to instructing technological and cultural factors to students and integrate them into the teaching and learning process. Besides, mental health care and learner autonomous skills should be taught to students, especially for freshmen to support them in the new learning environment. It is noteworthy that DTU should improve LMS system quality as well as have a suitable policy to help poor students approach eLearning effectively in the future.

This study shows a few limitations. Due to lockdown time, the researcher conducted this study via questionnaire and observation. At that time, most people, including her students, were very stressed, tired, and scared of Covid19, and they stayed at home, so she could not use other instruments such as interviews. Besides, she did not invite Chinese-major students in Foreign Languages Faculty at DTU to participate in this research to obtain a variety of perceptions and implementation of this field as the circumstance at that time did not allow her to fulfill that. Additionally, she did not invite her colleagues to join this research as participants because of the difficult situations in that period.

In further research, the author should investigate other major students' and EFL teachers' perceptions and practices of online teaching to have full insight into this field. What lecturers think may affect what they do, so their understanding of real teaching in the online platform is very important. More research methods should be added to make future studies more reliable.

REFERENCES

Al Kurdi, B., Alshurideh, M., Salloum, S. A., Obeidat, Z. M., & Al-Dweeri, R. M. (2020). An empirical investigation into examination of factors influencing university students' behavior towards ELearning acceptance using SEM approach. *Int. J. Interact. Mob. Technol.*, *14*(2), 19–41. doi:10.3991/ijim.v14i02.11115

Alzahrani, L., & Seth, K. P. (2021). Factors influencing students' satisfaction with continuous use of learning management systems during the COVID-19 pandemic: An empirical study. *Education and Information Technologies*, *26*(6), 6787–6805. doi:10.100710639-021-10492-5 PMID:33841029

Bertea, P. (2009). Measuring students' attitude toward E-learning: A case study. *The 5th International Scientific Conference: eLearning and Software for Education, Bucharest.*

Boca, G. D. (2021). Factors influencing students' behavior and attitude towards online education during COVID-19. Sustainability, 13(13), 7469. <jrn> Bui, H. P. (2022). Students' and teachers' perceptions of effective ESP teaching. [PubMed]. *Heliyon*, *10628*(9), e10628. doi:10.1016/j.heliyon.2022.e10628

Butler, K. C. (2012). A model of successful adaptation to online learning for college-bound native American high school students. *Multicultural Education & Technology Journal*, *6*(2), 60–76. doi:10.1108/17504971211236245

Cakrawati, L. M. (2017). Students' perceptions on the use of online learning platforms in EFL classroom. *English Language Teaching and Technology Journal*, *1*(1), 22–30. doi:10.17509/elt%20tech.v1i1.9428

Chakraborty, M., & Muyia, N. F. (2014). Strengthening student engagement: What do students want in online courses? *European Journal of Training and Development*, *38*(9), 782–802. doi:10.1108/EJTD-11-2013-0123

Creswell, J. W. (2014). *Research design: Qualitative, quantitative, and mixed methods approaches* (4th ed.). Sage.

Croft, N., Dalton, A., & Grant, M. (2010). Overcoming isolation in distance learning: Building a learning community through time and space. *The Journal for Education in the Built Environment*, *5*(1), 27–64. doi:10.11120/jebe.2010.05010027

Davis, F. D. (1993). User acceptance of information technology: System characteristics, user perceptions and behavioral impacts. *International Journal of Man-Machine Studies*, *38*(3), 475–487. doi:10.1006/imms.1993.1022

Estacio, R. R., & Raga, R. C. Jr. (2017). Analyzing students online learning behavior in blended courses using Moodle. *Asian Association of Open Universities Journal, 12*(1), 52–68. doi:10.1108/AAOUJ-01-2017-0016

Famularsih, S. (2020). Students' experiences in using online learning applications due to COVID-19 in English classroom. *Studies in Learning and Teaching, 1*(2), 112–121. doi:10.46627ilet.v1i2.40

Gao, L. X., & Zhang, L. J. (2020). Teacher learning in difficult times: Examining foreign language teachers' cognitions about online teaching to tide over COVID-19. *Frontiers in Psychology, 11*, 1–14. doi:10.3389/fpsyg.2020.549653 PMID:33071866

Gillett-Swan, J. (2017). The challenges of online learning: Supporting and engaging the isolated learner. *Journal of Learning Design, 10*(1), 20–30. doi:10.5204/jld. v9i3.293

Haron, N. N., Yasmin, H. Z., & Ibrahim, N. A. (2015). E-learning as a platform to learn English among ESL learners: Benefits and barriers. In Mahani, S., & Haliza, J. (Eds.), *Research in Language Teaching and Learning,* 79-106. UTM Press. https://www.researchgate.net/publication/306119651_ELearning _as_a_Platform_to_Learn_English_among_ESL_Learners_Benefits_ and_Barriers

Hung, B. P., & Nguyen, L. T. (2022). Scaffolding Language Learning in the Online Classroom. In: Sharma, R., Sharma, D. (eds) New Trends and Applications in Internet of Things (IoT) and Big Data Analytics. Intelligent Systems Reference Library, 221. Springer, Cham. doi:10.1007/978-3-030-99329-0_8

Hwee, J., Koh, L., & Chai, C. S. (2016). Teacher professional development for TPACK-21CL: Effects on teacher ICT integration and student outcomes. *Journal of Educational Computing Research, 55*(2), 1–25. doi:10.1177%2F0735633116656848

Kebritchi, M., Lipschuetz, A., & Santiague, L. (2017). Issues and challenges for teaching successful online courses in higher education: A literature review. *Journal of Educational Technology Systems, 46*(1), 4–29. doi:10.1177/0047239516661713

Le, T. N. A. (2021). *Interaction and evaluation of teaching and learning English online: Challenges and solution.* The 9th OPEN TESOL International Conference 2021: Language education in challenging times: Designing digital transformations, Ho Chi Minh City, Vietnam. https://opentesol.ou.edu.vn/2021proceedings.html

Lee, J. W. (2010). Online support service quality, online learning acceptance, and student satisfaction. *The Internet Higher Education, 13*(4), 277–283. doi:10.1016/j. iheduc.2010.08.002

Lin, C. L., Jin, Y. Q., Zhao, Q., Yu, S.-W., & Su, Y.-S. (2021). Factors influence students' switching behavior to online learning under COVID-19 pandemic: A push–pull–mooring model perspective. *The Asia-Pacific Education Researcher, 30*(3), 229–245. doi:10.100740299-021-00570-0

Liu, I.-F., Chen, M. C., Sun, Y. S., Wible, D., & Kuo, C.-H. (2010). Extending the TAM model to explore the factors that affect intention to use an online learning community. *Computers & Education, 54*(2), 600–610. doi:10.1016/j.compedu.2009.09.009

Mailizar, M., Burg, D., & Maulina, S. (2021). Examining university students' behavioural intention to use e-learning during the COVID-19 pandemic: An extended TAM model. *Education and Information Technologies, 26*(6), 7057–7077. doi:10.100710639-021-10557-5 PMID:33935579

Md Yunus, M., Ang, W. S., & Hashim, H. (2021). Factors affecting teaching English as a Second Language (TESL) postgraduate students' behavioural intention for online learning during the COVID-19 pandemic. *Sustainability, 13*(6), 3524. doi:10.3390u13063524

Mukhtar, K., Javed, K., Arooj, M., & Sethi, A. (2020). Advantages, limitations and recommendations for online learning during COVID-19 pandemic Era. *Pakistan Journal of Medical Sciences, 36*(COVID19-S4). doi:10.12669/pjms.36. COVID19-S4.2785 PMID:32582310

Mustafa, M. B. (2015). One size does not fit all: Students' perceptions about Edmodo at Al Ain University of Science & Technology. *Journal of Studies in Social Sciences, 13*(2), 135–160. https://core.ac.uk/download/pdf/229607465.pdf

Nartiningrum, N., & Nugroho, A. (2020). Online learning amidst global pandemic: EFL students' challenges, suggestions, and needed materials. *ENGLISH FRANCA: Academic Journal of English Language and Education, 4*(2), 115–140. doi:10.29240/ ef.v4i2.1494

Ngo, D. H. (2021). Perceptions of EFL tertiary students towards the correlation between e-learning and learning engagement during the COVID-19 pandemic. *International Journal of TESOL & Education, 1*(3). https://eoi.citefactor. org/10.11250/ijte.01.03.013

Okita, S. Y. (2012). Social interactions and learning. In Seel N. M. (Ed.), Encyclopedia of the Sciences of Learning, 182-211. Springer. doi:10.1007/978-1-4419-1428-6_1770

Pham, M. T., Luu, T. T. U., Mai, T. H. U., Thai, T. T. T., & Ngo, T. C. T. (2022). EFL students' challenges of online courses at Van Lang University during the COVID-19 pandemic. *International Journal of TESOL & Education, 2*(2), 1–26. doi:10.54855/ijte.22221

Singh, V., & Thurman, A. (2019). How many ways can we define online learning? A Systematic literature review of definitions of online learning (1988-2018). *American Journal of Distance Education, 33*(4), 289–306. doi:10.1080/08923647 .2019.1663082

Tran, T. P., & Nguyen, T. T. A. (2022). Online education at Saigon University during the COVID-19 pandemic: A survey on non-English major college students' attitudes towards learning English. *AsiaCALL Online Journal, 13*(2), 1–20. doi:10.54855/ acoj.221321

Zhang, P. (2021). Understanding digital learning behaviors: Moderating roles of goal setting behavior and social pressure in large-scale open online courses. *Frontiers in Psychology, 12*, 783610. doi:10.3389/fpsyg.2021.783610 PMID:34899535

Zia, A. (2020). Exploring factors influencing online classes due to social distancing in COVID-19 pandemic: A business students perspective. *International Journal of Information and Learning Technology, 37*(4), 197–211. doi:10.1108/ IJILT-05-2020-0089

ADDITIONAL READINGS

Amri, Z., & Alasmari, N. (2021). Self-efficacy of Saudi English majors after the emergent transition to online learning and online assessment during the COVID-19 pandemic. *International Journal of Higher Education, 10*(3), 127. doi:10.5430/ ijhe.v10n3p127

Baber, H. (2020). Determinants of students' perceived learning outcome and satisfaction in online learning during the pandemic of COVID-19. *Journal of Education and E-Learning Research, 7*(3), 285–292. doi:10.20448/journal.509.2020.73.285.292

Efriana, L. (2021). Problems of Online Learning during Covid-19 Pandemic in EFL Classroom and the Solution. *Journal of English Languange Teaching and Literature, 2*(1). https://jurnal.stkipmb.ac.id/index.php/jelita/article/view/7 4

Hadeel, A. S., & Ahmad, S. H. (2021). The use of YouTube in developing the speaking skills of Jordanian EFL university students. *Heliyon*, *7*(7), e07543. doi:10.1016/j. heliyon.2021.e07543 PMID:34307951

Okyar, H. (2022). University-level EFL students' views on learning English online: A qualitative study. *Education and Information Technologies*. doi:10.100710639-022-11155-9 PMID:35756361

Pham, N. T. (2020). Factors influencing interaction in an online English course in Vietnam. *VNU Journal of Foreign Studies*, *36*(3), 149–163.

Ravindran, L., Ikhram, R., & Bee, E. W. (2022). The impact of social media on the teaching and learning of EFL speaking skills during the COVID-19 pandemic. *Proceedings*, *82*(1), 38. doi:10.3390/proceedings2022082038

Tran, Q. H., & Nguyen, T. M. (2021). Determinants in student satisfaction with online learning: A survey study of second-year students at private universities in HCMC. *International Journal of TESOL & Education*, *2*(1), 63–80. doi:10.54855/ijte22215

Truong, N. K. V., & Le, Q. T. (2022). Utilizing YouTube to enhance English speaking skill: EFL tertiary students' practices and perceptions. *AsiaCALL Online Journal*, *13*(4), 7–31. doi:10.54855/acoj.221342

Xu, Q., Wu, J., & Peng, H. (2022). Chinese EFL university students' self-efficacy for online self-regulated learning: Dynamic features and influencing factors. *Frontiers in Psychology*, *13*, 912970. Advance online publication. doi:10.3389/fpsyg.2022.912970 PMID:35874382

KEY TERMS AND DEFINITIONS

Asynchronous: (of learning or teaching) Involving students working separately at different times, for example using recorded lessons or the Internet, rather than involving students and teacher taking part in a lesson at the same time.

Benefit: A helpful or good effect, or something intended to help.

Challenge: (the situation of or being faced with) Something that needs great mental or physical effort in order to be done successfully and therefore tests a person's ability.

EFL: English as a Foreign Language: the teaching of English to students whose first language is not English.

Online Education: This is defined as education being delivered in an online environment hrough the use of the Internet for teaching and learning. This includes online learning on the part of the students that is not dependent on their physical or virtual co-location. The teaching content is delivered online, and the instructors develop teaching modules that enhance learning and interactivity in the synchronous or asynchronous environment.

Perception: A belief or opinion is often held by many people and based on how things seem.

Practice: Something is usually or regularly done, often as a habit, tradition, or custom.

Quality: How good or bad something is.

Synchronous: (of learning or teaching) hHappening with a teacher and group of students who are all taking part in a lesson at the same time, either physically together in one place or using the Internet.

Technology: (the study and knowledge of) The practical, especially industrial, use of scientific discoveries.

Chapter 7
Facebook–Based Language Learning:
Vietnamese University EFL Students' Attitudes and Practices

Tham My Duong
Ho Chi Minh City University of Economics and Finance, Vietnam

Thao Quoc Tran
iD https://orcid.org/0000-0001-8063-8853
HUTECH University, Vietnam

ABSTRACT

The emergence of Facebook has benefited language educators and learners in different ESL/EFL contexts as Facebook can function as a learning management system (LMS), facilitating the English language teaching and learning (ELTL) process. Nonetheless, research on the Facebook-based language learning (FBLL) activities is scarce. This book chapter presents a study delving into tertiary EFL English majors' attitudes toward FBLL activities, their FBLL strategy use, and the correlation between the two variables mentioned above at an institution of higher education in Vietnam. This study adopted the postpositivist perspective for quantitative data collection from a cohort of 126 English majors answering closed-ended questionnaires. The SPSS software processed the data in terms of descriptive and inferential statistics. The findings revealed that tertiary English majors showed positive attitudes towards FBLL activities and employed FBLL strategies at a high level. Furthermore, the English majors' attitudes towards FBLL activities did not affect how they utilized their FBLL strategies. This book chapter suggests some practical pedagogical implications for both teachers and students aiming to leverage the quality of ELTL concerning the use of FBLL activities.

DOI: 10.4018/978-1-6684-7034-3.ch007

INTRODUCTION

Facebook, a new form of computer-mediated communication (CMC), has emerged and gained immense popularity (e.g., Boyd & Ellison, 2007; Nham & Nguyen, 2013). With its exclusive features, Facebook has proved to be one of the most powerful social networking sites (SNS) in the field of education as it embraces the interactivity of the Internet by allowing educators and learners to use various features for teaching and learning communication (e.g., Bosch, 2009; Kalelioğlu, 2017; Munoz & Towner, 2009). Accordingly, Facebook has been seen to have positive effects on general education and English language education. Furthermore, Facebook-based language learning (FBLL) activities have been applied in ESL/EFL teaching and learning in various contexts. In the context of Vietnam, Facebook has emerged as one of the educational tools which can help teachers and learners in both general education and foreign language acquisition (e.g., Le, 2018; Tran & Ngo, 2020). More and more teachers and learners tend to use Facebook and adapt it for educational purposes as it is free and contains different features of a learning management system (LMS). They can make use of educational features embedded in Facebook (e.g., posting and sharing the teaching and learning materials, discussing, watching video clips, and giving comments) to boost the teaching and learning quality.

It is seen that the study on the employment of Facebook as a language teaching and learning platform has attracted different researchers' attention (e.g., Camus et al., 2016; Godwin-Jones, 2008; Kalelioğlu, 2017; Le, 2018; Tran & Ngo, 2020; Ulla & Perales, 2021). Camus et al. (2016), for example, examined the effects of the Facebook based online discussion on student engagement and learning achievement. Another focus is EFL students' attitude towards Facebook-based activities for English language learning conducted by Tran and Ngo (2020) in the Vietnamese context. Recently, Ulla and Perales (2021) studied students' perspectives and experiences in using Facebook as a tool for online learning during the period of the COVID-19 pandemic in the Thai EFL context. Nevertheless, it is likely that there has been a scarcity of research on FBLL activities.

This current study highlights the application of Facebook as a learning platform to enhance EFL students' language proficiency, and it aims to scrutinize the English majors' attitudes towards FBLL activities, their employment of FBLL strategies, and their relationship between the attitudes towards FBLL activities and the use of FBLL strategies at an institution of higher education in Vietnam. This study endeavors to answer the research questions as follows:

1.   What are tertiary English majors' attitudes towards FBLL activities?
2.   To what extent do tertiary English majors employ FBLL strategies?

3.   Is there any significant correlation between tertiary English majors' attitudes towards FBLL activities and their use of FBLL strategies? If yes, what?

This study highlights some significant contributions. It is theoretically expected to contribute its part to the body of literature in terms of the use of FBLL activities, more specifically the attitudes towards FBLL activities, employment of FBLL activities, and relationship between attitudes towards FBLL activities and employment of FBLL activities. Practically, this study is hoped to provide the stakeholders (e.g., teachers, students, and administrators) with some pedagogical implications on the application of FBLL activities in an attempt to leverage the quality of English language teaching and learning (ELTL). Furthermore, the study's findings can be hopefully considered as references for policymakers and faculty members in similar EFL contexts to deploy the Facebook application in ELTL.

LITERATURE REVIEW

Facebook as a Learning Platform

Facebook, an online SNS, functions as a learning platform (Wang et al., 2012) providing teachers and students with various tools for teaching and learning communication (e.g., Bosch, 2009; Kalelioğlu, 2017; Munoz & Towner 2009). It can provide a closed group in which learning materials (e.g., lecture notes, presentations, articles, audio-visual materials, exercises, and website links) can be uploaded (Kalelioğlu, 2017), and Facebook group members can create and delete content. Besides, Facebook can provide asynchronous and synchronous communication (e.g., discussion, chat, message, conference call) to support the teaching and learning process, and a learning platform to work collaboratively (Kabilan et al., 2010).

Facebook can allow users to share materials and resources, discuss and work one-to-one and in groups using an instant messaging system and group conferencing. It can create different academic activities to develop learners' language skills. Different materials as well as exercises/assignments can be posted to enhance learners' vocabulary, grammar, and pronunciation as well as language skills. Learners can read reading materials and answer reading comprehension exercises to develop their vocabulary, grammar, and reading skill. They can start online voice chats with their classmates, teachers, foreign friends so that their speaking skill can be improved. Audio-visual materials can be uploaded to Facebook so that learners can listen and watch to improve their listening skills, and they can get involved in writing activities (e.g., writing comments, writing journals) (TeachThought, n.d.). Therefore, Facebook can serve as an alternative to LMS since it can feature pedagogical, social, and technological functions (Wang et al., 2012).

Attitudes Towards Facebook-Based Language Learning Activities

The term attitude can be understood as "a conceptual concept used to describe human behavior's course and persistence" (Baker, 1992, p. 10). It has three interrelated components, namely cognitive, affective and behavioral components (Wenden, 1991). The cognitive aspect involves mental activities and the brain (Isti & Istikharoh, 2019) in indicating knowledge and expectations (Schiffman & Kanuk, 2004). The affective component is relevant to the thoughts and emotions of a person towards objects of disposition (e.g., likes or dislikes, with or against) (Eagly & Chaiken, 1998). The behavioral component indicates an individual's tendencies, behaviors, or reactions to respond or behave towards a particular object (Jain, 2014). Within the scope of this study, the cognitive component refers to the perceptions and conceptions (belief or disbelief) about FBLL activities; the affective component focuses on the emotional response (likes or dislikes) towards FBLL activities; and the behavioral component involves actions or observable responses towards FBLL activities.

Research has shown that FBLL activities have great impacts on students' attitudes towards language learning (e.g., Bosch, 2009; Doğan & Gulbahar, 2018; Faryadi, 2017; Polok & Harężak, 2018; Ríos & Campos, 2015; Tran & Ngo, 2020; Wang & Chen, 2013). Facebook can be employed as an online learning tool not only to shape students' attitudes towards English language learning but also to help, improve and/ or reinforce students' English language learning (Kabilan et al., 2010). Students' attitudes towards FBLL activities can be positive or negative; however, their attitudes can have impacts on their language learning strategies (LLS) in general (e.g., Duong et al., 2021; Tran & Duong, 2021; Tran & Tran, 2020) and FBLL strategies in specific. Students with positive attitudes towards language learning are found to become more successful in developing their language proficiency than those with negative attitudes (Tran & Ngo, 2020).

Facebook-Based Language Learning Strategies

FBLL strategies can be derived from LLS which have been variously defined. Oxford (2011) delineates LLS as "the learner's goal-directed actions for improving language proficiency or achievement, completing a task, or making learning more efficient, more effective, and easier" (p. 167). Meanwhile, Griffiths (2013) describes LLS as activities that learners employ intentionally and regulate for the appropriateness of their language learning process. Within the scope of this study, FBLL strategies can be understood as LLS consciously used by learners to accomplish learning tasks when they are engaged in FBLL activities.

LLS have been differently classified. Rubin (1987) groups LLS into three main types: learning, communication, and social strategies. Similarly, O'Malley and Chamot (1990) classify LLS into three major groups: cognitive, metacognitive, and social/affective strategies. Meanwhile, Oxford (1990) divides LLS into six groups: memory, cognitive, compensation, metacognitive, affective, and social strategies. Nevertheless, Dörnyei (2005) groups LLS into five types: commitment, metacognitive, satiation, emotion, and environmental control strategies. Based on the purpose of this study and Oxford's (1990) strategy classification, FBLL strategies can be grouped into six types, namely Cognitive, metacognitive, memory, compensation, affective, and social strategies. (1) cognitive strategies are strategies for improving language skills; (2) metacognitive strategies are strategies for monitoring the usage of FBLL activities to improve language skills; (3) memory strategies are strategies for understanding and remembering the information in doing FBLL activities; (4) compensation strategies are strategies for compensating for their limitations in doing FBLL activities; (5) affective strategies are strategies for coping with the emotions in doing FBLL activities; (6) social strategies are strategies for seeking assistance from others in doing FBLL activities.

RESEARCH METHODOLOGY

Research Context and Participants

This study, which was designed based on the quantitative approach, adopted the postpositivist perspective to garner the data through the structured (closed-ended) questionnaire to delve into the participants' attitudes towards FBLL activities and their employment of FBLL strategies (Crowther & Lancaster, 2008). It was carried out at a multidisciplinary institution of higher education in Ho Chi Minh City, Vietnam. Among the different training programs offered by this institution, the English language training program is one of the accredited training programs taken by students. English majors have to learn different courses ranging from English language skills to English language theories within the first three years. They have to choose one of the specialized English courses (e.g., Business English, English for translation and interpretation, or English language teaching methodology) in the fourth year.

For the purpose of this study, the course of English Reading 2 lasting eight weeks was designed as a blended course, i.e., students had to attend the class in person, and they had to do extra exercises online. A Facebook group was created for students who were taking this course. Different materials (e.g., reading articles, video clips) were posted in this Facebook group twice a week, followed by exercises and discussions. Students were required to do exercises individually and in groups alternatively. During the course, the students were required to make a video clip in groups of four (the clip is the role play reflecting the content of the chosen reading topic that students had learned. The guided questions for each reading topic were prepared to ensure that students' video clips could reflect the learned reading topic) and post it in the Facebook group. Other groups then had to give comments on the posted clips. Besides, the students were encouraged to post questions or share relevant materials in the Facebook group. For the evaluation, the students had to do progress and achievement tests in class.

A cohort of 126 second-year English majors taking the course of English Reading 2 were conveniently sampled. Of 126 students, most of them (56%) self-reported that their level of English language proficiency was pre-intermediate, 37% of them believed that their English level was intermediate, and a small proportion (7%) of them thought that they had an advanced level of English language proficiency. Additionally, 53 (42.1%) out of 126 students allocated less than one hour daily doing FBLL activities, while 64 (50.8%) students spared more than three hours daily to do FBLL activities. Only 9 (7.1%) students spent from one to three hours daily doing FBLL activities. Among the research participants, only 21 (16.7%) reported that they were taking English extra-courses at English language centers.

Research Instruments

This study employed the questionnaire to garner the data. This closed-ended questionnaire includes two main sections: Section I collects participants' background information; Section II is the main questionnaire content addressing English majors' attitudes towards FBLL activities and their FBLL strategy use. The content for attitudes towards FBLL activities, which was adapted from Tran and Ngo's (2020) study, consists of 21 five-point scaled items (from *Strongly disagree* to *Strongly agree*) divided into three sub-parts: Cognitive attitudes (6 items), Affective attitudes (9 items) and Behavioral attitudes (6 items). The content for FBLL strategy use, which was designed based on the theoretical framework, includes 30 five-point scaled items (from *Always* to *Never*) split into six categories: Cognitive strategies (7 items), Metacognitive strategies (7 items), Memory strategies (4 items), Compensation strategies (3 items), Affective strategies (5 items), and Social strategies (4 items). To ensure the questionnaire's internal reliability, the confirmatory factor analysis was carried out for the questionnaire items of attitudes towards FBLL activities, while the exploratory factor analysis was to explore the questionnaire items of FBLL strategy use (Pallant, 2016). Additionally, the questionnaire in English was first designed, and then converted into Vietnamese for the purpose of assisting participants to answer the questionnaire without any language barriers. The Cronbach's alpha of attitudes towards FBLL activities and FBLL strategies use was .82 and .87 respectively, which means that the questionnaire's reliability was high.

Data Collection and Analysis Procedures

A pilot study had been conducted with a group of ten students who shared similar background information with the target research participants in the official data collection After the modification, 150 copies of the questionnaire were officially administered to English majors taking the course of English Reading 2; however, 126 copies of the questionnaire with valid answers were received. Regarding data analysis, the direct approach was adapted for quantitative data analysis (Nykiel, 2007), using the SPSS software (version 23) in terms of mean (M), standard deviation (SD), and Pearson's correlation coefficient (r). The interval scale was interpreted as 1.00-1.80 for *Strongly disagree/Always*; 1.81-2.6 for *Disagree/Seldom*; 2.61-3.40 for *Neutral/Sometimes*; 3.41-4.20 for *Agree/Often*; 4.21-5.00 for *Strongly agree/Never* (Kan, 2009). The intra-rating method was conducted for data analysis double-checks. i.e., the random data analysis was re-analyzed by the researchers one month after the first data analysis.

RESULTS AND DISCUSSION

Results

Tertiary English Majors' Attitudes Towards FBLL Activities

Table 1 reveals the results for English majors' attitudes towards FBLL activities. It is seen that the overall mean score was 3.48 (SD=.35) out of 5.00. Specifically, the mean score of affective attitudes was 3.49 (SD=.40), followed by that of behavioral attitudes (M=3.48; SD=.56) and that of cognitive attitudes (M= 3.46; SD=.45). This can be understood that English majors had positive attitudes towards FBLL activities. They were aware of the benefits of FBLL activities in improving their English language proficiency and had good affection for FBLL activities. Additionally, they were likely to take part in FBLL activities to foster their English language proficiency.

Table 1. Tertiary English majors' attitudes towards FBLL activities

	N=126	
	M	**SD**
Cognitive attitudes	3.46	.45
Affective attitudes	3.49	.40
Behavioral attitudes	3.48	.56
Average	**3.48**	**.35**

Cognitive Attitudes Towards FBLL Activities

As seen in Table 2, English majors believed in the potentials of FBLL activities in boosting their English language proficiency. Regarding micro language skills, they supposed that "[t]aking part in FBLL activities (e.g., reading materials, watching video clips) [could] give [them] chances to learn more vocabulary" (item A1: M=3.47; SD=.73), and "[t]aking part in FBLL activities (e.g., reading materials, watching video clips, giving comments, writing posts, sending messages) [could] help [them] to improve [their] grammar" (item A2: M=3.48; SD=.65). With respect to macro language skills, English majors reckoned that they could improve their reading skills by "reading materials posted in the Facebook group" (item A3: M=3.49; SD=.84), their listening skills by "watching video clips posted in the Facebook group" (item 4: M=3.48; SD=.61), their speaking skill by "making a video clip with [their] group members" (item A5: M=3.40; SD=.76), and their writing skill by "giving comments or writing posts in the Facebook group" (item A6: M=3.51; SD=.83.).

Affective Attitudes Towards FBLL Activities

With respect to the affective attitudes towards FBLL activities, the results in Table 3 show that English majors had good affection in FBLL activities. During the course of Reading using FBLL activities, they found it "interesting to take part in FBLL activities (e.g., reading materials, watching video clips, giving comments, writing posts, sending messages)" (item A7: M=3.46; SD=.70), "enjoyable to take part in FBLL activities (e.g., reading materials, watching video clips, giving comments, writing posts, sending messages)" (item A8: M=3.52; SD=.81), and "necessary to take part in FBLL activities (e.g., reading materials, watching video clips, giving comments, writing posts, sending messages)" (item A9: M=3.55; SD=.83). Additionally, they believed that they felt confident about their "vocabulary when taking part in FBLL activities (e.g., reading materials, watching video clips)" (item A10: M=3.44; SD=.75) and "grammar when taking part in FBLL activities (e.g.,

Table 2. Tertiary English majors' cognitive attitudes towards FBLL activities

	N=126	
	M	**SD**
A1. Taking part in FBLL activities (e.g., reading materials, watching video clips) can give me chances to learn more vocabulary.	3.47	.72
A2. Taking part in FBLL activities (e.g., reading materials, watching video clips, giving comments, writing posts, sending messages) can help me to improve my grammar.	3.48	.65
A3. Reading materials posted in the Facebook group can help me to improve my reading skill.	3.49	.84
A4. Watching video clips posted in the Facebook group can help me improve my listening skills.	3.48	.61
A5. Making a video clip with my group members can help me improve my speaking skill.	3.40	.76
A6. Giving comments or writing posts in the Facebook group can provide me more chances to enhance my writing skills.	3.51	.83

reading materials, watching video clips, giving comments, writing posts, sending messages)" (item A11: M=3.63; SD=.79). As for their macro language skills, they also had confidence in their "reading skill when reading materials posted in the Facebook group" (item A12: M=3.53; SD=.75), "listening skill when watching video clips posted in the Facebook group" (item A13: M=3.45; SD=.72), "speaking skill when making a video clip in groups" (item A14: M=3.47; SD=.76), and "writing skill when giving comments or writing posts in the Facebook group" (item A15: M=3.43; SD=.70).

Behavioral Attitudes Towards FBLL Activities

Regarding the behavioral attitudes towards FBLL activities (see Table 4), English majors reported that they tried to allocate more time to different FBLL activities during the course of English Reading 2. In order to improve their vocabulary and grammar, they took part in different FBLL activities such as reading materials, watching video clips "to improve [their] vocabulary" (item A16: M=3.40; SD=.79) and reading materials, watching video clips, giving comments, writing posts, sending messages "to improve [their] grammar" (item A17: M=3.49; SD=.74). In an attempt to enhance their major language skills, they spared more time to "[give] comments or [write] posts in the Facebook group to improve [their] speaking skill" (item A21: M=3.72; SD=.84), [watch] video clips posted in the Facebook group to improve [their] listening skill' (item A19: M=3.46; SD=.78), "[read] materials posted in the Facebook group to improve [their] reading skill" (item A18: M=3.44; SD=.87),

Table 3. Tertiary English majors' affective attitudes towards FBLL activities

During the course of Reading, …	N=126	
	M	**SD**
A7. I find it interesting to take part in FBLL activities (e.g., reading materials, watching video clips, giving comments, writing posts, sending messages).	3.46	.70
A8. I find it enjoyable to take part in FBLL activities (e.g., reading materials, watching video clips, giving comments, writing posts, sending messages).	3.52	.81
A9. I find it necessary to take part in FBLL activities (e.g., reading materials, watching video clips, giving comments, writing posts, sending messages).	3.55	.83
A10. I feel confident about my vocabulary when taking part in FBLL activities (e.g., reading materials, watching video clips).	3.44	.75
A11. I feel confident about my grammar when taking part in FBLL activities (e.g., reading materials, watching video clips, giving comments, writing posts, sending messages).	3.63	.79
A12. I feel confident about my reading skill when reading materials posted in the Facebook group.	3.53	.75
A13. I feel confident about my listening skill when watching video clips posted in the Facebook group.	3.45	.72
A14. I feel confident about my speaking skill when making a video clip in groups.	3.47	.76
A15. I feel confident about my writing skill when giving comments or writing posts in the Facebook group.	3.43	.70

and "[rehearse] with [their] group members before making a video clip to improve [their] speaking skill" (item A20: M=.3.42; SD=.80).

Tertiary English Majors' Employment of FBLL Strategies

The findings in Table 5 reveal that the total mean score of the English majors' employment of FBLL strategies was 3.95 (SD=.45) out of 5.00. Regarding the different groups of FBLL strategies, English majors utilized the compensation strategies the most (M=4.05; SD=.53), followed by cognitive strategies (M=4.03; SD=.75). Additionally, the mean scores of affective strategies, metacognitive strategies, and memory strategies were 3.97 (SD=.68), 3.94 (SD=.84) and 3.90 (SD=.41), respectively. The least employed group of FBLL strategies was social strategies (M=3.74; SD=.71). Such findings imply that English majors often employed FBLL strategies. They tended to use compensation and cognitive strategies more than other groups of FBLL strategies in fulfilling their FBLL activities.

Table 4. Tertiary English majors' behavioral attitudes towards FBLL activities

	N=126	
	M	**SD**
A16. I try to spend more time taking part in different FBLL activities (e.g., reading materials, watching video clips) to improve my vocabulary.	3.40	.79
A17. I try to spend more time taking part in different FBLL activities (e.g., reading materials, watching video clips, giving comments, writing posts, sending messages) to improve my grammar.	3.49	.74
A18. I try to spend more time reading materials posted in the Facebook group to improve my reading skill.	3.44	.87
A19. I try to spend more time watching video clips posted in the Facebook group to improve my listening skill.	3.46	.78
A20. I try to spend more time rehearsing with my group members before making a video clip to improve my speaking skill.	3.42	.80
A21. I try to spend more time giving comments or writing posts in the Facebook group to improve my writing skill.	3.72	.84

Table 5. Tertiary English majors' employment of FBLL strategies

	N=126	
	M	**SD**
Cognitive strategies	4.03	.75
Metacognitive strategies	3.94	.84
Memory strategies	3.90	.41
Compensation strategies	4.05	.53
Affective strategies	3.97	.68
Social strategies	3.74	.71
Average	**3.95**	**.45**

FBLL Cognitive Strategies

In respect of the FBLL cognitive strategies in Table 6, the English majors always "[read] the reading materials posted in the Facebook group" (item B1: M=4.36; SD=.62) and "[chatted] with [their] friends in the Facebook group" (item B7: M=4.22; SD=.81). Nonetheless, they often "[discussed] the requirement of FBLL activities orally with [their] friends in the Facebook group" (item B6: M=4.14; SD=.67) and "[gave] comments to the posts in the Facebook group" (item B4; M=4.11; SD=.80). What is more, they often "[watched] the video clips" (item B2; M=3.88; SD=.77),

Table 6. Tertiary English majors' employment of FBLL cognitive strategies

	N=126	
	M	**SD**
B1. I read the reading materials posted in the Facebook group.	4.36	.62
B2. I watch the video clips posted in the Facebook group.	3.88	.77
B3. I read comments posted in the Facebook group.	3.83	.88
B4. I give comments to the posts in the Facebook group.	4.11	.80
B5. I do the reading exercises posted in the Facebook group.	3.71	.79
B6. I discuss the requirement of FBLL activities orally with my friends in the Facebook group.	4.14	.67
B7. I chat with my friends in the Facebook group.	4.22	.81

"[read] comments" (item B3; M=3.83; SD=.88), and "[did] the reading exercises posted in the Facebook group" (item B5; M=3.71; SD=.79).

FBLL Metacognitive Strategies

With regard to the FBLL metacognitive strategies, Table 7 reveals that the English majors always "[set] clear goals for doing FBLL activities" (item B9: M=4.25; SD=.78). They often "[set] time for different FBLL activities" (item B11: M=4.17; SD=.69), "[paid] attention to the requirements of FBLL activities" (item B8: M=4.08; SD=.64), and "[set] a clear schedule for doing FBLL activities" (item B10: M=4.03; SD=.63). Besides, they also often "[took] notes of the pointed-out mistakes in the Facebook group so that [they could] avoid them" (item B14: M=3.90; SD=.65), "[tried] to find different ways to get engaged in FBLL activities" (item B12: M=3.63; SD=.59) and "[tried] to improve [their] language skills thanks to FBLL activities" (item B13: M=3.53; SD=.60).

FBLL Memory Strategies

Regarding the FBLL memory strategies as presented in Table 8, English majors tried to review what they had learned by "[rewatching] the video clips" (item B16: M=4.06; SD=.74), "[re-doing] the reading exercises" (item B17: M=4.02; SD=.64), "[taking] notes of the comments" (item B18: M=3.87; SD=.71) and "[re-reading] the reading materials" (item B15: M=3.84; SD=.68).

Table 7. Tertiary English majors' employment of FBLL metacognitive strategies

	N=126	
	M	**SD**
B8. I pay attention to the requirements of FBLL activities.	4.08	.64
B9. I set clear goals for doing FBLL activities.	4.25	.78
B10. I set a clear schedule for doing FBLL activities.	4.03	.63
B11. I set time for different FBLL activities.	4.17	.69
B12. I try to find different ways to get engaged in FBLL activities.	3.63	.59
B13. I try to improve my language skills thanks to FBLL activities.	3.53	.60
B14. I take notes of the pointed-out mistakes in the Facebook group so that I can avoid them.	3.90	.65

Table 8. Tertiary English majors' employment of FBLL memory strategies

	N=126	
	M	**SD**
B15. I re-read the reading materials posted in the Facebook group to review what I should remember.	3.84	.68
B16. I re-watch the video clips posted in the Facebook group to review what I have learned.	4.06	.74
B17. I re-do the reading exercises posted in the Facebook group to remember the learned reading skills.	4.02	.64
B18. I take notes of the comments posted in the Facebook group to remember what I have learned.	3.87	.71

FBLL Compensation Strategies

The results of FBLL compensation strategies in Table 9 unravel that when English majors did not understand the requirements of FBLL activities, they often "[tried] to figure out what to do" (item B20: M=4.13; SD=.71), "[posted] questions in the Facebook group" (item B19: M=4.08; SD=.76), and "[re-checked] materials posted in the Facebook group" (item B21: M=3.97; SD=.81).

FBLL Affective Strategies

As regards FBLL affective strategies in Table 10, the English majors reported that they often "[encouraged themselves] to get engaged in FBLL activities" (item B23:

Table 9. Tertiary English majors' employment of FBLL compensation strategies

	N=126	
	M	**SD**
B19. I post questions in the Facebook group when I do not understand the requirements of FBLL activities.	4.08	.76
B20. I try to figure out what to do when I do not understand the requirements of FBLL activities.	4.13	.71
B21. I re-check materials posted in the Facebook group when I do not understand the requirements of FBLL activities.	3.97	.81

M=4.11; SD=.68), "[relaxed] when [they were] overloaded with the FBLL activities" (item B22: M=4.07; SD=.59), "[rewarded themselves] when [they did] the FBLL activities well (item B24: M=3.96; SD=.51), and "[talked] to [their] friends how [they felt] about the FBLL activities" (item B25: M=3.94; SD=.76).

Table 10. Tertiary English majors' employment of FBLL affective strategies

	N=126	
	M	**SD**
B22. I relax when I am overloaded with the FBLL activities.	4.07	.59
B23. I encourage myself to get engaged in FBLL activities.	4.11	.68
B24. I reward myself when I do the FBLL activities well.	3.96	.51
B25. I talk to my friends how I feel about the FBLL activities.	3.94	.76
B26. I try to calm myself down when I am confused with the FBLL activities.	3.80	.70

FBLL Social Strategies

In respect of FBLL social strategies shown in Table 11, the English majors often "[asked their] friends to clarify the requirements of FBLL activities" (item B27: M=4.09; SD=.65) and "[consulted their] lecturer for assistance" (item B30: M=3.80; SD=.65) when they could not figure out requirements of FBLL activities. In addition, they also often "[asked their] friends to check what [they intended] to post in the Facebook group" (item B28: M=3.60; SD=.59) and "[discussed] with [their] friends what [they intended] to post in the Facebook group" (item B29: M=3.49; SD=.50).

Table 11. Tertiary English majors' employment of FBLL social strategies

	N=126	
	M	**SD**
B27. I ask my friends to clarify the requirements of FBLL activities when I cannot figure out them.	4.09	.65
B28. I ask my friends to check what I intend to post in the Facebook group.	3.60	.59
B29. I discuss with my friends what I should post in the Facebook group.	3.49	.50
B30. I consult my lecturer for assistance when I cannot figure the requirements of FBLL activities.	3.80	.65

The Correlation Between Tertiary English Majors' Attitudes Towards FBLL Activities and Their Employment of FBLL Strategies

Table 12 depicts that no significant correlation between the English majors' attitudes towards FBLL activities and their employment of FBLL strategies ($r=-.011; p=.900$) was observed. This means that the English majors' attitudes towards FBLL activities did not affect the ways they employed FBLL strategies.

Table 12. The correlation between tertiary English majors' attitudes towards FBLL activities and their employment of FBLL strategies

		Attitudes towards FBLL activities
	Pearson's *r*	-.011
Employment of FBLL strategies	Sig. (2-tailed)	.900
	N	126

Nevertheless, as found in Table 13, only tertiary English majors' affective attitudes towards FBLL activities were found to be negatively correlated with their employment of FBLL metacognitive strategies ($r=-.201; p=.024$). Meanwhile, other components of attitudes towards FBLL activities and categories of FBLL strategies were not significantly correlated. That is, the higher English majors' affective attitudes towards FBLL activities were, the less frequently they employed FBLL metacognitive strategies.

Table 13. The correlation between components of attitude towards FBLL activities and categories of FBLL strategies

		Cognitive attitudes	Affective attitudes	Behavioral attitudes
FBLL cognitive strategies	Pearson's *r*	.077	-.038	.094
	Sig. (2-tailed)	.389	.674	.293
	N	126	126	126
FBLL metacognitive strategies	Pearson's *r*	-.018	**-.201***	.005
	Sig. (2-tailed)	.840	.024	.953
	N	126	126	126
FBLL memory strategies	Pearson's *r*	.104	.018	-.077
	Sig. (2-tailed)	.247	.840	.391
	N	126	126	126
FBLL compensation strategies	Pearson's *r*	.020	-.150	-.019
	Sig. (2-tailed)	.824	.095	.836
	N	126	126	126
FBLL affective strategies	Pearson's *r*	-.074	-.044	.045
	Sig. (2-tailed)	.413	.622	.620
	N	126	126	126
FBLL social strategies	Pearson's *r*	.157	-.037	.106
	Sig. (2-tailed)	.079	.685	.235
	N	126	126	126

DISCUSSION

This present study which aimed to find out tertiary English majors' attitudes towards FBLL activities, their FBLL strategy employment, and the correlation between the two mentioned components has unraveled some major findings. First, it was found out that the tertiary English majors in this research context had positive attitudes towards FBLL activities. They were aware of the benefits of FBLL activities in enhancing their English language knowledge and skills, had beliefs in FBLL activities which could create interest, joy, and confidence for students to get involved in English language learning, and had intentions to get engaged in FBLL activities for their English language improvement. This could be due to one of the facts that Facebook activities are common among young people who could exploit the Facebook activities for a wide range of purposes, which could be attributed to shaping students' attitudes towards FBLL activities. Although FBLL activities were very new to students in this study, they had good attitudes towards the application of Facebook for language learning and teaching activities. This finding further confirms that FBLL activities could be beneficial to learners in their English language learning and empower them to improve their English language proficiency (e.g., Bosch, 2009; Kalelioğlu, 2017). This finding was partially corroborated with that of the study carried out by Tran and Ngo (2020) who have found that students showed their positive attitudes towards FBLL activities, but the difference lies in one of the attitude components (behavioral attitudes) which was neutral. This dissimilarity could be because of the characteristics of the research participants in the two studies. The research participants in this current study were English majors, while those in Tran and Ngo (2020) were non-English majors. Furthermore, the FBLL activities could be relatively unfamiliar to students within this study, and they had to conform to the course regulations when they did FBLL activities; therefore, they may encounter some difficulties in fulfilling the FBLL activities. That could explain why the level of attitudes towards FBLL activities, albeit being positive, was not high.

The second main finding was that English majors tended to deploy their FBLL strategies at a high level. On the one hand, they seemed to employ FBLL compensation, cognitive, and affective strategies more than other types of FBLL ones. They employed FBLL compensation strategies to repair what they could not understand when doing FBLL activities, cognitive strategies to improve their language skills, and affective strategies to encourage their FBLL activity engagement. As found, English majors in this study always read the reading materials posted in the Facebook group, chatted with their friends in the Facebook group, and set clear goals for doing FBLL activities. Since the course was a combination of the onsite and online platform in which students had to learn face-to-face in class, and had extra materials and exercises in the Facebook group. On the other hand, although Facebook is one of the common social networking sites, English majors had a tendency to utilize the FBLL social strategies the least in comparison to other groups of FBLL strategies. It seemed that the students in this study taking the FBLL course of English Reading 2 were highly aware of what to do so as to accomplish the course. As a matter of fact, more than 50% of the research participants allocated more than three hours daily to do FBLL activities, and it was reported that their English language proficiency was sufficiently good enough to get engaged in FBLL activities. This may imply that Facebook could be a platform useful for ESL/EFL teaching and learning purposes if it was appropriately and meticulously designed. This finding is supported by scholars (e.g., Bosch, 2009; Godwin-Jones, 2008; Kalelioğlu, 2017; Manan et al., 2012; Ulla & Perales, 2021) who have confirmed that Facebook is one of the useful tools for improving ESL/EFL students' language skills as students can get engaged in various activities such as reading, writing, watching, chatting, learning grammar and vocabulary.

The last finding was that the English majors' attitudes towards FBLL activities did not affect their employment of FBLL strategies; nonetheless, English majors' employment of FBLL metacognitive strategies was negatively affected by their affective attitudes towards FBLL activities. As discussed in the first finding that the level of attitudes towards FBLL activities was not significantly high (M=3.48; SD=.35), the level of FBLL strategies used by the English majors was high (M=3.95 out of 5.00). This was because of the fact that students had to get engaged in FBLL activities as one of the course requirements. That is why they often employed different types of FBLL strategies to fulfil the tasks (e.g., reading materials, watching video clips, giving comments, writing posts). Additionally, Facebook is a social media that can function as a learning platform (e.g., Chugh & Ruhi, 2018; Ulla & Perales, 2021), so the English majors could make use of Facebook to accomplish their learning tasks regardless of their attitudes towards FBLL activities.

CONCLUSION

This study has confirmed that Facebook, one form of CMC, can be a useful learning tool for ELTL. Via Facebook engagement, tertiary English majors could shape their attitudes towards FBLL activities, and they could get actively involved in employing FBLL strategies at a high level. Nonetheless, the level of students' attitudes towards FBLL activities did not significantly affect the extent to which they employed the FBLL strategies except for the negative correlation between their behavioral attitudes towards FBLL activities and their employment of FBLL metacognitive strategies. It is, therefore, recommended that teachers can make use of the Facebook features for ELTL. Although the Facebook application within this study was the form of a blended course of English Reading 2, it has been proved that FBLL activities were influential in shaping students' attitudes towards English language learning and engaging them in using strategies to improve the quality of English language learning. It is advised that teachers should shed light on the benefits of FBLL activities in ELTL so that students will be able to adjust their attitudes towards FBLL activities; teachers should select appropriate Facebook features (e.g., posting and sharing materials, giving comments, discussion and so on) functioning as an LMS and choose extra materials relevant to students' interests and compatible with their English language skills. Furthermore, teachers should instruct students on how to utilize the strategies when using Facebook for English language learning. When students are able to understand the purposes of FBLL strategies and know how to use them effectively and appropriately, they can fulfill and achieve their learning goals effectively.

With respect to students, it is suggested that students should be proactively engaged in FBLL activities since Facebook can function as a bridge to connect students' academic learning activities with their daily life. Students should shift their language learning style to autonomous language learning so that they can accommodate themselves to the new learning environment. As for administrators, it is important to develop incentive policies to encourage teachers to embed the application of information and communication technologies (ICT) in ELTL. Besides, the design of ELTL curriculum should indicate the explicit use of technological tools (including Facebook) to facilitate the ELTL process.

REFERENCES

Baker, C. (1992). *Attitudes and language*. Multilingual Matters.

Bosch, T. E. (2009). Using online social networking for teaching and learning: Facebook use at the University of Cape Town. *Communication, 35*(2), 185–200. doi:10.1080/02500160903250648

Boyd, D. M., & Ellison, N. B. (2007). Social network sites: Definition, history, and scholarship. *Journal of Computer-Mediated Communication, 13*(1), 210–230. https://bit.ly/2uo0IAl. doi:10.1111/j.1083-6101.2007.00393.x

Camus, M., Hurt, N. E., Larson, L. R., & Prevost, L. (2016). Facebook as an online teaching tool: Effects on student participation, learning, and overall course performance. *College Teaching, 64*(2), 84–94. doi:10.1080/87567555.2015.1099093

Chugh, R., & Ruhi, U. (2018). Social media in higher education: A literature review of Facebook. *Education and Information Technologies, 23*(2), 605–616. doi:10.100710639-017-9621-2

Crowther, D., & Lancaster, G. (2008). *Research methods: A concise introduction to research in Management and Business consultancy*. Butterworth-Heinemann.

Doğan, D., & Gulbahar, Y. (2018). Using Facebook as social learning environment. *Informatics in Education, 17*(2), 207–228. doi:10.15388/infedu.2018.11

Dörnyei, Z. (2005). *The psychology of the language learner: Individual differences in second language acquisition*. Lawrence Erlbaum.

Duong, T. M., Tran, T. Q., & Nguyen, T. T. P. (2021). Non-English majored students' use of English vocabulary learning strategies with technology-enhanced language learning tools. [AJUE]. *Asian Journal of University Education, 17*(4), 455–463. doi:10.24191/ajue.v17i4.16252

Eagly, A., & Chaiken, S. (1998). *Attitude structure and function. Handbook of social psychology*. McGrow Company.

Faryadi, Q. (2017). Effectiveness of Facebook in English language learning: A case study. *Open Access Library Journal*, *4*(11), 1–11. doi:10.4236/oalib.1104017

Godwin-Jones, R. (2008). Mobile computing technologies: Lighter, faster, smarter. *Language Learning & Technology*, *12*(3), 3–9. 10125/44150

Griffiths, M. D. (2013). Social networking addiction: Emerging themes and issues. *Journal of Addiction Research & Therapy*, *4*(05), 1–2. doi:10.4172/2155-6105.1000e118

Isti, M., & Istikharoh, L. (2019). EFL students' attitude toward learning English. *Journal of Sains Social dan Humaniora, 3*(2), 95–105. doi:10.5539/ass.v8n2p119

Jain, V. (2014). 3D model of attitude. *International Journal of Advanced Research in Management and Social Sciences, 3*(3), 1–12. https://docplayer.net/22666307-3d- model-of-attitude.html

Kabilan, M. K., Ahmad, N., & Abidin, M. J. Z. (2010). Facebook: An online environment for learning of English in institutions of higher education? *Internet and Higher Education, 13*(4), 179–187. doi:10.1016/j.iheduc.2010.07.003

Kalelioğlu, F. (2017). Using Facebook as a learning management system: Experiences of pre- service teachers. *Informatics in Education-An International Journal, 16*(1), 83–101. doi:10.15388/infedu.2017.05

Kan, A. (2009). *Statistical procedures on measurement results. In the H. Atilgan. Assessment and evaluation in education*. Ani Publications.

Le, V. T. (2018). *Social media in learning English in Viet Nam* [Unpublished doctoral dissertation, The University of Canterbury].

Manan, N., Alias, A., & Pandian, A. (2012). Utilizing a social networking website as an ESL pedagogical tool in a blended learning environment: An exploratory study. *International Journal of Social Sciences & Education, 2*(1), 1–9. https://rb.gy/48j3wt

Munoz, C., & Towner, T. (2009). *Opening Facebook: How to use Facebook in the college classroom*. Paper presented at the Proceedings of society for information technology & teacher education international conference. Research Gate.

Nham, P. T., & Nguyen, T. T. (2013). The impact of online social networking on students'study. *VNU Journal of Education Research, 29*(1), 1–13.https://js.vnu.edu.vn/ER/article/view/486

Nykiel, R. (2007). *Handbook of marketing research methodologies for hospitality and tourism*. Haworth Press. doi:10.4324/9780203448557

O'Malley, J. M., & Chamot, A. U. (1990). *Learning strategies in second language acquisition*. Cambridge University Press. doi:10.1017/CBO9781139524490

Oxford, R. L. (1990). *Language learning strategies: What every teacher should know*. Heinle & Heinle.

Oxford, R. L. (2011). *Teaching and researching language learning strategies*. Pearson Longman.

Pallant, J. (2016). *SPSS survival manual: A step by step guide to data analysis using IBM SPSS* (6th ed.). Allen & Unwin.

Polok, K., & Harężak, J. (2018). Facebook as a beneficial tool while used in learning second language environment. *Open Access Library Journal*, *5*(07), 1–13. doi:10.4236/oalib.1104732

Ríos, A. A., & Campos, J. L. E. (2015). The role of Facebook in foreign language learning. *Revista de Lenguas ModeRnas*, *23*(23), 253–262. doi:10.15517/rlm. v0i23.22349

Rubin, D. B. (1987). *Multiple imputation for nonresponse in surveys*. John Wiley & Sons Inc. doi:10.1002/9780470316696

Schiffman, L. G., & Kanuk, L. L. (2004). Consumer Behavior, 8th International edition. Prentice Hall, Upper Saddle.

TeachThought. (n.d.). *100 ways to use Facebook in education by categories*. TeachThought. http://www.teachthought.com/technology/100-ways-to-use-facebook-in-education-by-category/

Tran, T. Q., & Duong, H. (2021). Tertiary non-English majors' attitudes towards autonomous technology-based language learning. *Advances in Social Science, Education and Humanities Research*, *533*, 141–148. doi:10.2991/assehr.k.210226.018

Tran, T. Q., & Ngo, D. X. (2020). *Attitudes towards Facebook-based activities for English language learning among non-English majors*. Proceedings of the International conference 2020: Language for global competence: Finding authentic voices and embracing meaningful practices (pp. 624–643). Ho Chi Minh City: Publishing House of Economics.

Tran, T. Q., & Tran, T. N. P. (2020). Attitudes toward the use of project-based learning: A case study of Vietnamese High school students. *Journal of Language and Education*, 6(3), 140–152. doi:10.17323/jle.2020.10109

Ulla, M. B., & Perales, W. F. (2021). Facebook as an integrated online learning supportapplication during the COVID19 pandemic: Thai university students' experiences and perspectives. *Heliyon*, 7(11), 1–8. doi:10.1016/j.heliyon.2021. e08317 PMID:34746477

Wang, C., & Chen, C. (2013). Effects of Facebook tutoring on learning English as a second language. *IADIS International Conference e-Learning*, 135–142. https:// files.eric.ed.gov/fulltext/ED562299.pdf

Wang, Q., Woo, H. L., Quek, C. L., Yang, Y., & Liu, M. (2012). Using the Facebook group as a learning management system: An exploratory study. *British Journal of Educational Technology*, 43(3), 428–438. doi:10.1111/j.1467-8535.2011.01195.x

Wenden, A. (1991). *Learner strategies for learner autonomy*. Prentice Hall.

ADDITIONAL READINGS

Adinata, I. B. P. A., Ramendra, D. P., & Mahendrayana, G. (2020). The uses of code mixing on Facebook by the students in English language education department. *Indonesian Journal of Educational Research and Review*, 3(3), 179–186. doi:10.23887/ijerr.v3i3.27662

Alm, A. (2015). "Facebook" for informal language learning: Perspectives from tertiary language students. *The EuroCALL Review*, 23(2), 3–18. https://files.eric. ed.gov/fulltext/EJ1082622.pdf

Barrot, J. S. (2018). Facebook as a learning environment for language teaching and learning: A critical analysis of the literature from 2010 to 2017. *Journal of Computer Assisted Learning*, 34(6), 863–875. doi:10.1111/jcal.12295

Buga, R., Căpeneaţă, I., Chirasnel, C., & Popa, A. (2014). Facebook in foreign language teaching–A tool to improve communication competences. *Procedia: Social and Behavioral Sciences*, 128, 93–98. doi:10.1016/j.sbspro.2014.03.124

Duong, Q. P., & Pham, T. N. (2022). Moving beyond four walls and forming a learning community for speaking practice under the auspices of Facebook. *E-Learning and Digital Media*, 19(1), 1–18. doi:10.1177/20427530211028067

Niu, L. (2019). Using Facebook for academic purposes: Current literature and directions for future research. *Journal of Educational Computing Research*, *56*(8), 1384–1406. doi:10.1177/0735633117745161

Terantino, J. (2011). Students' perceptions of language learning with Facebook: An exploratory study of writing-based activities. *Frontiers of Language and Teaching*, *2*, 230–249. shorturl.at/hkrzG

Tran, T. Q. (2020). EFL students' attitudes towards learner autonomy in English vocabulary learning. *English Language Teaching Educational Journal*, *3*(2), 86–94. doi:10.12928/eltej.v3i2.2361

Tran, T. Q., Duong, T. M., & Huynh, N. T. T. (2019). Attitudes toward the use of TELL tools in English language learning among Vietnamese tertiary English majors. *VNU Journal of Social Sciences and Humanities*, *5*(5), 581–594. doi:10.1172/vjossh.v5i5.498

Ulla, M. B., & Perales, W. F. (2021). Facebook as an integrated online learning support application during the COVID19 pandemic: Thai university students' experiences and perspectives. *Heliyon*, *7*(11), e08317. doi:10.1016/j.heliyon.2021.e08317 PMID:34746477

KEY TERMS AND DEFINITIONS

Attitude: This refers to one's behavior's course and persistence, and it consists of three main components, namely cognitive, affective and behavioral attitudes.

English majors: This refers to students majoring in English at an educational institution.

Facebook: This refers to a social networking platform where one can share information and have synchronous and asynchronous interaction with others.

Facebook-based language learning activities: This refers to activities which involve learners in sharing information and doing exercises on the Facebook group in an attempt to enhance their English language proficiency.

Facebook-based language learning strategies: This refers to language learning strategies consciously used by learners to accomplish the learning tasks when they are engaged in Facebook-based language learning activities.

Learning management system: (LMS): This refers to a learning platform used to plan, implement and assess a specific learning process.

Chapter 8
Explicit Lexical Collocation Instruction in Online Teaching

Cao Yen Ngoc
Victory Language Center, Tra Vinh University, Vietnam

Chau Thi Hoang Hoa
International Collabouration Office, Tra Vinh University, Vietnam

ABSTRACT

This present study aims to investigate the impact of explicit lexical collocation instruction on Vietnamese students' vocabulary use and their attitudes on the explicit lexical collocation instruction during online teaching due to COVID-19. Participants were 47 EFL students, divided into the experimental group and the control group learning online via Google Meet. Results collected from two writing tests (pre and posttest) and semi-structured interview demonstrated that the vocabulary use in students' writing performance in the experimental group had a significant improvement after receiving the explicit lexical collocation instruction online and students had positive attitude toward the explicit lexical collocation instruction. It is recommended that the application of explicit lexical collocation instruction at the beginning of a writing period in EFL teaching. It is recommended that EFL teachers can integrate lexical collocation instruction into teaching writing online to improve students' English proficiency in terms of quality of vocabulary use and writing performance.

INTRODUCTION

When mentioning studying a foreign language, the four language skills: listening,

DOI: 10.4018/978-1-6684-7034-3.ch008

speaking, writing, and reading were immediately reminisced. To advance these skills, a great wealth of vocabulary must be retained. Vocabulary is essential for the development of both receptive and productive skills. In addition, the linguist, David Wilkins (1972), declared that "without grammar very little can be conveyed, without vocabulary nothing can be conveyed" (pp. 111-112). It can be explained that if foreign learners maintain some words, a simple sentence can be created but without vocabulary, the learners cannot communicate in this language. Teaching vocabulary, in contrast to teaching grammar, has been ignored and has mostly been instructed by integrating it with other language skills. In opposition, grammar has always been instructed separately which led to supposing an inferior role for vocabulary. Fortunately, thanks to the assumption related to the important role of vocabulary and the shifts of sands in language teaching, vocabulary instruction has been upgraded the position. However, while acquiring vocabulary, students face considerable difficulty involving using appropriate words in various contexts, especially in writing skill which is described as a complicated skill by most Vietnamese teachers (Nguyen, 2009). It is extremely hard for students to guess the meanings and determine which words can be combined. This challenge can be seen in the example of Bui (2021, p99) about some common combinations employed by Vietnamese students that "feel headache or feel stomachache, drink some medicine, a fast lunch, problems happen, She has yellow hair, We are meeting many difficulties, She smiled with me, I very like music, I often go to eat in a restaurant near my school" which are incorrect and came from the translation directly from Vietnamese to English. Thus, to assist students in "expressing precisely what they want to say" (Lewis, 2000, p. 49) as well as producing natural target language (Siik, 2006), introducing collocation in English language learning class should be accelerated.

There are numerous international studies carried out to highlight the importance of collocational knowledge in language learning, namely Hsu (2010), Chang (2011) and Nizonkiza (2017) as well as most collocation studies in Vietnam were concerned about collocation in general and for English major students and the non - English major students seemed to be ignored. Therefore, the concept of "collocation" especially "lexical collocation" seems to be a relatively novel concept in the Vietnamese language learning and teaching context. Little literature involving the influence of a specific kind of collocation on a feature of writing skill was explored. In the current study, the effect of explicit lexical collocation instruction on vocabulary use in non – English major students' writing performance during online education due to COVID-19 was examined. What is more, to ensure the quality of the findings collected from the writing tests, this research was an attempt to explore students' attitudes toward the explicit instruction of lexical collocation regarding their vocabulary use in writing performance during online learning. To this end, two research questions are raised below:

RQ1. To what extent, does explicit lexical collocation instruction affect Vietnamese students' vocabulary use in their writing on online teaching?

RQ2. What are students' attitudes toward the explicit instruction of lexical collocation regarding their vocabulary use in writing on online teaching?

LITERATURE REVIEW

Definition of Collocation

The literature on the definitions of collocation varied among many scholars. In particular, the term "collocation" originally comes from the Latin s' word "collocure" indicating to arrange or to set in order (Muller, 2008). According to Firth (1957), who was known as a father of linguistics signifying the elucidation of collocation declared that collocation is delineated to be easily recognized because one will "know a word by the company it keeps" (p. 11) and "words are mutually expectant and mutually comprehended" (p. 12). Lewis (1993) elucidated collocation as the words that are habitual co – occurrence. Correspondingly, O'Dell and McCarthy (2017) used the term collocation to indicate lexical items that were combined by a semantic and syntactic word sounding natural to the natives. Sharing the same point of view, Lewis (2000) defined collocation as "words which are statistically much more likely to appear together than random chance suggests" by Lewis (2000 p. 29).

Classifications of Collocation

The classification of collocation has been carried out by many researchers in various ways. Lewis (1997) divided collocation into four groups, such as strong and weak; frequent and infrequent collocations. According to Wei (1999), collocation fell into three groups, namely lexical collocations, grammatical collocations and idiomatic expressions. Based on syntactic features, Benson et al (1997) offered a classification comprising lexical collocations and grammatical collocations which was employed in this current study due to its logical and comprehensive features as well as being a considerable reference for many corresponding Bahns (1993), Hsueh and Chiu (2008), and Wang and Good (2007). As presented above, the concentration of this study was lexical collocation so it was dealt with in detail in this study and concreted in the next part. Lexical collocation is defined as the combination of word units namely nouns, adjectives, verbs, and adverbs. According to Benson et al (1997), lexical collocations were distinguished between several structural types of lexical collocations marked as L1 to L6 (see Table 1).

Table 1. Lexical collocations by Benson et al (1997, p. 30 - 35)

Type	Pattern	Examples
L1	Verb + Noun	cancel an appointment
L2	Adjective + Noun	strong tea
L3	Noun + Verb	bombs explode
L4	Quantifier + Noun	a swarm of bees
L5	Adverb + Adjective	closely acquainted
L6	Verb + Adverb	run rapidly

The Relationship Between The Development of Lexical Collocation Knowledge and Vocabulary Use in Students' Writing Performance

Five subscales are worth assessing to identify a good writing paper: content, organization, vocabulary, language use, and mechanics (Heaton, 1988). As presented above, collocation knowledge, especially lexical collocation, confers students an opportunity to identify which couple of words could be in a company as well as avoid them from producing wrong word choices in their writing performance.

Various attempts have been made to discover the effect of collocation instruction on students' vocabulary used in writing performance. According to Carter and McCarthy (1988), "collocations teach students expectations about what sort of language can follow from what has preceded. Students will not have to go about reconstructing the language each time they want to say something but instead can use these collocations as pre - packaged building blocks" (p.75). In addition, McCarthy (1990, as cited in Jaya, Wijaya & Kurniawan, 2019) stated that "knowledge of collocation appropriation is part of the native speaker's competence and can be problematic for learners in cases where collocability is language - specific and does not seem solely determined by universal semantic constraints" (p. 2). In other words, when mastering the collocation knowledge, students can gain a benefit corresponding to producing the language appropriately and naturally like a native speaker. Additionally, O'Dell and McCarthy (2017) presented their view related to learning collocations that collocations have definite benefits regarding expressing their productive skills naturally and concisely, widening their lexical knowledge, and upgrading their vocabulary source. A conclusion can be made that collocation knowledge brings to students an opportunity to distinguish the appropriate combination of these words as well as obstruct them from creating wrong couples of words and making wrong word choice errors in their pieces of writing.

For example, Ghonsooli et al. (2008) conducted a study to examine the impact of teaching collocation on upper-intermediate learners' writing skills. Literature on explicit lexical collocation instruction in offline teaching is abundant but in online education, its effects have not been discovered. The finding showed that teaching collocation positively affected writing fluency and vocabulary components. Besides, Trisviana et al (2019) declared that there is a correlation between the lexical collocation knowledge of students and their level of English proficiency that can be explained by attaining a good level of collocation, students can use their range of vocabulary so that words can be combined correctly and applied in the appropriate context to express the meaning precisely. Also, Mousavi and Darani (2018) investigated the influences of collocation as language chunks on the writing skills of Iranian female intermediate English as Foreign Language (EFL) learners as well as the students' attitude towards writing after learning collocations with thirty female Iranian students. Oxford Placement Test, writing pretest and posttest and an attitude questionnaire was provided to explore students' writing proficiency and their attitude on explicit instruction of collocation. The primary outcome of this research showed that thanks to giving the collocation instruction, students outperformed in their writing performance compared to the pretest after the intervention as well as had a positive attitude toward it.

Explicit Lexical Collocation Instruction

Schmitt (2000) delineated two approaches to acquiring vocabulary: explicit and incidental learning. Based on the idea of Schmitt, the focus of explicit learning was what is going to be learned while incidental learning can be acquired when the learners apply the language to communicate. To distribute the idea of Lewis (2001), students are taught lexical collocation explicitly. The advocates of explicit learning believe that there are many benefits that students can attain when they put their attention directly to the learning objectives (Boers et al., 2006; Bruce et al., 2009Norris & Ortega, 2000, as cited in Mahvelati, 2019). Lewis (2000) stated that explicit lexical collocation instruction not only assists students in widening their collocation knowledge, noticing and practicing them, but also boosts their language acquisition through conscious learning. Particularly, the result of Rassaei and Karrbor 's study (2012) showed that explicit teaching is described as the most effective method for students to attain collocation knowledge. Additionally, as Zaferanieh and Behrooznia (2011) noted, explicit teaching is more beneficial than implicit teaching in the evolution of students' collocation knowledge.

In this present study, the collocational grid of Channel (1981) was employed to introduce the combinations of collocations and the input enhancement with typographical techniques was applied to explore the meaning and use of collocations during online teaching in the local education context.

Collocational Grid

Due to using collocation being certain about, students find it difficult to employ various kinds of collocation. To solve this problem, a collocation grid of Channel (1981) was introduced to support students' exposure to different kinds of collocation related to a previously restricted word.

Table 2. Collocation grid (Channel,1981, p.120)

	Woman	Man	Child	Dog	Bird	Flower	Weather	Landscapes	View	House	Furniture	Bed	Picture	Dress	Present	Voice
handsome	+	+	+													
pretty	+	+	+	+	+	+		+	+	+		+				
charming	+	+	+	+	+											
lovely	+	+	+	+	+	+		+	+	+		+	+	+	+	+

When it comes to introducing collocation, this technique is considered an effective way. Qader (2018) declared that collocation grid "only provide information on the forms of collocations and do not include their usage". In other words, the collocation grid provides students with information about the words combined and the forms of these words while the information about the meaning and how to implement these words are absent. To illustrate, according to Table 2, students can recognize that the word "pretty" can combine with the words "dog" and "flower," but it cannot combine with the words "voice" and "picture". Hence, in this current study, the collocational grid is employed as a lead-in instrument for students to get acquainted with lexical collocation in association with input enhancement.

Input Enhancement

In this technique, reading text with bolding or highlighting the intended collocation is employed to provide input for students. Khanchobani's view (2012) stated that visual input enhancement using typographical techniques namely underlining, bolding or highlighting is an effective tool to draw the students' attention to the formal aspects of language. Alanen (1995), Doughty (1991), Shook (1994), and Williams (1999) give their agreement about the positive influence of visual input enhancement on gravitating the students to the intended forms in the assigned texts (as cited in Park, Choi & Lee, 2012). Vu and Peters (2021, p.17) claimed that "through reading - only learners had a chance to see the form of the target collocations in the texts, which could help them recognize and memorize them". Besides the input provided by reading, reading-while-listening is another way to foster students' attention and long-term intention by identifying the form by reading and the sound by listening (Vu & Peters, 2021). By applying the reading text, students can recognize not only the forms of vocabulary but also guess their meaning from the context as well as identify how to use these words.

METHODOLOGY

Participants

Forty – seven sophomores of a four-year bachelor training coming from two EFL online classes were invited to participate in the current study. All of the participants are Vietnamese so they share a similar social, cultural, demographical and economic context. One class was the controlled group and the other was the experimental group. Both studied online via Google Meet during the COVID-19 breakout.

Instruments

Tests (Pretest - Posttest)

The writing achievement tests (pre and posttest) were employed to examine the collocation knowledge of students and the level of vocabulary use in their writing performance. The tests comprised two main parts, namely Part One, "Multiple choice" including 10 items in which students were required to elect an appropriate word to fill in the blank, and Part Two, a writing task. Notably, the collocations in Part One and the writing task in Part Two shared the same topic. The requirement of the writing task in Part Two was to write a letter or an email from 100 to 120 words about a given topic. As presented above, the minor component assessed to examine the impacts of lexical collocation instruction, defined as a factor affecting the ways students choose and combine the words in this paper was vocabulary. When examining numerous writing rubrics, there were two prominent options involving the writing assessment scale of B1 Preliminary for School - Cambridge and the writing rubric of Heaton. Although the two writing assessment scales contained vocabulary, the assessment criteria differed. After obtaining an intensive observation, due to the writing task collected from the Preliminary English Test (PET) as well as holding satisfied criteria related to vocabulary assessment, vocabulary subscale in the writing rubric of B1 Preliminary for School was employed to assess the vocabulary use in tertiary students' writing performance.

Online Semi- Structure Interview

The interview was used to shed light on the students' attitude toward the explicit lexical collocation instruction regarding their vocabulary use in writing performance. Wilkingson and Birmingham (2003) proposed three prominent kinds of interviews such as unstructured interview, semi-structured interview and structured interview. However, in this current research, the semi-structured interview was employed because "semi–structured interviews have predefined areas for discussion" (Wilkingson & Birmingham, 2003, p. 45). There are two main parts in the interviews corresponding to investigating students' attitudes on the instruction of explicit lexical collocation and examining the students' attitudes on the uses of lexical collocation in their writing performance. The participants in these interviews were classified into three groups based on their scores in the experimental group: four students with the lowest scores, four with the average scores, and four with the highest scores to synthesize which level of students was preferable for the treatment. Moreover, all of the interviews were carried out in Vietnamese in which students could confidently share their attitudes without language barriers to fulfill the insightful data. The interview was conducted online due to social distance. Each interview lasted about 20 to 30 minutes at the most convenient time for students.

Course Specification

In the first week, the revision related to the format of a letter or email was given to experimental and control groups under the teacher's guidance, and then students in both groups are invited to do the pretest.

Two groups were instructed on the different treatments from the second to the sixth week. In particular, the lessons of the experimental group commenced with learning lexical collocation. First, students were asked to look at the collocation grid and recognize which word could be combined. After that, students were given a reading text and asked to complete the tasks. Based on the questions related to the topic, the teacher asked students to brainstorm their ideas under the attention involving the relevant collocations in the warm-up stage, which can be used to answer these questions. Through these activities, students are expected to expand their vocabulary sources to boost their writing performance.

On the other hand, students in the control group were instructed with activities presented in the coursebook. First, to lead in the topic, the teacher asked students to match or fill in the blank in which reading texts were consistent with the writing topic. Next, the vocabulary instruction only with isolated words such as adjectives, verbs and linking words conveyed to the students. To create fairness of time for learning and extra resources, the teacher gave the writing instruction comprising

Table 3. Teaching scenario and two conditions

Experimental condition Teaching writing with the integration of explicit lexical collocation instruction	*Control condition* Teaching writing with the integration of implicit lexical collocation instruction
Warm–up: Explicit lexical collocations instruction To stimulate the student's curiosity regarding how to combine the words, the teacher introduced collocation grids. The teacher delivered the reading text to the students with two tasks that assisted students in exploring the meaning and the usage of these lexical collocations.	*Warm–up: Implicit lexical collocation instruction* Students were asked to complete the tasks regarding to the ideas of writing topics and how to apply these ideas. The teacher introduced isolated words.
Brainstorming: Giving setting and developing ideas with collocations in the warm-up stage Questions were raised corresponding to the topic and students were asked to answer these questions accompanied by the teacher's reminder that the collocations in the warm-up stage were pertinent and can be applied to answer these questions.	*Brainstorming: Giving setting and developing ideas by themselves* Questions were raised corresponding to the topic and students were asked to answer these questions.
Planning Students exchanged the information and discussed with other pairs	
Mind – map: Organize the ideas after discussing the lexical collocations in the warm-up stage Students imagined the situation about the writing task's requirement and then used their ideas mentioned above as well as developed more ideas to accomplish the mind - map by the collocations as much as possible.	*Mind–map: Organize the ideas after discussing.* Students imagined the situation related to the writing task's requirement and then employed their ideas mentioned above as well as developed more ideas.
Drafting After accomplishing the mind - map, students were asked to write an email/letter (100 words) according to the information mentioned above individually	
Peer feedback Students were asked to exchange their letters/emails with their partners to receive feedback from their friends.	
Editing After receiving the feedback, students revised their writing	
Final draft Students constructed the final draft accompanied by necessary changes at home	
Evaluation and teacher's feedback Students submitted the final draft on LMS to receive the teacher's feedback.	

eight stages, namely brainstorming, planning, mind - map, drafting, peer feedback, editing, and final draft which corresponded to writing instruction employed for the experimental group as illustrated in Table 3.

Data Analysis Procedure

First, the scores gathered from the pretest of both the experimental group and the control group were analyzed statically by an Independent Sample T-test to find whether the levels of English writing proficiency in the two groups were the same. Then, after six weeks of giving the explicit lexical collocation instruction to the students in the experimental group, the scores of the posttest in each group were run by the General Linear Model repeated measures to examine whether the intervention program had an impact on the improvement of the quality of vocabulary use in experimental group's writings.

The answers from the semi-structured interviews of the three groups were transcribed and translated into English. The Thematic analysis with a six-phrase guide, suggested by Braun and Clarke (2006, as cited in Maguire & Delahunt, 2017) was employed in this current study. Initially, the data was got familiar by reading and re-reading the transcripts of the interview. Secondly, the data were generated in the initial code and arranged meaningfully and systematically. After that, the excerpt was grouped into a theme used to select significant or interesting supporting data about the research question (Maguire & Delahunt, 2017). In this step, according to the broad themes, the sub–themes were identified. The next step of this procedure was reviewing and revising themes and sub-themes coherently based on the following questions:

"Do the themes make sense?

Does the data support the themes?

Am I trying to fit too much into a theme?

If themes overlap, are they really separate themes?

Are there themes within themes (subthemes)?

Are there other themes within the data?" (Maguire & Delahunt, 2017, p. 8)

Next, themes were defined which were aimed to "identify the 'essence' of what each theme is about" (Braun & Clarke, 2006, p.92, as cited in Maguire & Delahunt, 2017, p.11). The end-point of this procedure was to conduct a report based on these data.

FINDINGS

Research Question 1: To What Extent, Does Explicit Lexical Collocation Instruction Affect Vietnamese Students' Vocabulary Use in Their Writing on Online Teaching?

To observe the influence of explicit lexical collocation instruction described as a device on students' vocabulary use improvement in their writing performance, the researcher made the following comparisons. To begin with, the students in the experimental group achieved higher total mean scores (M = 3.0400) of the pretest than those in the control group (M = 3.000). An Independent samples T-test was used to examine whether the mean scores of the pretest on the participants' vocabulary use in their writing performance in the experimental group and the control group were statistically different. The outputs are displayed in Table 4 and the distinction between the two mean scores was not significantly different (df = 48; p = 0.839). It is therefore apparent that there is a homogeneity of the experimental and control group's levels of vocabulary use in their writing performance before the treatment.

Table 4. The result of comparing the mean scores of vocabulary use in students' writing performance in the pretest within each group by Independent samples T-test

	Group	**Mean**	**t**	**Df**	**Sig (2- tailed)**
Pretest	Experimental group (N = 25)	3.0400	0.204	48	0.839
	Control group (N = 25)	3.0000	0.204		

In the same way, an Independent sample T-test was to examine whether the means scores of participants in the experimental group (M = 3.6818) and that of the control group (M = 3.1600) were different with regard to vocabulary use after given treatment. With the result illustrated in Table 5, the two–tailed probability of the data of these two groups p = 0.024 and 0.030 respectively were lower than 0.05. As a result, it signified that the two mean scores were statistically different. It means that after receiving the explicit lexical collocation instruction, the students in the experimental group achieved a higher level of vocabulary use than those in the control group.

Table 5. The result of comparing the mean scores of vocabulary use in students' writing performance in the posttest within each group by Independent samples T-test

	Group	**Mean**	**t**	**Df**	**Sig (2- tailed)**
Posttest	Experimental group (N = 22)	3.6818	2.269	32.964	0.024 0.030
	Control group (N = 25)	3.1600			

In order to explore students' vocabulary use in writing in the control group before and after the intervention, a General Linear Model Repeated Model was run. Table 6 indicates the results obtained. It was observed that the mean of vocabulary use

Table 6. The result of comparing the mean scores of vocabulary used in the control group's writing performance in the pretest and posttest by General Linear Model Repeated Model

	Mean	F	Df	Sig
Pretest (N = 25)	3.0000	2.087	1	0.161
Posttest (N = 25)	3.1600			

score for the control group in the pretest was 3.000 and in the posttest was 3.160. Regarding the significance level (p < 0.05), it resulted that the difference between students' vocabulary use in the control group's writing of pre and posttest was not determined ($p = 0.161$). Hence, it indicates that students' vocabulary use in the control group was not improved.

Table 7. The result of comparing the mean scores of vocabulary use in experimental group's writing performance in the pretest and posttest by General Linear Model Repeated Model

	Mean	F	Df	Sig
Pretest (N = 22)	3.0400	6.176	1	0.021
Posttest (N = 22)	3.6818			

A General Linear Model Repeated Model was run to examine whether the intervention had an impact on students' level of vocabulary use in their writing performance. From the pretest to the posttest, there was an upward trend in vocabulary use in students' writing performance of the experimental group demonstrated through the growth of mean scores from 3.0400 to 3.6818 with p - value = 0.021 as shown in Table 7. An assumption could be drawn that the vocabulary use in students' writing performance before and after the intervention was enhanced as well as the explicit lexical collocation instruction exerted a positive impact on the Vietnamese student's vocabulary use in their writing performance. More specifically, the input enhancement technique facilitated the development of lexical collocation knowledge of participants in the experimental group involving exploring the form, meaning and use of lexical collocation through the reading text.

Research Question 2: What Are Students' Attitudes Toward the Explicit Instruction of Lexical Collocation Regarding Their Vocabulary Use in Writing on Online Teaching?

Students' Attitudes on the Explicit Lexical Collocation Instruction

A difference in lexical collocation acquisition among the three interviewed groups was pointed out that the students in the average and highest score group acquired lexical collocation knowledge effortlessly while the lowest score group encountered many difficulties. Nevertheless, the interviewees had positive attitudes toward the impact of the explicit lexical collocation instruction on their vocabulary use in writing performance. More specially, all of the participants assented that explicit lexical collocation learning supported them in advancing the level of vocabulary use in their writing performance. Most of the participants showed their agreement on the use of collocation grid and input enhancement in learning lexical collocation. Most of the respondents from the three groups emphasized and acknowledged that the collocational grid assisted them in identifying the acceptable combinations among the words easily. Markedly, two students of the average group figured out the additional drawback of collocation grid involving the challenge of exploring the rule of collocation grid without teacher's instruction. In addition, numerous students from three groups highlighted that collocation grid fulfilled well its function regarding knowing how the words combined appropriately but there was an absence of information related to knowing the meaning and the way to use these collocations.

What needs to be highlighted is that participants cannot deny the pivotal role of reading texts in learning lexical collocation. Resulting from the reading text and the tasks designed based on them, students not only discovered the meaning but also noticed the form and the context to use these collocations appropriately. Numerous respondents share their agreement involving the sufficient role of tasks designed according to the reading text in comprehending the meaning and the way to apply these lexical collocations. Conversely, interviewees from the lowest score groups dealt with many obstacles related to their hesitation in exploring the meanings and the use of them. More specifically, due to the limitation of vocabulary sources, they got into trouble with guessing the diverse meaning of these lexical collocations in the context and being confused about the way to use these collocations appropriately.

In the nutshell, according to the evidence gathered from the interviews, it can be acknowledged that the students in the average and highest score groups found lexical collocation knowledge more effortlessly to acquire than those in the lowest score group. Moreover, the level of challenge in learning lexical collocation encountered by the participants is inversely proportional to their English proficiency.

Students' Attitudes on the Explicit Lexical Collocation Instruction Regarding Vocabulary Use in Their Writing Performance

When it comes to the question of how research participants perceived the explicit lexical collocation regarding vocabulary use in their writing performance, all of the participants in the interviews strongly agreed with the belief that explicit lexical collocation learning helped them improve the vocabulary used in writing performance. This idea was also presented in the answer of participants in the average group, as they believed that applying lexical collocation instruction has certainly diversified their vocabulary use which restricted the repetitive word errors as well as developing the ideas conveniently. In addition, students in the average and high score groups declared that thanks to explicit lexical collocation instruction, students could exposure to advanced English knowledge which upgrades their writing product. Notably, a student in the highest score group highlighted that using lexical collocations in my writing products confers an advantage regarding updating the vocabulary source employed for a long time. Taking into consideration the opinion and implication of this participant, it could be admitted that lexical collocation learning correlates to reinforce the vocabulary source, especially for students in the highest score group who found learning lexical collocation intuitively appealing and easily comprehending as practical implications. Adding to this point of view, participants in this current research pointed out the positive influence of explicit collocation on other aspects of writing performance. To illustrate, two students from the average group claimed the coherent and cohesion features as one of the strong points of learning lexical collocations and employing them in their pieces of writing.

However, there existed some challenges that students in the lowest score group not being well-equipped with a sufficient amount of English language knowledge might find acquiring lexical collocation by collocation grid and reading texts and applying them in their writing performance quite overwhelming for their comprehension. Conversely, thanks to obtaining assured English proficiency renders fewer problems, students in the high score group perceived vigorously the explicit instruction of lexical collocation regarding their vocabulary use.

With regards to the perception of explicit lexical collocation instruction on vocabulary use, participants in this study admitted the sufficient role of explicit lexical collocation instruction in facilitating the quality of vocabulary use. Additionally, great recognition was conveyed for the importance of learning lexical collocations in developing not only vocabulary use but also other features like grammar and fluency as well as other language skills.

DISCUSSION

The findings of this study turned out to be as desirable. In light of indicating the influence of explicit lexical collocation instruction on the Vietnamese students' vocabulary use, there was a significant enhancement of their vocabulary level in the pieces of writing. This result was determined by the progression of mean scores from the pretest to the posttest in the experimental group after intervention Mousavi and Darani (2018) emphasized the considerable assistance of lexical collocation instruction in vocabulary use in their writing performance. Thanks to possessing a host of collocation knowledge, students could easily imagine what they want to write about and convey their ideas into the paper immediately with appropriate vocabulary and meaningful structures. This was also illustrated by Trisviana et al (2019) who discovered the correlation between the students' lexical collocation knowledge and their language skill. They assumed that students who obtain a good level of collocation knowledge, they can write effortlessly accompanied by accurate and natural word choices. Furthermore, it is worth noting that among different levels of English proficiency, there was a significant difference in lexical collocation acquisition and use. In other words, the level of lexical collocation use in their writing performance is proportional to their English proficiency. It could be signified that under the given treatment, participants showed a better performance related to the quality of vocabulary use that rejected the findings of Chang (2011) which indicated that collocation instruction had no considerable impact on EFL students' writing performance.

When it comes to lexical collocation learning, participants acknowledged the success of explicit learning. A finding highlighted the effectiveness of learning lexical collocation explicitly through the collocation grid and reading text. Many students in this present study reflected that by using learning lexical collocation explicitly through employing collocation grid and reading text, they can not only know the words but also comprehend the meaning as well as the way to apply these collocations appropriately. According to Lewis (2000), with a view to imparting knowledge efficiently, explicit lexical collocation instruction provides knowledge to students through noticing and practising the knowledge as well as promotes the students' language acquisition through conscious learning. This result is consistent with the findings of Zaferanieh and Behrooznia (2011) and Rassaei and Karrbor (2012) noted that explicit teaching is the most effective method in the evolution of students' collocation knowledge.

With regards to the attitude toward explicit lexical collocation instruction in the quality of vocabulary use in their writing performance, students concurred with the pivotal role of explicit lexical collocation instruction in stimulating vocabulary use in their pieces of writing. This claim has been supported by the study of Mousavi and

Darani (2018) that by not knowing many collocations and meaningful structures to delineate their ideas, students identify writing as a complicated and stressful task, however, after learning collocations, they can write easier and become more interested in writing. This could be seen on the account of learning collocation, the students' attitudes toward writing adjusted from a negative attitude to a positive one. Students perceived positively the explicit collocation instruction regarding the development of vocabulary use. Moreover, the importance of learning lexical collocations was recognized which pertains to giving a boost to not only the vocabulary use in their writing performance but also other receptive and productive skills. Parallel with the results of the current study, Trisviana et al (2019) revealed that a student who acquires numerous lexical collocation knowledge could get better comprehension when they read a text as well as they can easily apply them in their speaking and writing.

IMPLICATIONS AND CONCLUSION

According to quantitative and qualitative findings, the hypothesis of this present study was confirmed. The findings of this study accentuated the contribution of explicit lexical collocation instruction to students' vocabulary use in their writing on online teaching. If the students achieved a good level of lexical collocation, it can help the students to build up a stable vocabulary source then students would get better results in their writing, especially vocabulary use. It can be concluded that explicit lexical collocation instruction in online teaching had a durable influence on students' vocabulary use.

On the whole, this current study contributed to the series of empirical research studies investigating the effects of explicit lexical collocation instruction on students' vocabulary use while studying online. This study determined that such influences exist. The findings of this paper, therefore, can transfer a great benefit to language educators, namely teachers, instructional designers, material developers, as well as language learners. This study bears some implications. First, based on the fact that students' mastery of lexical collocation has a positive effect on vocabulary use in their writing, English teachers should put attention to integrating lexical collocation into their writing teaching material because lexical collocations do not receive the attention they deserve in textbooks. In addition, lexical collocation instruction should be instructed at the beginning of a writing period as a scaffolding step to equip the potent vocabulary foundation that served the writing process. Online teaching did not cause any difficulties in the implementation of explicit instruction of collocation.

The second implication was carried out that it is favorable for teachers to pay attention to the student's English proficiency to mitigate the negative influences and to convey pertinent support to accommodate the students to become efficient language learners. In doing so, teachers should implement appropriate pedagogical methods as well as design appropriate tasks for students' English levels.

The result of this study can be essential for language learners by knowing that mastery of lexical collocation developed their writing performance, especially vocabulary use. It is a must for them to enhance their collocation knowledge with the regard to assisting them to combine and use the words completely and accurately in their pieces of writing.

This study was limited to only one writing component, that is, vocabulary use. Future research can examine the efficacy of explicit lexical collocation on other writing components. Moreover, future empirical research can be conducted in a longer period of time to examine the long-term effect of the employed treatment through delayed posttests to determine its impacts over time.

REFERENCES

Bahns, J., & Eldaw, M. (1993). Should we teach EFL students collocations? *System*, *21*(1), 101–114. doi:10.1016/0346-251X(93)90010-E

Benson, M., Benson, E., & Ilson, R. (1997). *The BBI dictionary of English word combinations* (2nd ed.). John Benjamins. doi:10.1075/z.bbi1(2nd)

Bui, L. T. (2021). The role of collocations in the English teaching and learning. *International Journal of TESOL & Education*, *1*(2), 99–109. https://i-jte.org/index.php/journal/article/view/26

Carter, R., & McCarthy, M. (1988). *Vocabulary and language teaching*. Longman.

Chang, L. (2011). Integrating collocation instruction into a writing class: A case study of Taiwanese EFL students' writing production. *Journal of Kao - Tech University - Humanities and Social Sciences*, 281 - 304.

Channel, J. (1981). Applying semantic theory to vocabulary teaching. *ELT Journal*, 115 - 122. doi:10.1093/elt/XXXV.2.115

Channel, J. (1981). Applying semantic theory to vocabulary teaching. ELT Journal, 115 - 122. doi:10.1093/elt/XXXV.2.115

Firth, J. R. (1957). *Papers in linguistics, 1934 - 1951*. Oxford University Press.

Ghonsooli, B., Pishghadam, R., & Mahjoobi, F. (2008). The impact of collocational instruction on the writing skill of Iranian EFL learners: A case of product and process study. *Iranian EFL Jounal*, 36 - 59. https://www.researchgate.net/publication/277841864_The_impac t_of_colocational_instrucion_on_the_writing_skill_of_Iranian _EFL_learners#fullTextFileContent

Heaton, J. B. (1988). *Writing English language tests.* Longman.

Hsu, T. J., & Chiu, C. (2008). Lexical collocations and their relation to speaking proficiency. *Asian EFL journal, 10*(1), 181 - 204. https:// www.asian-efl-journal.com/main-editions-new/lexical-collocations-and-their-relation-to-speaking-proficiency-of-c ollege-efl-learners-in-taiwan/index.htm

Hsu, Y. J. (2010). The effect of collocation instruction on reading comprehension and vocabulary learning of Taiwanese college English majors. *Asian EFL Journal, 12*(1), 47 - 87. https://www.asian-efl-journal.com/main-editions-new/the-effe cts-of-collocation-instruction-on-the-reading-comprehension-and-vocabulary-learning-of-college-english-majors/index.htm

Jaya, H. P., Wijaya, A., & Kurniawan, D. (2019). Correlation between the ability of using English collocation and academic achievements of students of faculty of teacher training and education Universitas Sriwijaya. *Holistics Journal, 11*(2), 908–917. doi:10.20319/pijss.2019.52.908917

Khanchobani, A. (2012). Input enhancement and EFL learners" collocation acquisition. *International Journal of Academic Research, 4*(1), 96–101.

Lewis, M. (1993). The Lexical Approach. The state of ELT and a way forward. Commercial Colour Press Lewis, M. (Eds.). (2000). *Teaching collocation: Further developments in lexical approach.* Oxford University Press.

Lewis, M. (1997). *Implementing the lexical approach: Putting theory into practice.* Language Teaching Publications.

Maguire, M., & Delahunt, B. (2017). Doing a thematic analysis: A practical, Step - by – Step guide for learning and teaching scholar. *Journal of Teaching and Learning in Higher Education, 3*, 1–13. http://ojs.aishe.org/index.php/aishe -j/article/view/335

Mahvelati, E. H. (2019). Explicit and implicit collocation teaching methods: Empirical research and issues. *Advances in Language and Literary Studies, 10*(3), 105. http://www.journals.aiac.org.au/index.php/alls/article/view/ 5545. doi:10.7575//aiac.alls.v.10n.3p.105

Mousavi, S. M., & Heidari Darani, L. (2018). Effect of collocations on Iranian EFL learners' writing: Attitude in focus. *Global Journal of Foreign Language Teaching*, *8*(4), 131–145. doi:10.18844/gjflt.v8i4.3568

Müller, Y. (2008). *Collocation - A linguistic view and didactic aspects*. GRIN Verlag.

Nguyen, H. H. T. (2009). Teaching EFL writing in Vietnam: Problems and solutions -a discussion from the outlook of applied linguistics. *VNU Journal of Science, 25,* 61 - 66. https://js.vnu.edu.vn/FS/article/view/2236

Nizonkiza, D. (2017). Improving academic literacy by teaching collocations. *Stellenbosch Papers in Linguistics, 47*(0), 153 - 179. https://journals.co.za/doi/abs/10.5774/47-0-267

O'Dell, F., & McCarthy, M. (2017). *English collocations in use advanced*. Cambridge University Press.

Park, H., Choi, S., & Lee, M. (2012). Visual input enhancement, attention, grammar learning, & reading comprehension: An eye movement study. *English Teaching*, *67*(4), 241–265. http://journal.kate.or.kr/wp-content/uploads/2015/01/kate_67_4_11.pdf. doi:10.15858/engtea.67.4.201212.241

Qader, D. S. (2018). The role of teaching lexical collocations in raising EFL learners' speaking fluency. *Journal of Literature. Language and Linguistics (Taipei)*, *46*, 42–53. https://www.iiste.org/Journals/index.php/JLLL/article/view/43454

Rassaei, E., & Karbor, T. (2012). The effects of three types of attention drawing techniques on the acquisition of English collocations. *International Journal of Research Studies in Language Learning*, *2*(1), 15–28. doi:10.5861/ijrsll.2012.117

Schmitt, N. (2000). *Vocabulary in language teaching*. Cambridge University Press.

Siik, S. (2006). *The teaching of lexical collocations and its effects on the quality of essays and knowledge of collocations among students of program Persedian Ijazah Sarjana Muda of Instut Perguruan Batu Lintang, Kuchiing*. [Unpublished Master Thesis, University Teknologi Malasia, Malaysia].

Trisviana, W., Afriazi, R., & Hati, G. M. (2019). The correlation between students' mastery on lexical collocation and their reading comprehension. *Journal of English Education and Teaching*, *3*(1), 53–65. doi:10.33369/jeet.3.1.53-65

Vu, D. V., & Peters, E. (2021). Incidental learning of collocations from meaningful input. *Studies in Second Language Acquisition*, 1 -23. doi:10.1017/S0272263121000462

Wang, J. T., & Good, R. L. (2007, November). The repetition of collocations in EFL textbooks: A corpus study [Paper presentation]. *The Sixteenth International Symposium and Book Fair on English Teaching, Taipei.* https://eric.ed.gov/?id=ED502758

Wei, Y. (1999). *Teaching Collocations for Productive Vocabulary Development.* Paper presented at the Annual Meeting of the Teachers of English to Speakers of Other Languages. New York, USA.

Wilkins, D. A. (1972). *Linguistics in Language Teaching.* Edward Arnold.

Wilkinson, D., & Birmingham, P. (2003). *Using research instruments: A guide for researchers.* Psychology Press.

Zaferanieh, E., & Behrooznia, S. (2011). On the impacts of four collocation instructional methods: Web - based concordancing vs. traditional method, explicit vs. implicit instruction. *Studies in Literature and Language*, *3*(3), 120–126. http://cscanada.net/index.php/sll/article/view/j.sll.1923156 320110303.110

ADDITIONAL READINGS

Adelian, M., Nemati, A., & Fumani, M. R. F. Q. (2015). The effect of Iranian advanced EFL learners' knowledge of collocation on their writing ability. *Theory and Practice in Language Studies*, *5*(5), 974–980. https://www.semanticscholar.org/paper/The-Effect-of-Iranian-Advanced-EFL-Learners%E2%80%99-of-on-Adelian-Nemati/5eed75e9 67b5db4885c1123f3927bb48abdf6073. doi:10.17507/tpls.0505.12

Bahardoust, M. (2012). Lexical collocations in writing production of EFL learners: A study of L2 collocation learning. *Iranian EFL Journal*, *8*(2), 185–200.

Choi, S. (2017). Processing and learning of enhanced English collocations: An eye movement study. *Language Teaching Research*, *21*(3), 403–426. doi:10.1177/1362168816653271

Farrokh, P. (2012). Raising awareness of Collocation in ESL/ EFL Classrooms. *Journal of Studies in Education*, *2*(3). https://www.macrothink.org/journal/index.php/jse/article/vie w/1615/1525. doi:10.5296/jse.v2i3.1615

Firth, J. R. (1957). *Papers in linguistics.* Oxford University Press.

Mounya, A. (2010). *Teaching lexical collocations to raise proficiency in foreign language writing: A case study of first year students at Guelma University, Algeria.* [Master"s dissertation, Guelma University]. https://bu.umc.edu.dz/theses/anglais/ABD1089.pdf

Mutlu & Kaslioglu. (2016). Turkish EFL teachers" and learners" perceptions of collocations. *Sakarya University Journal of Education.* (pp. 81-99). http://dergipark.gov.tr/download/article-file/262852

O'Dell, F., & McCarthy, M. (2017). *English collocations in use advanced.* Cambridge University Press.

Obeidat, A., & Sepora, T. (2019, September). Collocation translation errors from Arabic into English: A case study of naguib mahfouz's novel "Awlad Haratina". *International Journal of Humanities. Philosophy and Language, 2*(7), 129–138. doi:10.35631/ijhpl.270011

Pym, A. (2014). *Exploring translation theories* (2nd ed.). Routledge.

Qader, D. S. (2018). The role of teaching lexical collocations in raising EFL learners' speaking fluency. *Journal of Literature. Language and Linguistics (Taipei), 46*, 42–53.

Rao, V. C. (2018). The importance of collocations in teaching of vocabulary. *Journal For Research Scholars And Professionals English Language Teaching, 2*(7), 1–8.

Rao, V. C. (2018). The importance of collocations in teaching of vocabulary. *Journal For Research Scholars And Professionals English Language Teaching, 2*(7), 1–8.

Rassaei, E., & Karbor, T. (2012). The effects of three types of attention drawing techniques on the acquisition of English collocations. *International Journal of Research Studies in Language Learning, 2*(1), 15–28. doi:10.5861/ijrsll.2012.117

Siengsanoh, B. (2021). Lexical collocational use by Thai EFL learners in writing. *LEARN Journal: Language Education and Acquisition Research Network, 14*(2), 171-193. https://files.eric.ed.gov/fulltext/EJ1310912.pdf

Szudarski, P., & Carter, R. (2016). The role of input flood and input enhancement in EFL learners' acquisition of collocations. *International Journal of Applied Linguistics, 26*(2), 245–265. https://onlinelibrary.wiley.com/doi/abs/10.1111/ijal.12092. doi:10.1111/ijal.12092

Toomer, M., & Elgort, I. (2019). The development of implicit and explicit knowledge of collocations: A conceptual replication and extension of Sonbul and Schmitt (2013). *Language Learning, 69*(3), 405–439. doi:10.1111/lang.12335

Ünver, M. M. (2018). Lexical collocations: Issues in teaching and ways to raise awareness. *European Journal of English Language Teaching*, *3*(4), 144–124. https://halshs.archives-ouvertes.fr/halshs-00371418. doi:10.5281/zenodo.1344700va

Vasiljevic, Z. (2014). Teaching collocations in a second language: Why, what and how? *ELTA Journal*, *2*(2), 48–73.

Webb, S., & Chang, A. C. S. (2020). How does mode of input affect the incidental learning of collocations? *Studies in Second Language Acquisition*. doi:10.1017/S0272263120000297

KEY TERMS AND DEFINITIONS

Explicit Instruction: This can be described as a direct instruction in which teacher guided students how to do and what to do.

Lexical Collocation: This is the combination of word units namely nouns, adjectives, verbs, and adverbs

Online Teaching: This is the process of educating others employing virtual platforms and taking place via the internet.

Vocabulary Use: This is related to the words that students select to express their ideas in writing or speaking skills communicatively.

Writing Performance: This refers to the achievement of writing in which students use vocabulary and structures to clarify and organize their thoughts coherently and fluently.

Chapter 9
Video–Mediated Dialogic Reflection for Teacher Professional Development:
A Case in Vietnam

Khoa Do

 https://orcid.org/0000-0002-0016-8511
University of Warwick, UK

ABSTRACT

This study investigates the nature of discourse, the advantages and disadvantages of Video-mediated Cooperative Development (VMCD) and the potential of the technique being applied as a means of Continuing Professional Development in Vietnam. The study is a qualitative case study, and the data encompass recordings of the VMCD sessions, recordings of the interviews, recordings of the online lessons, and the participants' drawings. Overall, the nature of discourse in VMCD meetings is similar to that of face-to-face Group Development (GD), with Attending suffering the most due to problems which emerge from online interaction. Additionally, even though they experienced some difficulties being the Understanders, the participants generally enjoyed the VMCD sessions, acknowledging its novelty, its relevance, and the fact that it conveys a sense of community. It is also reported that solutions or the 'moments of enlightenment' can come during the VMCD sessions, or later on when the teachers have their own time to reflect. The results have implications for the possible modifications of future versions of VMCD, or any VMCD-integrated teacher training programmes.

DOI: 10.4018/978-1-6684-7034-3.ch009

INTRODUCTION

This chapter offers an insight into the use of Computer-mediated Communication in the field of Second Language Teacher Education (SLTE), the field whose coverage entails preparation, training, and education of second language teachers. An introductory outline of different approaches SLTA, which further encompasses different approaches to SLTA and Julian Edge's Cooperative Development and its variants, is provided in the first part. The second part of the chapter presents findings from Do's (2022) study about Video-mediated Cooperative Development (VMCD) and how the elements of CMC are shown in teacher exchanges.

A BRIEF OVERVIEW OF SECOND LANGUAGE TEACHER EDUCATION

There are several choices accessible to in-service English language teachers for pursuing Continuing Professional Development (CPD). Those entail formal ones like conferences, workshops, or seminars to informal ones like mentoring, written journals, teacher study groups, or peer-coaching (Bailey et al., 2001; Wedell, 2017) The following paragraphs set out some efforts at categorizing second language teacher education (SLTE).

Wallace (1991) outlines three models which have historically existed in CPD: *the craft model, the applied science model,* and *the reflective model.* The craft model, which emphasizes the experiential component of learning, has inexperienced teachers observe and replicate more experienced ones. The applied science model, on the other hand, highlights the importance of imparting scientific knowledge through research from researchers to teachers, which represents intellectual learning. Finally, the reflective model encompasses teachers reflecting on what they previously know, and the model acts as a 'compromise solution' (Wallace, 1991, p.17) to both experience and science. Borg (2016) uses the term the training-transmission model to refer to training generated externally from teachers; Freeman (2009) refers to it as the input-application approach and Farrell (2021) uses the phrase knowledge transmission approach to refer to a similar concept. In contrast to this external model is reflective practice (RP), which deals with teachers' internal growth.

Figure 1 below provides a visual summary of the aforementioned concepts.

Across SLTE literature, there has been increasing criticism towards the external models, for they might not reflect what actually happens in the classroom (Farrell, 2021; Timperley, 2011), and there exists the problem of the theory/practice gap (Edge, 2011; Farrell, 2021; Schön, 1992). Reflective practice, on the other hand, is generally favored, playing a central position in teachers' professional development

Figure 1. Second language teacher education models

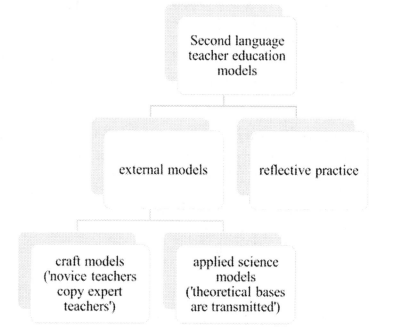

(Mann & Walsh, 2013, 2017; Walsh & Mann, 2015), attaining the 'status of orthodoxy' (Mann & Walsh, 2017, p.5). Popular reflective practice activities include critical friends group, team teaching, peer coaching, action research, written journal, and Cooperative Development (CD) (Farrell, 2019; Walsh & Mann, 2015)).

COOPERATIVE DEVELOPMENT: A CONVERSATIONAL APPROACH TO SLTE

Although reflective practice has been considered central in SLTE, it has been stereotypically deemed an individual process; a great number of reflective paradigms present reflection as a process carried out alone rather than a collaborative one. Another inherent challenge of reflective practice is that the written forms of reflection (i.e., reflective journals, diaries) generally dominates its dialogic counterparts (see Walsh & Mann, 2015). In reality, working together and picking up knowledge from one another is very much aligned with Dewey's original definition of reflection, which placed an emphasis on communication and collaboration.

Cooperative Development (CD), a CPD framework developed by Julian Edge (see Edge, 1992a, 2002), is a strong example of reflective practice that places a strong emphasis on active, two-sided conversations. 'We learn by speaking, by working to put our thoughts together so that someone else can understand them' (Edge, 2002, p.19). In CD, teachers play the roles of the Speaker and the Understander for a certain amount of time or until the participants decide to stop. The Understander listens and supports the process without passing judgment. On the one hand, the Speaker makes an effort to organize his views by speaking in a non-defensive manner. The third role, the Observer, is sometimes needed to do what its name implies: observe the process and make thoughtful notes. Non-judgmental discourse lies at the heart of CD, and this is achievable under three attitudinal elements: Respect, Empathy and Sincerity. Respect refers to the unconditional acceptance of what the Speaker has to say; Empathy reflects the ability to see things from the Speaker's perspective, and Sincerity manifests itself in the ability to offer genuineness when showing Respect and Empathy. The following table summarizes the speech acts that an Understander is supposed to perform in a CD session.

CD can be either conducted in pairs, or in a group (in which case it is also called Group Development - GD).

COMPUTER-MEDIATED COOPERATIVE DEVELOPMENT (CMCD)

Research into CMCD hitherto encompasses CD by emails (EMCD) (Cowie, 2002), Instant Message CD (IMCD) (Boon, 2007, 2011, 2019) and Video-mediated Cooperative Development (VMCD) (Do, 2022; Webb et al., 2022). Overall, the exchanges among teachers in EMCD are asynchronous, whilst with IMCD, the interaction can be quasi-synchronous and entails an element of social presence (Edge, 2006). Research into VMCD appears scarce; there is one recent study by (Webb et al., 2022) which explores the clime between a loose version to a fixed version of two case studies of paired VMCD which were conducted on a videoconferencing platform (Microsoft Teams). This section presents the first two modes of CD; VMCD is presented in a separate section as it is the main session of this chapter.

First proposed by Neil Cowie, EMCD is an act of dialogic reflection maintained by email exchanges. Cowie (2002) sets out some important features of EMCD, including (1) asynchronous feature of feedback, (2) seamless recycling of previous topics, thanks to the cut-paste function, and (3) the convenience of maintaining discussions of themes and topics in a parallel fashion. Unlike EMCD which is asynchronous by nature, IMCD locates itself into a real-time domain. It takes advantage of Instant Messaging applications to provide instantaneous responses whilst secure of elements of social exchanges. The key difference between IMCD and its in-person version is that the Speaker and the Understander do not see one

Table 1. The Understander's moves

Discovery		
	Attending	The Understander adopts attentive listening and non-threatening physical stances. Attending is a 'prerequisite to all the other skills, and its quality informs the quality of the entire exchange.' (Edge & Attia, 2014, p.67)
	Reflecting	The Understander mirrors what the Speaker has just said. Two possible scenarios might happen during Reflecting. If the Understander's reflection is accurate, then the Speaker can continue knowing that their ideas have been fully grasped. If there are hiccups in the reflections, the Speaker is provided an opportunity to revise his articulation.
	Focusing	The Understander invites the Speaker for an in-depth discussion on a particular point of their talk. For instance, the Understander might want to ask which one the Speaker wants to Focus on among his/her/their three ideas X, Y and Z.
Exploration		
Relating	**Thematizing**	The Understander draws the Speaker's attention to possible links between ideas previously mentioned.
	Challenging	The Understander draws the Speaker's attention to possible contradictions and ask him/her/them for elaboration.
	Disclosing	The Understander shares his/her/their experiences on the matter in question. The Speaker might want to use what the Understander Disclose as a point of comparison or contrast.
Action		
	Goal setting	The Understander helps the Speaker develop a possible goal.
	Trialing	The Understander helps the Speaker consider the course of action needed to achieve the Goal.
	Planning	This is the final stage of a CD session when 'administrative arrangements for continuity are made' (Edge, 1992b, p.69)

another face to face, and this, as Boon (2007) claimed, can help 'isolated teachers interact with a potential global community of online participants from the comfort of their own workspace' (See Boon (2015)).

VIDEO-MEDIATED COOPERATIVE DEVELOPMENT

First mentioned in Webb et al (2022) and further studied by Do (2022), VMCD steps into the fourth generation of CD in the modern era. It takes advantage of video-conferencing applications such as Zoom, Google Meet or Microsoft Teams as a medium for the CD sessions to take place. The following part is adapted from the first part of Do's case study on a series of CD conducted online by a group of six teachers.

Table 2. Participants' background information

Participant	Male/ Female (M/F)	Educational background	Work experience (years)	Exposure to CPD	CD before?
Mike	M	Undergraduate student of Teaching English as a Foreign Language (TEFL)	3	moderate	No
Phineas	M	Undergraduate student of TEFL (pending graduation)	2.5	very high	No
Rebecca	F	BA in Business Administration-Marketing	4	moderate	No
Derrick	M	BA in Hospitality	4	moderate	No
Lee	M	BA in Economics and Finance CELTA	3	moderate	No

The Context of The Study

The study was based at a language school in Ho Chi Minh city, Vietnam. This is a medium-sized language school which specializes in IELTS preparation. The majority of the teachers here fall into the 21-to-35 bracket, many of whom have bachelor's degrees or at least certificates in Teaching English to Speakers of other Languages (TESOL). All of the in-service professional development opportunities that have hitherto taken place at this language school are a series of seminars and workshops, which are conducted on a termly basis (3-5 sessions per term). None of them have engaged in CD before.

The number of participants for the project was 5, including 4 males and 1 female. The table below serves as a summary of their background, with their names pseudonymized as (Teacher) Mike, Phineas, Rebecca, Derrick and Lee. The use of 'Teacher' in the transcripts of the subsequent chapters is understood as a common honorific often used when teachers refer to one another. For example, my colleague might refer to me as 'Teacher Khoa' in a unit meeting.

With regard to the participants' former CPD experience (other than CD), four over five of the participants shared similar experiences. Mike, Rebecca, Derrick, and Lee engaged in CPD to a moderate extent, and their involvement revolved around the compulsory INSET workshops conducted by Vietop (the language school they work for) and the series of IELTS teacher training provided by the International Development Program (IDP) Australia and the British Council (BC). Phineas, by contrast, actively sought different CPD opportunities. He has so far participated in a great degree of workshops, webinars and MOOCs provided by different organizations

Table 3. Speakers' topics

Dates (dd/mm/yyyy)	Speaker	Topic/Critical incident	Notes
21/06/2022	Mike	profit-oriented versus educational aspects of a language school	later narrowed down to how to use debates and videos in the classroom
23/06/2022	Phineas	teacher's depression when the students came very late	
28/06/2022	Rebecca	making use of breakout rooms	
30/06/2022	Derrick	students' short-termed memory problems	
05/07/2022	Lee	Vietnamese-English structural confusion	

and institutions. He even had a blackboard with all of the upcoming TESOL events and courses he found interesting.

The VMCD sessions were conducted two times a week, on Tuesday and Thursday at 7 AM (BST – UK time)/1 PM (GMT+7 – Vietnam time). For each session, one participant played the role of the Speaker, and the others and the researcher were the Understanders. The procedures of the meetings were adapted from the GD procedures by Edge (2002) and Mann (2002). The following table summarizes the dates and the original topics.

Due to the relatively short training time with only one induction and a handbook, during each session the slide of moves was shared on the screen as a reminder (Figure 2 below).

Research Design and Approach

This study is qualitative in its nature. The major reason why the quantitative or mixed-method approach were not chosen is the sample size: as CD is a relatively novel concept in Vietnam, the number of Vietnamese practitioners of CD before this study was close to zero, so it is impossible to collect statistically meaningful quantitative data. In addition, this research is also a case study. Rose et al. (2019) outline three major types of a case study:

1. An intriguing case, when the unusual characteristic of the case is the primary focus.
2. A typical case, where the phenomenon is the focus, rather than the case itself.
3. A multiple case where the phenomenon is also the primary focus, and for variability to be captured, multiple perspectives are added to the case.

Figure 2. A snapshot of a VMCD session

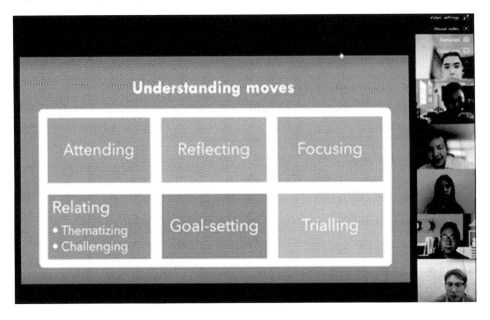

This study falls into the third category, as the research's focus is on the 'phenomenon' of VMCD, yet the participants, despite being relatively young and working in the same language school, their work experience, educational backgrounds (i.e. whether they have former teacher training or not), and CPD experience greatly vary from one another, which offers different insightful perspectives to the phenomenon in question.

Finally, the study, despite being inherently descriptive as a case study, also encompasses elements of intervention. The participants of the study received initial training on how CD worked, and the CD process also works in a similar fashion to action research or a reflective practice cycle (see Farrell, 2021; Wallace, 1991) as it also aims at identifying an issue and developing a course of action to target the issue.

For this study six strands of data were collected, including (1) initial questionnaire results of teacher CPD background and experience, (2) recordings of the VMCD sessions and the induction, (3) 15-minute immediately-after-CD interviews, (4) 40-minute interviews 3 weeks after the CD session, (5) recordings of the online lessons and (6) visual representations how the participants perceive CD. The array of different types of data and the fact that two interviews were conducted at two different points offer data and time triangulation, which helped provide a well-rounded picture to a particular point of analysis and enhance the overall validity of the research (Cohen et al., 2018). The following table summarizes the correspondence between the data and the research questions.

Table 4. Research questions and corresponding data

Research questions	Corresponding data
RQ1: What is the nature of discourse in a Video-mediated Cooperative Development session?	2, 3, 4, 6
RQ2: What are the teachers' general attitudes, challenges and difficulties towards Cooperative Development versus other means of CPD?	1, 2, 3, 4, 6
RQ3: To what extent has Cooperative Development encouraged teachers to make positive changes in their teaching?	3, 4, 5

The Nature of Discourse of a VMCD Session

Shares of Talk

Table 5 below gives information about the percentage of talk for each VMCD participant. From an overall perspective, it is evident that the Speakers (labelled yellow in the table) always took up the highest percentages, ranging from 38% to 50%. The participants were generally satisfied with the amount of talk, as they all said in the interview that as Speakers, they were given sufficient space for development.

As the Understander, the researcher (Alan), who also acted as the host of the session, ranked first since he took extra turns to introduce the Speakers, commence each part, and make mediating moves when necessary. However, it can be seen that his rates dropped gradually over time as his 'mediator' role was somewhat reduced due to the increased familiarity of the participants. Lee did not contribute a lot in the first session, at only 3%, but the percentage soared more than fivefold to 16% in the second session and remained consistent at 7-8%. He briefly described his sense of unpreparedness in the first session:

Interview Extract 1: Lee

Table 5. Percentages of talk in 5 VMCD sessions.

	Mike %	Phineas %	Derrick %	Lee %	Rebecca %	Alan %
Session 1	50%	4%	9%	3%	3%	31%
Session 2	9%	41%	9%	16%	4%	21%
Session 3	10%	4%	17%	7%	38%	24%
Session 4	5%	8%	52%	8%	1%	26%
Session 5	8%	7%	14%	50%	3%	18%

I was unable to stay focused in the first session, as I wasn't accustomed to what we were expected to do, the moves, the procedures and everything. Things got better over the next sessions when I started to think I got the hang of the moves.

Rebecca contributed the fewest talk as Understanders, accounting for around 1-4%, whilst Derrick constituted a sizable share of talk in most sessions, ranking second only to the host-mediator Alan. Mike and Phineas, despite each having one low, amounted to a consistent rate of 9-10%.

The Nature of Multiparty VMCD

Overall, VMCD meetings share similarities to other video-mediated forms of meetings, such as Zoom-enabled lessons, business meetings or virtual teacher education exchanges. On the plus side, virtual exchanges open up more opportunities to connect people from different time zones and locations (Loranc et al., 2021; Milić et al., 2020). Such an affordance helps bring together teachers residing in two locations, Ho Chi Minh city and Coventry, for regular meetings every week. Enhanced connectivity also helps facilitate the establishment of communities of practice (Dooly & Sadler, 2013), which is particularly the case with the 'CD Players' group in the dissertation. However, it also suffers difficulties regarding internet connection and latency (Seuren et al., 2021), gaze direction (Bannink & van Dam, 2021), and fractured ecologies (Luff et al., 2003). The subsequent part delineates the nature of discourse in a VMCD session through the lens of the Understander's moves.

1. Attending and setting up rapport

Four non-verbal, paralinguistic cues that constitute positive Attending are examined, including sitting postures, eye gazes, backchanneling, and smiles. Regarding the postures, since the participants suffer fractured ecologies (see Luff et al., 2003) as we were all seated in different rooms, coupled with the limited webcam angles, it was generally difficult to see a person's full physical stance. Throughout the sessions, there were moments when only the heads and parts of the shoulders of the participants were shown, with Figure 3 being a typical example. In addition, the power of human eye gazes in turn delegating and rapport building was rendered more or less null, as it is virtually impossible to know who was looking at whom, and to let the other(s) know that we were looking at them.

Unable to use eye gazes as a tool of turn delegating the sessions largely depend on the combo of the hand-raising functionality and address terms. The typical motifs that happened throughout the sessions were [Speaker talk] + silence + 'Teacher X, please'. The frequent use of address terms 'Teacher + name' appears to have

Figure 3. VMCD postures

increased the formality (and possibly awkwardness) of the talk, which to a certain extent causes harm to the attempt to build rapport among members. In the excerpt below, Mike (the Speaker) was trying to untangle his struggle as a novice teacher facing harsh reality (lines 1-7), which was followed by 5 seconds of silence (line 7) and the eventual selection of Derrick as the next speaker. Although this part is not directly related to Attending, I decided to mention it here because firstly, it is the direct result of the lack of eye gaze, and secondly, it influences the process of setting up initial trust and rapport, which is also what a good Attending move attempts to do.

Excerpt 1

Another problem that hinders effective Attending is the fact that participants were distracted by other businesses that happened on-screen. Mike and Derrick mentioned the distractors they had during the VMCD sessions:

01	MIKE:	hmmm talking about changes (.) to a certain extent(.) I
02		think(.) at a certain level(.) the problem must be on
03		me if the students refused to do what I want to(.) as
04		things keep moving forward but the problem's still
05		there(.) but perhaps as a novice teacher, my knowledge
06		is not good enough to interpret the problem and find
07		the solution (.) a way to go about it(5.0) ((Derrick))
08		Teacher Derrick(.) do you have any questions?(2.0)
09		So previously what I heard is that(.) what I have so
10	DERRICK:	far understood is that(.) you are still looking for the
11		appropriate teaching style right?

Interview Extract 2: Mike

At certain times, I had to admit that some of what everybody was saying fell on deaf ears as I was looking at several tabs on my computer screen. (…) It would have been different if it had been a face-to-face session – I would not have had anything to distract me.

Interview Extract 3: Derrick

Just like other online meetings, I got that kind of distractors from people sending me messages. During one of the CD sessions, I got a message from the office and it caused me to be temporarily negligent.

Two identifiable distractors are mentioned in the two extracts, including what the Understanders could see on their computer screen, and their text messages.

Backchannelling during the CD sessions was generally scarce. In fact, there were not a lot of instances where interruptions and overlapping occurred, as people often had their mics off when the other was speaking, and that functionality might have brought in its wake the reluctance to interrupt for some healthy signs of approval such as 'uh-huh' or 'that's true'. Excerpt 2 is one of the rare moments when backchanneling happened.

Excerpt 2

01 02 03	MIKE:	that's right, and it helps in brainstorming (.) and it also improves students' attitude (.) or like (.) stance. This also helps reduce the situation of (.) sitting on the fence(2.0)
04 05 06 07	ALAN:	uh huh (.) I see (1.0) occasionally the Understander might want to have his or her mic on and say 'uh huh I see' to show they are still listening (.) and not looking at Facebook. ((giggles and grins from everybody))

In this Excerpt, the host Alan gently reminded the other members to occasionally have their mics on for some 'uh huh' and 'I see'.

Smiles and grins appear to be the only feature that has helped improved Attending in VMCD, now that the power of the other elements was greatly constricted. Visual

Figure 4. Smiling faces

2 was taken from the same session as Visual 1 but in a more light-hearted moment when some members grinned and laughed at Mike's absent-mindedness.

Focusing

Normally the Speaker is the one setting his Focus (Mann, 2002). In the VMCD sessions, the Speakers were given some time to think about the critical incident they would like to discuss before coming to the meetings, but they were not obliged, and not supposed to, spend an extended amount of time at home thinking about it. There were cases when the Speakers were not very sure about the area of the topic they would like to talk about, and this happened to the youngest teacher Mike. Mike had a very broad topic of profits versus education in a language school, and through the session, it has been narrowed down to how to use English-speaking videos in a sustainable manner.

Excerpt 3

01 02 03 04 05 06	ALAN:	can I just squeeze in for a moment? Erm:: it seems that we have started to narrow down (.)the topic from a very general one (.)of the dilemma between business and English language learning to a smaller-scale classroom topic(.) like(.) er: how to bring watching videos, listening to TED talks, or Masterchef in a sustainable manner? So is that the case?

Context-Exploring

During the process of analyzing the VMCD data, a repeating set of patterns in the way the Understanders interacted with the Speaker was noticed. This particular move happened when the Understanders find that the context the Speaker has provided is insufficient for them to comprehend, and that more information should be added to better Understanding. As it is not listed as a standard CD move, for ease of analysis, I temporarily call it Context-Exploring. Context-Exploring might be in two forms. The first form is the inquiry for extra information, for example, the class size, the student's age, or whether it is an online or an online lesson (Excerpts 4 and 5). I believe this can lead to better understanding, as the Speaker did not always provide a clear enough picture for the Understanders to look into, and those questions help unravelling parts of that picture, which in turn fosters Empathy.

Excerpt 4

01	ALAN:	Is it a big class? How many people?

Excerpt 5

01 02	LEE:	Did that problem just happen in the Speaking class? Did it happen in the other classes as well?

The second form is the question of whether the Speaker has done something before. It is often in the form of 'Did you + [action]'?

Excerpt 6

01	LEE:	Did you provide instructions before you started the breakout rooms?

Lee might ask this question with a view to Context-Exploring: as it is difficult to visualize how a situation like breakout rooms happens, the answer to that question might provide a clearer piece of the breakout-room picture - the one about how instruction to breakout rooms were given. However, it can also be viewed as 'advice in disguise' – an attempt to provide a suggestion. In this context, I believe Lee was truly wanting to know more about the context rather than proposing a camouflaged suggestion, as throughout the session he has developed great Respect and Sincerity towards other members. In addition, perhaps it might sound more like a suggestion if it read 'Did you provide clear instructions before you started the breakout rooms?' (The implied version of 'your breakout rooms didn't work because your instructions weren't clear').

Reflecting

Reflecting is one of the most commonly used verbal moves throughout the researched VMCD sessions. It has worked as a mirror which granted the Speaker a chance to re-examine their thoughts from a different perspective (See Edge 1992a, 2002). Reflecting happened at different points; it can be either 2-3 minutes right after the Speaker initiated the talk, in the middle of the Speaker-articulation stage, or right in the Goal Setting and Trialing. Excerpt 6 shows Reflecting being done very early on, right after the speaker-articulation stage (lines 1-27). Alan took the chance when there was a small technical error (lines 28-34) to make a small joke (line 35) and briefly Reflect on Mike's long talk (lines 37-39).

Excerpt 7

01 02 03 04 05 06 07 08 09 10 11 12 13 14 15 16 17 18 19 20 21 22 23 24 25 26 27 28 29 30 31	MIKE: DERRICK: MIKE: DERRICK:	Thank you very much for attending my session, dear Teachers(.) Hmm (.) As I am the youngest here, and I have to start first(.) to be to be honest, hmm, eh, I have a rough overview of what my critical incident entails, but it has come eh, been influenced by different dimensions. Because it's a problem I encountered during the process of converting and (..) what I learned and the reality. Hmm… let's first talk about my background, what I study – I'm studying Teaching English as a Foreign Language (.) so so normally when I learn TEFL (.) eh when the professors teach us English language teaching methodology(.) they consider communicative competence very highly – hmm how to put it(.) like students are not supposed to know the language, but to use it and apply it in life. However when we look at the market er: education has been heavily commercialized and exam-oriented. Hmm uh uh Like IELTS, for example. Here I don't mean that it's a bad thing, as it's business. But what I'm concerned about is how to balance between uh uh uh how to put it(.) teaching for eh eh eh (2.0) sustainably so that students can actually use the language and at the same time achieving business interests, IELTS band scores. And that's one of my biggest concern. And when I keep thinking about this(.) gradually it has affected the way I teach. In my former workplace which I wouldn't want to mention the name, its teaching styles focus heavily on grades and templates, and it I keep on teaching like that(.) gradua—lly I have become a teaching machine which puts everything into templates and formulaic expressions – everybody can do that. Sorry, Was I out for a moment? Was I out? ((waving hysterically)) yeah yeah, it was laggy for like 1 min. can't hear, can't hear anything=
32	MIKE:	=I was still talking – sorry for the noises=
33 34	DERRICK:	=My connection was unstable for a while, Teacher Alan – I think Mike just froze.
35	ALAN:	let's blame the school's wifi
36		((giggle)) ((smiley faces with mics off))
37 38 39	ALAN:	ok ok, let's take this chance to reflect – so if I understand it correctly (.) your current problem is (.)to balance between preparation for the test and teaching English, is it correct?

Reflecting also occurred in the middle of the Speaker-articulation stage. Excerpt 7 below was taken around 20 minutes after the Speaker initiated the conversation. Phineas attempted to Reflect on Mike's philosophical struggle between the money-making versus educational domains of a language school, and between learning to the test and learning for knowledge. Mike interrupted midway through the Reflecting to confirm that it was accurate, and then did that a second time when Phineas finished his Reflection.

Excerpt 8

01	PHINEAS:	ermm Teacher Mike may I ask erm: let me reflect what you have
02		said so far erm:: so now (.) what you are concerned about now
03		is how to balance the work ethics(.) how to balance the work
04		ethics and the business at the language school(.) there::::
05		should be a balance between the two aspects(.) as students are
06		supposed to gain the knowledge and use it(.) but at the same
07		time the students also need to attain the desired band
08		scores(.) is that right?=
09	MIKE:	= yes that's right=
10	PHINEAS:	=but but you are also wondering how to achieve the best of
11		both worlds(.) which means giving students some sort of
12		benefits(.) and because of the time constraint you
13		have yet to have any concrete plans(.) the innovative ideas are still
14		somewhat primitive and you just use those as they are(.) like
15		when you feel that particular method is good you use it(.) Is
16		that correct(.) Teacher Mike?
17	MIKE:	Yeah that's correct(.) To elaborate that further for
18		example(.) when I ask a question in my class, but if I have
19		too many questions, it might become tedious, but yeah there
20		are things that are useful, and there are also things that
21		aren't very useful. For instance, like, for instance, high-
22		intermediate classes find debates pretty useful, and hopefully
23		I can (..) apply it more in the future.

Reflecting also occurred during Goal setting and Trialing. In Excerpt 8 below, Alan Reflected on Phineas' series of solutions he had previously come up with during Goal setting and Trialing.

Excerpt 9

01	ALAN:	Okay. Let me reflect on the series of solutions you've just
02		come up with (.) You you seem to have some solutions to the
03		main problem and its associated sub-problems (.)the first one
04		is (.)improving your accents and the way you deliver the
05		lessons in a way that make those lessons sound more
06		attractive=
07	PHINEAS:	=yeah=
08	ALAN:	=the second one is informing the office of students who are
09		absent without notice (.) right?=
10	PHINEAS	=yeah=

The Reflecting move Alan made in this situation is particularly interesting, as it was an attempt to Reflect Phineas' talk particles that spread over more than 20 minutes of Speaker-Understanders exchanges. In other words, reflecting in this regard acted as a way of summarizing the Speaker's talk segments, compressing them into a verbal 'morsel' and ultimately gave it back to the Speaker. Three over five participants acknowledged the importance of Reflecting-Summarizing, and it is interesting to note that those three were the older teachers (Derrick, Lee and Rebecca). Rebecca stated that Reflecting in this sense helped her problem as a whole; Lee also shared a similar opinion as he highlighted the importance of seeing his ideas in systematic manner.

Interview Extract 4: Rebecca

What I've been uttering was kinda scattered all over. I didn't even know what I was saying, so when everybody helped me summarise, I got to see for myself what my problem actually was

Interview Extract 5: Lee

At the beginning of the talk I had this jumble of chaotic thoughts, and then I got them sorted out in a clear-cut, systematic manner.

Alan appeared to Reflect in this summative sense a lot compared to the other members; this might be because that he was slightly more experienced in CD.

Relating

Relating is two-pronged by nature; it helps link (Thematizing) or show the contradictions (Challenging) between two points (see Mann, 2002; Edge, 2009). In the dataset, they were the lesser used ones – the participants did not perform a lot of Thematizing and Challenging, with Challenging happening even less. Excerpts 9 and 10 are two examples of Thematizing and Challenging (in that order), both performed by Lee. In Excerpt 9, Lee invited the Speaker to think about the possible connection between the teacher's mental state and the effectiveness of a lesson, whilst in Excerpt 10, he raised a question about whether the Speaker meant learning English to prepare for a test does not help students improve their communication skill. This can also be seen as Reflecting, as Lee appeared to rephrase what he understood from the Speaker's articulation.

Excerpt 10

01 02 03	LEE:	Erm:may I say something (.)teacher Phineas? So you have mentioned the teacher's mood and then the effectiveness (.) of a lesson (.) are they connected in any way?

Excerpt 11

01 02 03 04 05 06	LEE:	ah Teacher Mike, I would like you to clarify this one point – you said the traditional ways of teaching is to teach the knowledge (..) and students are gonna use that to communicate, but from the test-taking perspective, the focus is on the scores, so, does that mean that (..) test preparation is not relevant to the students' ability to communicate?

Later in the interview, Lee and Mike stated that Thematizing and Challenging are the hardest moves for them to master because (1) the brain somehow makes assumptions and eventually ignores any connections, and (2) trying to stay focused and at the same time attempting to look for possible connections is mentally exhausting.

Interview Extract 6: Lee

Yeah I think the hardest is probably Thematizing and Challenging. I don't know really, maybe it's just me. I tried several times testing out how they worked, but it's just too hard to find that point A and point B from the Speaker's talk to make connections. Or maybe because my head is already making connections you know. I made assumptions that there were obvious connections, so I didn't ask.

Interview Extract 7: Mike

It is already very difficult trying to stay focused and really understand the Speaker, not to mention connecting the dots and seeing the linkable or contradictory relations.

Goal Setting and Trialing

Goal setting and Trialing are the two moves that help the Speaker develop possible courses of action towards the issue in question. In the VMCD meetings, we often called this 'trying to get the solutions out'. We did not 'get the solutions out' in all of the VMCD sessions. The one with the strongest and most concrete courses of action was Rebecca trying to make her online breakout rooms more exciting, and she was very happy about being able to formulate her own solutions in the speaker-review stage.

Figure 5. Lee's drawing

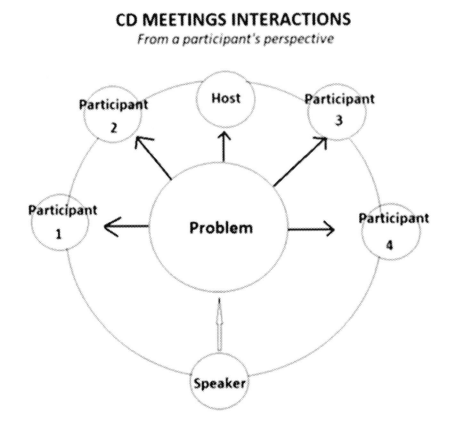

CD MEETINGS INTERACTIONS
From a participant's perspective

Excerpt 12

01		What should I do now? Should I(.) summarise everything or
02	REBECCA:	should I say anything else?
03		Just say what you want to say (.)maybe some moments of
04	ALAN:	enlightenment?
		There are a lot of moments of enlightenment really (.) I
05		really need to note them all down so that I can remember and
06	REBECCA:	use them later.

Mike, after having his topic narrowed down from 'profits versus education' to how to use videos and debates effectively, appear to develop some courses of action as well, and Derrick also had some solutions on how to help his student in a one-on-one class remember the grammar and vocabulary in a sustainable manner.

Figure 6. Derrick's drawing

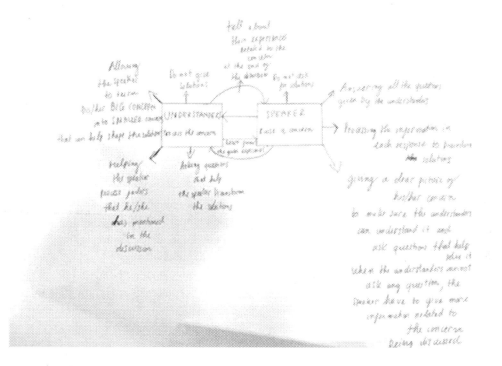

Figure 7. Rebecca's drawing.

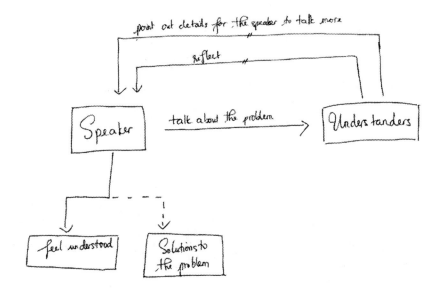

Figure 8. Phineas' drawing

Figure 9. Mike's drawing

Lee did not seem to develop a concrete 'solution', and this might be due to time constraints, and also his topic (English-Vietnamese structural differences and how those can hinder the learner's competence) is a relatively complicated.

Interaction From the Participants' Perspectives

Besides the analysis of CD interaction and nature of discourse from the conversations, I believe it is useful to also know how the participants mentally visualize the interactions, and one of the useful data is visuals ('I did not realize this until you asked me to draw it out' - Lee). Explanations of the visual were provided in the interview.

For Lee (see Figure 5 below), CD is about a Speaker putting a problem on the table (the big red oval-shaped one), spreading it out to the participants, and the Understanders ('participants' and 'host' in the photo) helped by picking up the problem and breaking it down into smaller chunks by using moves. It is interesting to note that he also mentioned besides the Speaker-Understander interaction, there also exists an Understander-Understander interaction (the big circle around the participants), albeit to a much lesser extent. This might work in a similar fashion to a focus group interview: one Understander's stream of talk might help the other Understander formulate an Understanding move.

From Derrick's diagram, it is clear that he shared a similar perception of CD interaction: the Speaker raise a concern, and the Understanders process the concern by breaking down big concerns into smaller concerns. He also provide a summary of CD key ideas, and he was one of the two who mentioned the Resonances ('talk about their experiences related to the concern at the end of the discussion').

Rebecca used a different description of CD as not 'breaking the big problem into smaller chunks', but 'reflecting and pinpointing the details. She summarizes the six core moves of CD into two branches: Reflecting and 'pointing out details for the Speaker to talk more'. By putting Reflecting in a separate branch, she highlights the importance of the move for the personal growth of the Speaker, and the 'details' in the latter branch refer to the connections drawn as a result of Thematizing and Challenging. The results of the two branches of discourse might lead to the Speaker feeling understood and/or possibly (represented by the intermittent arrow) coming up with a solution.

The two youngest teachers, Mike and Phineas, provide cartoonish visual representations (which is fascinating as it provides us with an insight into how a particular professional development activity can resonate with the teacher's personal experiences). Mike's CD involves a team of Speaker-Understander Power Rangers, with the Speaker playing the lead (the 'red ranger' in the series), fighting the Problem monster. Whilst it does not mention in detail what happens in a CD session, it highlights the power of teamwork and collegiality. It is interesting to notice that he perceives the Speaker's topic as a monster, and this might be due to the fact that he is going through a number of uncertainties as a novice teacher.

Mike's drawing, the only one with colors, delineates an interesting relationship between the Speaker and the Understanders ('Listeners') with the Understanders providing essential conditions (rain and sunlight) for the Speaker (the flowers and the tree) to grow. It is also worth noting that he mentioned the knowledge and experience of the Understanders as the conditions for growth, which is unusual as practitioners of CD tend to avoid offering their own experiences except during the Understander-resonance stage. Mike also confirmed in the interview that he learned most from the Resonances, which is understandable for a novice teacher who might need early-career advice.

Teachers' Attitudes Towards VMCD

Advantages of VMCD

Compared with the traditional top-down, external models of CPD, what CD appears to stand out owing to the following features: (1) 2-sided instructiveness (2) a sense of self-development via articulation, (3) feeling heard without judgment, (4) becoming a better listener, (5) more opportunities for professional bonding, (6) the use of hand-raising functionality and (7) flexibility with regard to scheduling (the last two points are VMCD-specific).

Firstly, the participants praised CD for its novel and unique way of interaction. Mike further commented on this unique feature compared to going to workshops.

Interview Extract 8: Mike

A lot of workshops and seminars are one-sided, with the presenter doing all the speaking and the audience doing all the listening, and in many cases, the audience misses huge parts of what the presenter was saying. CD, on the other hand, involves two-sided interaction where both parties are required to continuously listen and speak.

The second advantage of CD is that it helps the practitioners untangle and recognize the threads of unexamined thoughts in an organized way. Derrick said that CD has helped him 'activate' his brain, and Lee, in both interviews, highlighted the fact that he could see his problem reordered in a systematic way.

The third and fourth advantages refer to the features of non-judgmental discourse. As Speakers, all of the participants mentioned that they enjoyed the feeling of being heard as opposed to being advised. As Understanders, practicing the Understander's moves help them contemplate on their daily articulation.

Interview Extract 9.1: Mike

CD makes me re-think about my everyday way of articulation. I used to have a bad habit of giving advice nobody asked for – it can be tactless in a sense.

Interview Extract 9.2: Mike

I have become more patient when I listen to my students. I practice full listening – waiting until my students finish what they have to say so that I can fully understand them, and thereafter formulate my responses.

Both Interview extracts 9a and 9b were from Mike, the former of which was conducted right after his session as the Speaker, and the latter of which was taken 3 weeks after that session. It is clear from his statements that besides the potential of coming up with a solution to the issue discussed, CD can also help shape the teacher's general communication towards a non-judgmental, empathetic way. Improved communicative competence appears to be the positive 'side effect' of CD that the participants greatly enjoyed.

In addition, the teachers also commented both in the interviews and in the VMCD sessions that they got to realize they share similar classroom issues. Lee exclaimed that talking to others gave him a sign of relief that he is not 'alone in this world of teaching'. Those 'you are not alone' moments generally congregated in high density during Understander-resonance stage, but it also sometimes happened in the middle of the Speaker-articulation, for example in Excerpt 11 below.

Excerpt 12

01	((Rebecca))	
02	Phineas:	Teacher Rebecca please
03 04 05 06	Rebecca:	Erm before I say anything, I would like to share that you are not alone (.)I greatly resonated with you on this (.) I got that situation when the students came to class very late and I was pretty upset too…

During this Phineas' session as the Speaker, Rebecca cyber-raised her hand to interrupt and eventually Disclosed that she was also very upset when the students came late. This, alongside her similar Resonance that happened later on in this session, has built up a great sense of Empathy which built up the ambiance of the session.

The fifth advantage of CD is the fact that it has created more opportunities for the teachers to meet and have a chat.

Interview Extract 10: Derrick

I've never met Mike before. I rarely meet Rebecca [NB: they do not work in the same branch]. The CD meetings have given me a chance to get to know people more, and what's better, to have a sense of community, to feel that we belong to a group of the CD Players. When we have that opportunity to meet in real life, as part of the CD Players we will already have something in common to strike up a conversation.

Derrick's quote in Interview extract 8 resonated with other teachers' opinions about CD sessions working as an additional gathering zone for colleagues to meet up. Mike also highlighted the fact that most of the talk's teachers share with each other are short and often at lunchtime; we hardly ever have a chance to have an extended, more-than-30-minute in-depth talk.

The sixth and final advantages specific to the online nature of VMCD are the flexibility in scheduling and the use of hand-raising functionality. Online CD significantly opens up more time frames; Phineas mentioned that many of his evening classes end at 7-8 PM, and he would love to attend the VMCD meetings at around 10 PM, calling it his 'ideal' time for a meeting. Additionally, the participants also enjoyed the use of hand-raising functionality as a tool of turn delegating, as they believe it helped turn-taking more organized.

All of the participants stated that they generally enjoyed VMCD, and that they will definitely attend VMCD again in the future. In addition, it is also suggested that CD could be a great addition to other CPD activities such as classroom observation.

Disadvantages and Difficulties

Three general setbacks identifiable from the data include (1) the feeling of uncertainties, especially when no solutions are made and (2) the uncomfortable 'low' moments during the sessions, and (3) the technological problems.

Firstly, the two younger teachers – Mike and Phineas – commented that they felt something is missing from CD. Whilst they refrained from directly criticizing VMCD, their responses hinted that they were not generally satisfied with the lack of a concrete, tangible, and 'practical' solution to their issues ('I want something practical; something I can use right away in my class's – Phineas). Mike also commented that CD is like the 'blueprint of possible solutions', and that the feedback from expert teachers following a classroom observation 'offers a more realistic, on-the-spot feeling' compared to CD.

Secondly, Phineas also noticed that there were some lows during the sessions where people stayed silent for an extended period of time. It is not rare to see 7-second period of silence, and the longest recorded was 13 seconds (Excerpt 12).

Excerpt 14

01 02 03 04 05 06 07 08 09 10 11 12 13	Mike:	I think it's a yes (.) but but it's quite difficult in the current context, but do you know why? Erm I think if, for example, a student(.) for example (.) with great English might not be interested in learning (.) it's not worth mentioning if they do like it (.) but if they like it and they are also good at it(.) and they are willing to make efforts because they know it's good for them(.) the long-term benefits will include the benefits of test prep(.) like(.) once you get the hang of English that sort of thing(.) so the test's not only about the scores(.) but it also reflects their actual ability(.) and they don't have that fear of tests(.) as they know it reflects their command of English(.) and when they're excellent(.) those tests of English are just a piece of cake(.) Yeah
14		(13.0)
15 16 17 18 19	Mike:	that's the thing I've always:: wanted to find out, but as I'm quite new, and I'm still testing out at different environments, so I'm trying to find an appropriate teaching style erm:: to cater for the needs of the society, and I think I'm doing the right thing too.
20		(4.0)
21 22	Derrick	So:: (2.0) what have you done erm::: to find what you want to find::, like, what have you achieved?

From this excerpt, we can see that for 13 seconds, no one self-selected as the next speaker, which prompted Mike to continue his talk with some more details. However, again the talk was disrupted for another 4 seconds of silence before Derrick self-selected and started a new move.

The third disadvantage is technologically related: the teachers commented on the unstable Internet connection that happened at times, the noticeable latency, and the noises from different backgrounds which prompted them to have their mics off. The participants also faced problems related to fractured ecologies, which were largely covered in the Attending section (3.1.2.1) of this dissertation.

The five Understanders reported that they faced the following difficulties and challenges upon doing CD: (1) the deteriorated level of concentration, (2) the feeling of restraint upon having to abstain from giving suggestions, (3) difficulties in formulating the moves, and (4) the struggle upon receiving multiple, continuous moves from different Understanders.

Firstly, the Understanders found it relatively difficult to maintain the same degree of focus throughout the 50-minute long session. One reason for this is the on-screen distractions. Phineas also noted that sitting at the table for 50 minutes straight tired him out, and Derrick also agreed with this idea, further adding that prolonged sitting can cause harm to his neck and back. Another important factor affecting the practitioner's concentration is the relevance of the topic, as is discussed by Lee in Interview extract 10 below:

Interview Extract 11: Lee

The most important thing that affects my concentration is whether the topic the Speaker brings to the table is something I have knowledge of or have experienced before. If I share similar concerns and interests, I'll probably pay more attention. If I can't fully relate to the topic, maybe because I have never experienced it before, my attention span will probably drop somewhere in the middle even though I've tried to stay focused.

Interestingly, all of the five participants shared a similar concern with Lee. Phineas said as a novice teacher, he felt left out on several occasions, and Derrick suggested that the Understanders be informed of the topic several days beforehand to get their minds prepared.

The second difficulty is the result of repressing the urge to give advice. Lee mentioned that sometimes he got this very interesting suggestion he believed would be very helpful to the Speaker, but he had to keep it dormant until the Resonance stage. Phineas were also conscious of the fact that asking whether someone has done a particular classroom activity is considered a suggestion and therefore goes against the 'rule' of CD.

The third difficulty lies in mentally formulating the moves. Thematizing and Challenging are generally the harder moves to make (see 4.1.2.5), and Rebecca also mentioned another struggle she faced when she had the habit of listening until the very end to make sure that she fully understood the Speaker, yet for some reason she ended up not saying anything at all, thereby giving up her turn to the other Understanders.

The final difficulty is related to the situation when there were multiple moves formulated in close proximity to one another. Lee noted that occasionally while he was trying to let the first Understander-Speaker encounter to sink in, he had to face another Understanding move, sometimes about a completely different thing, closing in on him.

CONCLUSION

Boon (2015) outlines three generations of CD, which commence at the original pair CD (first generation), continue with GD and EMCD (second generation), and with the advent of technology, IMCD (third generation). In the wake of the post-COVID era and the unprecedented development of videoconferencing technology, the present study, alongside Webb et al (2022), steps onto the fourth generation of CD: VMCD. In addition, by studying novice teachers, the research project offers another angle into GD as it sheds light on how early-career teachers responded to the model. The study also contributes to the repertoire of reflective practice studies in Vietnam a new, dialogically inspired form of reflective practice, alongside more familiar domains such as action research or lesson study. Finally, in the spirit of this book, this project also hopes to add in a humble contribution to the existing literature of multidisciplinary applications of CMC in the field of education.

REFERENCES

Bailey, K. M., Curtis, A., & Nunan, D. (2001). *Pursuing Professional Development: The Self as Course*. Heinle ELT.

Bannink, A., & van Dam, J. (2021). Teaching via Zoom: Emergent Discourse Practices and Complex Footings in the Online/Offline Classroom Interface. *Languages*, *6*(3), 148. doi:10.3390/languages6030148

Boon, A. (2007). Building bridges: Instant messenger cooperative development. *Language Teaching*, *31*(12), 9–13. https://jalt-publications.org/files/pdf/the_language_teacher /12_2007tlt.pdf#page=10

Boon, A. (2011). Developing Instant Messenger Cooperative Development. *Bulletin of Toyo Gakuen University*, *19*, 109–120. doi:10.24547/00000239

Boon, A. (2019). Facilitating reflective practice via Instant Messenger Cooperative Development. *Indonesian Journal of English Language Teaching*, *14*(1), 35–54. doi:10.25170/ijelt.v14i1.1417

Borg, S. (2016). Researching Language Teacher Education. In B. Paltridge & A. Phakiti (Eds.), *Research Methods in Applied Linguistics - a practical course* (pp. 541–555). Bloomsbury.

Cohen, L., Manion, L., & Morrison, K. (2018). *Research Methods in Education* (8th ed.). Routledge.

Cowie, N. (2002). CD by email. In Edge. J (Ed), Continuing Cooperative Development (pp. 225–229). University of Michigan Press.

Do, K. (2022). *Video-mediated Cooperative Development in Vietnam: a case study* [Unpublished Master Thesis, University of Warwick].

Dooly, M., & Sadler, R. (2013). Filling in the gaps: Linking theory and practice through telecollaboration in teacher education. *ReCALL, 25*(1), 4–29. doi:10.1017/S0958344012000237

Edge, J. (1992a). *Cooperative Development*. Longman Group UK Limited.

Edge, J. (1992b). Co-operative Development. *ELT Journal, 46*(1), 62–70. doi:10.1093/elt/46.1.62

Edge, J. (2002). *Continuing Cooperative Development - A Discourse Framework for Individuals and Colleagues*. The University of Michigan Press. doi:10.3998/mpub.8915

Edge, J. (2006). Computer-mediated cooperative development: Non-judgemental discourse in online environments. *Language Teaching Research, 10*(2), 205–227. doi:10.1191/1362168806lr192oa

Edge, J. (2011). *The reflexive Teacher Educator in TESOL: Roots and Wings*. Routledge. doi:10.4324/9780203832899

Farrell, T. S. C. (2019). Reflective Practice in L2 teacher education. In S. Walsh & S. Mann (Eds.), *The Routledge Handbook of English Language Teacher Education*. Routledge. doi:10.4324/9781315659824-5

Farrell, T. S. C. (2021). *TESOL Teacher Education - A Reflective Approach*. Edinburgh University Press. doi:10.1515/9781474474443

Freeman, D. (2009). The scope of second language teacher education. In A. Burns & J. C. Richards (Eds.), *The Cambridge Guide to Language Teacher Education* (pp. 11–19). Cambridge University Press.

Loranc, B., Hilliker, S. M., & Lenkaitis, C. A. (2021). Virtual exchanges in language teacher education: Facilitating reflection on teaching practice through the use of video. *TESOL Journal*, *12*(2), 1–15. doi:10.1002/tesj.580

Luff, P., Heath, C., Kuzuoka, H., Hindmarsh, J., Yamazaki, K., & Oyama, S. (2003). Fractured Ecologies: Creating Environments for Collaboration. *Human-Computer Interaction*, *18*(1–2), 51–84. doi:10.1207/S15327051HCI1812_3

Mann, S., & Walsh, S. (2013). RP or 'RIP': A critical perspective on reflective practice. *Applied Linguistics Review*, *4*(2), 291–315. doi:10.1515/applirev-2013-0013

Mann, S., & Walsh, S. (2017). *Reflective Practice in English Language Teaching*. Routledge. doi:10.4324/9781315733395

Milić, J., Ehrler, B., Molina, C., Saliba, M., & Bisquert, J. (2020). Online Meetings in Times of Global Crisis: Toward Sustainable Conferencing. *ACS Energy Letters*, *5*(6), 2024–2026. doi:10.1021/acsenergylett.0c01070 PMID:34192148

Schön, D. A. (1992). The Theory of Inquiry: Dewey's Legacy to Education. *Curriculum Inquiry*, *22*(2), 119–139. doi:10.1080/03626784.1992.11076093

Seuren, L. M., Wherton, J., Greenhalgh, T., & Shaw, S. E. (2021). Whose turn is it anyway? Latency and the organization of turn-taking in video-mediated interaction. *Journal of Pragmatics*, *172*, 63–78. doi:10.1016/j.pragma.2020.11.005 PMID:33519050

Timperley, H. (2011). *Realizing the Power of Professional Learning*. Open University Press.

Wallace, M. J. (1991). *Training Foreign Language Teachers - A reflective approach*. Cambridge University Press.

Walsh, S., & Mann, S. (2015). Doing reflective practice: A data-led way forward. *ELT Journal*, *69*(4), 351–362. doi:10.1093/elt/ccv018

Webb, K., Mann, S., & Shafie, K. A. (2022). Using Computer-Mediated Cooperative Development in a Virtual Reflective Environment Among English Language Teachers. In Z. Tajeddin & A. Watanabe (Eds.), *Teacher Reflection: Policies, Practices and Impacts* (pp. 224–237). Multilingual Matters. doi:10.21832/9781788921022-019

Wedell, M. (2017). Teacher education planning handbook: Working together to support teachers' continuing professional development. In M. Wedell (Ed.), *Teacher Education Planning Handbook: Working together to support teachers' continuing professional development*. British Council.

ADDITIONAL READINGS

Barfield, A. (2002). Refreshing the hearts that other approaches don't reach. In J. Edge (Ed.), *Continuing cooperative development: A discourse framework for individuals as colleagues* (pp. 237–243). University of Michigan Press.

Bibila, S. (2011). Professional Development in the era of Hermes. *TESOL Journal*, *2*(1), 91–102. doi:10.5054/tj.2011.220144

Croes, E. A., Antheunis, M. L., Schouten, A. P., & Krahmer, E. J. (2019). Social attraction in video-mediated communication: The role of nonverbal affiliative behavior. *Journal of Social and Personal Relationships*, *36*(4), 1210–1232. doi:10.1177/0265407518757382 PMID:30886451

Farrell, T. S. (2012). Reflecting on Reflective Practice:(Re) Visiting Dewey and Schon. *TESOL Journal*, *3*(1), 7–16. doi:10.1002/tesj.10

Farrell, T. S. (2018). Operationalizing reflective practice in second language teacher education. *Journal of Second Language Teacher Education*, *1*(1), 1–20.

Liddicoat, A. J. (2021). *An introduction to conversation analysis*. Bloomsbury Publishing.

O'Conaill, B., Whittaker, S., & Wilbur, S. (1993). Conversations over video conferences: An evaluation of the spoken aspects of video-mediated communication. *Human-Computer Interaction*, *8*(4), 389–428. doi:10.120715327051hci0804_4

Rogers, C. R. (1995). *On becoming a person: A therapist's view of psychotherapy*. Houghton Mifflin Harcourt.

Schmid, P. F. (1998). On becoming a person-centered approach': A person-centred understanding of the person. *Person-centred therapy: A European perspective*, 38-52

van der Kleij, R., Maarten Schraagen, J., Werkhoven, P., & De Dreu, C. K. (2009). How conversations change over time in face-to-face and video-mediated communication. *Small Group Research*, *40*(4), 355–381. doi:10.1177/1046496409333724

KEY TERMS AND EXPLANATIONS

Computer-Mediated Communication: Communication with the assistance and mediation of computers

Cooperative Development (CD): A dialogic reflection framework developed by Edge (1992). Practitioners of CD play the roles of the Speaker and the Understander to help each other self-explore their problems.

Reflective Practice: The act of teachers reflecting on their teaching practices with a view to identifying possible problems and working towards a course of action (see Schön (1992)).

Second Language Teacher Education: The field whose coverage entails preparation, training and education of second language teachers.

Video-Mediated Communication: Communication with the assistance and mediation of videos. In this chapter, it is referred to as communication with the assistance of video-conferencing applications such as Microsoft Teams, Google Meet or Zoom.

Chapter 10

CMC Users' Positive and Negative Emotions:
Features of Social Media Platforms and Users' Strategies

Hong Quan Bui
Ho Chi Minh City University of Education, Vietnam

Thanh Tra Tran
Ho Chi Minh City Open University, Vietnam

ABSTRACT

The relationship between social media and users' emotions is a prevalent research topic in computer-mediated communication. This study aims to explore the features of social media platforms that affect users' emotional self-expression and the strategies employed by users to express their emotions in text-based communication. Semi-structured interviews with ten regular users of major social media platforms, namely Facebook and Zalo, were conducted to collect data. Results showed that online network density and size, visual properties, and content display were three features that influenced users' emotional expressions. To convey their emotions on social media platforms, users employed both verbal strategies (e.g., using affect terms and verbosity) and non-verbal strategies (e.g., using punctuations and emoticons). Some participants used punctuation marks, especially question marks, and emoticons to express positive emotions, while others used verbosity to express negative feelings. Implications for using social media platforms are discussed.

DOI: 10.4018/978-1-6684-7034-3.ch010

INTRODUCTION

Computer-mediated communication (CMC) encompasses various forms of human communication through electronic gadgets (Adrianson, 2001, Hung et al., 2022). With the advent of social media platforms as a form of CMC, people can share their emotions with a broad audience with a single touch on a smartphone (Kross & Chandhok, 2020). Sharing information, including emotional self-expression, has been considered the most frequent purpose for joining social networking sites (Li et al., 2020; Lin & Lu, 2011). Social media is becoming an essential medium for users to exchange and receive information, particularly in seeking support for health and crises (Jin et al., 2016; Li et al., 2020). The relationship between emotions and online social networks is a prevalent topic of researchers (Dhingra & Parashar, 2022; Galen, 2017; Kross & Chandhok., 2020; Pantti & Tikka, 2014). In the past, researchers argued that it was challenging to express emotions in CMC because non-verbal cues such as facial expressions and body movements were absent (Rice & Love, 1987; Sproull & Kiesler, 1986; Short et al., 1976). However, other researchers have found that online communication seems to boost the expression of emotions rather than deter people from disclosing their feelings (Choi & Toma, 2014; Derks et al., 2008b; Serrano-Puche, 2016). Additionally, many studies have been conducted to investigate users' emotional communication on social networking sites. Some studies focused on examining a particular emotion, such as annoyance (Livingstone et al., 2014), envy or jealousy (Christofides & Desmarais, 2009), hope (Fürst, 2014), and resentment (Risi, 2014). Also, a few studies explored users' perceptions of the appropriateness of expressing positive or negative emotions (Lin et al., 2014; Waterloo et al., 2018).

The empirical results of how users disclose their emotions on social networking sites vary. For instance, some research indicated that social media users express more positive emotions than negative emotions (Galen, 2017; Valenzuela et al., 2009; Waterloo et al., 2018). In contrast, Kalpidou et al. (2011), Kross et al. (2020), and Bazarova (2012) argued that some social media had been used to express negative emotions. To the authors' best knowledge, although many studies on emotional self-expression have been conducted, more research is needed to explore social media features that affect users' emotional self-expression. Lin et al. (2017) pointed out that the number of online friends can influence users' emotional self-expression. Users with many online friends tend to share positive content to express emotions (Lin et al., 2017).

People use various strategies to convey emotions in social media communication, including using more punctuation or emotion icons (emoticons). Hancock et al. (2007) suggested that people can use more punctuation to emphasize their feelings. Derks et al. (2008b) stated that users could deal with the restriction of CMC by using emoticons, defined as typographical symbols that replace facial expressions in CMC (Walther & D'Addario, 2001).

Previous studies have mainly adopted quantitative methods to investigate the relationship between users' expressions of emotion and social networking sites such as Facebook (Choi & Toma, 2014; Lin et al., 2014; Stieglitz & Dang-Xuan, 2013; Young et al., 2017). Therefore, this study employs the qualitative method to examine factors affecting users' emotional self-expression and their strategies to express emotions on social networking sites to fill the mentioned gap. The results will contribute to the literature on emotions in computer-mediated communication and provide implications for social media platforms to improve users' experience. This qualitative research investigates ten Vietnamese adults' experience using social media and how they express their emotions in online environments such as Facebook and Zalo. Zalo is a Vietnamese social networking site developed by the VNG company.

LITERATURE REVIEW

Social Media Communication

Since the appearance of social media platforms, human life has changed significantly, especially in communication; many studies are exploring this change. *Social media platforms* are the technology that enables users' sharing and interaction in an online community (Nxumalo & Chiweshe, 2019). Notably, Kapoor et al. (2018) defined *social media* as various user-driven platforms that promote the information dissemination of exciting material, the creation of conversation, and communication to a larger audience. It is a digital area developed by and for the people that creates a favorable atmosphere for interactions and networking at various levels (for instance, personal, professional, business, marketing, political, and societal) (Kapoor et al., 2018).

There has been various research on social media platforms with different aspects. Social media studies focus primarily on behavior and consequences for users (Graciyal & Viswam, 2021; Matook et al., 2015; Turel & Serenko, 2012), concerns and risks to users (Griffiths & Light, 2008; Vishwanath, A. 2015); the role of social media as an aid to users (Spagnoletti et al., 2015; Yan et al., 2015). Social media communication can benefit and adversely affect users' emotional well-being. According to Matook et al. (2015), users' active or passive interaction with social media can influence the relationship between online social networks and reported loneliness. Griffiths and Light (2008) also researched media convergence, which occurs when a gaming website integrates social media features, putting a young audience in danger of scams. According to an Australian study, many users may view themselves as low-risk targets or be unaware of the potential risks of disclosing personal information on social media sites (Tow et al., 2010). According to Vishwanath (2015), Facebook users can fall victim to phishing attacks on two different levels: they can be directly solicited for personal information by phishers, or they can fall victim to phishers using fake profiles. Studies also show that social media platforms serve as a tool for users to seek community support. According to Yan et al. (2015), social media can help patients reach out to others with the same health issue but different medical treatments. As a result, patients can compare their health conditions to others. Yan and Tan (2014) present a partially observed Markov decision process model to gather enough data to support the idea that providing emotional support to patients significantly impacts their health. In their 2014 study, Kallinikos and Tempini explore the benefits and drawbacks of using a sizable unsupervised social network based on patient self-reporting to collect and analyze patient health data.

Facebook and Zalo are the two most recognizable social media platforms in Vietnam. Especially among young adults and adolescents, who consider these sites as platforms for self-expression and social networking. Facebook is a social networking site founded in 2004 by Mark Zuckerberg, an American computer programmer, for easy connection and sharing with family and friends online. Zalo is a Vietnamese social media platform developed from a messaging application. Facebook is currently one of the market's most popular social media platforms (Andreassen & Pallesen, 2013). Existing evidence suggests that Facebook can help people gratify their personal and social needs, such as forming and maintaining bonds (Ellison et al., 2007; Valenzuela et al., 2009).

Additionally, some researchers advocating for Facebook as a teaching and learning tool presented evidence demonstrating how Facebook can be beneficial for college students and help increase teacher-student and student-student interactions (Muñoz & Towner, 2011). Another study by Valkenburg et al. (2006) suggested that positive feedback on their Facebook profiles or posts may improve adolescents' self-image, self-esteem, and overall well-being. However, an independent research paper applying meta-analysis and a systematic review has shown a discernible link between Facebook usage and mental health problems (Frost & Rickwood, 2017). They discovered that Facebook use is also associated with addiction, anxiety, depression, body image, and other mental health disorders. A research paper by Satici and Uysal (2015) exploring the excessive use of Facebook and an individual's subjective happiness and flourishing levels also revealed a noticeable correlation between problematic usage of Facebook and an individual's lower well-being. Facebook users can also negatively impact young women's body image and mood (Fardouly et al., 2015). With over 10 million new photographs uploaded to Facebook every hour (Mayer-Schönberger & Cukier, 2013), the site enables young women to make appearance-based social comparisons, which can cause body image concerns to arise. Mayer-Schönberger and Cukier (2013) concluded that Facebook usage could negatively impact women's body image and mood, potentially leading to young women developing a greater desire to alter their appearance.

Due to the widespread of Facebook, more study is required to assess the possible effects that Facebook may have on its users and the interferences that are necessary when Facebook usage becomes detrimental. Facebook is often cited in current research. In contrast, Zalo, a social media platform launched in 2012 by a technology startup in Viet Nam, has yet to attract the attention of researchers.

Expressions of Emotion On Social Media Platforms

Individuals can communicate with others through social networking sites (e.g., Facebook, Twitter) in various ways, such as posting comments and status updates, chatting, or privately messaging (Kuss & Griffiths, 2011). With a gentle swipe on the screen, people can easily find personal inner thoughts shared publicly through comments or status updates (Kross & Chandhok, 2020). Results of studies on expressing emotions in CMC have been mixed (Lin et al., 2014; Young et al., 2017). Researchers argued whether social networking sites are associated with positive or negative expressions of emotion.

Regarding positive expressions of emotion, Waterloo et al. (2018) examined the norms of emotional expression on four popular social media platforms: WhatsApp, Facebook, Twitter, and Instagram. The researcher discovered that positive expressions were generally considered more acceptable than negative ones. Lin et al. (2014) explained that sharing positive events to express emotions is related to the need for impression management. The need for impression management is defined as the motivation to build a favorable image that is socially accepted (Leary et al., 2011; Martin et al., 2000). Barasch and Berger (2014) found that when individuals communicate with a broad audience, they have a greater desire to convey a favorable image. Additionally, frequent displays of positive emotions favorable to the public may help maintain an ideal social image (Barzarova et al., 2013). Similarly, Dhingra and Parashar (2022) found that students experience positive emotions more than negative emotions since they use Facebook or Instagram to seek emotional support and new friends. According to Galen (2017), the emotions we convey in status updates match the feelings we experience in our daily emotional lives. The researcher found that correlations are slightly more robust for positive and negative emotions.

On the other hand, some researchers discovered that social media platforms such as Facebook, Twitter, and YouTube are related to users' negative emotional self-expression (Moreno et al., 2011; Naveed et al., 2011; Sagioglou & Greitemeyer, 2014). After assessing 200 students' Facebook profiles, Moreno et al. (2011) concluded that students, who often receive supportive responses from friends, had developed a tendency to express their depression more frequently on Facebook. Naveed et al. (2011) explored the interestingness of Twitter users by analyzing several large sets of Twitter posts. The researcher found that Tweets related to negative emotions, such as unpleasant or annoying, are shared more frequently. This result indicates that Twitter users perceive that negative emotional expression is appropriate. Sagioglou and Greitemeyer (2014) stated that Facebook users might express dissatisfaction since they perceive using social media as a waste of time.

Overall, the empirical findings of studies on expressing emotions on social media platforms are inconsistent, emphasizing the need to conduct more research in the related field (Lin et al., 2014). Looking into features of social networking sites influencing users' emotional self-expression can provide new marketing insight for companies to improve their brands and products. Stieglitz and Dang-Xuan (2013) discovered that creating content that stimulates positive emotions can maximize the effectiveness of advertisements. Specifically, content that triggers emotions is more likely to be shared (Stieglitz & Dang-Xuan, 2013). Investigating users' strategies to convey emotions can improve the interface design of social media platforms, which supports CMC (Hancock et al., 2007).

Features of Social Media Platforms Affecting Users' Emotional Self-Expression

As mentioned, research focusing on features of social networking sites that impact users' emotional disclosures is scarce. To the authors' best knowledge, only a few studies examine the effect of social media features on how users express their emotions (Kramer et al., 2014; Lin et al., 2014; Livingstone et al., 2014; Sheldon & Bryant, 2016). This section reviews the mentioned studies to describe some social media features influencing how users disclose their emotions.

According to Lin et al. (2014), two factors affecting users' emotional disclosures are social network size and density. The first factor is related to the total number of online friends (network size), which correlates with the need to portray oneself as an ideal model on Facebook. The second factor concerns the total number of close connections, including best friends or family (network density), connecting with the need for emotional expression. After conducting three separate studies, Lin et al. (2014) revealed that users with a more significant number of online friends perceive updating a positive status to express emotions as more appropriate and crucial for maintaining a favorable image on social media. Furthermore, expressing negative emotions constantly on social media may lead to the impression that one cannot control their emotions, which harms the desired image of oneself to other users (Gross et al., 2006). Relating to the second feature, users with denser online networks or who had more connections with close friends and knew each other showed a significant need to express positive and negative emotions (Lin et al., 2014). Park et al. (2012) explained that people with a dense online network share their emotions more frequently to maintain interpersonal connections and gain more social support. Lin et al. (2014) conclude that social media platforms should provide appropriate functions based on users' needs to ensure their satisfaction.

Waterloo et al. (2018) discovered a correlation between the visual display of Instagram and positive emotional expressions, which relates to the work of Sheldon and Bryant (2016). After investigating the motives of 239 college students for using Instagram, Sheldon and Bryant (2016) discovered that thanks to the attractive visual properties of Instagram, functioning as a virtual photo album, people are more likely to use Instagram as a tool to gain popularity and create an ideal social image. This ties in with the existing research of Lin et al. (2014), pointing out that expressing positive emotions can fulfill one's need for impression management. The work of Sheldon and Bryant (2016) pointed out that the user interfaces design of Instagram, such as photo filters, links to trending tags, and ways to explore posts, allows users to create self-promotional content, which relates to positive emotional expression (Waterloo et al., 2018). In the same vein, Gruzd (2013) analyzed 46,097 status of Twitter users and opened an online survey for 100 participants to examine whether the user interface design affects users' emotions. The researcher found that some users preferred to share positive events to express their feelings because most of their shared information is set as public mode automatically, and they are afraid of being judged for expressing negative emotions.

Social media platforms' content can also influence users' emotional expression (Kramer et al., 2014; Livingstone et al., 2014). Kramer et al. (2014) conducted a controversial study that manipulated the content on the newsfeed of 690,000 Facebook users. The researcher sent a user group that perceived positive news, while another group was given negative ones. One of the results showed that users exposed to more harmful content in their feeds are more likely to update a negative status (and vice versa). Livingstone et al. (2014) conducted an open-ended survey with approximately 10000 children (9-16 years old) to examine users' emotional expressions on YouTube. The researcher found that users express shock and disgust because of harmful content such as violent videos or pornographic.

This section has reviewed three features of social networking sites: the network size and density, the visual properties or user interface, and the content display. The network size, density, and visual properties can facilitate users to express positive emotions (Lin et al., 2014; Sheldon & Bryant, 2016). In contrast, social media's harmful content negatively impacts users' emotional expression (Kramer et al., 2014; Livingstone et al., 2014).

Strategies to Convey Emotions in Social Media Communication.

The absence of facial expressions or body movements might lead one to believe that CMC is cold and inhibits the expression of emotions (Serrano-Puche, 2016). It is interesting to notice that people can employ various strategies to replace the lack of nonverbal cues, such as body movements or facial expressions in CMC, including using more punctuations, affects terms, and verbosity (Hancock et al., 2007). Besides, using emoticons can be a helpful strategy to express feelings in CMC (Derks et al., 2008a; Lo,2008). Emoticons are typographical symbols that replace facial expressions in CMC (Walther & D'Addario, 2001).

Hancock et al. (2007) explored how people convey their emotions in CMC by analyzing 80 text-based dyadic conversations of undergraduate students. The results showed that students employed four strategies to express feelings. The first strategy is to agree more or disagree with partners to express likes or dislikes. Based on the first strategy, Hancock et al. (2007) suggested that people can convey their emotions in CMC by using a disagreement. The second strategy involved using negative affect terms to express emotions (Tran & Bui, 2023). The third strategy is to use more punctuation, especially the exclamation mark, which can function as the tone of voice in CMC (Hancock, 2004). The final strategy is to increase the number of words in each message to express a positive emotional state. The study of Hancock et al. (2007) suggested that users can employ verbal strategies (e.g., using affect terms or verbosity) and non-verbal strategies (e.g., using punctuation) to express positive and negative emotions.

According to Walther and D'Addario (2001), emoticons can play an important role in deciphering text communications in CMC. Emoticons can be used as a communication tool to assist Internet users in expressing their emotions (Lo, 2008). After analyzing instant messages from 137 Internet users, Lo (2008) concluded that emoticons could reduce the ambiguity of messages and provide better comprehension. Emoticons help users to interpret emotion, attitude, and attention intents more effectively (Lo, 2008). Furthermore, Derks et al. (2008a) examined how emoticons are used to exchange emotional interactions. The study investigated 925 participants' experiences using emotions to respond to short Internet chats. Derks et al. (2008a) concluded that Internet users send texts with emoticons to express emotions and express a sense of humor. Similarly, Thompson et al. (2016) stated that emoticons can create fun interaction in CMC. Moreover, Internet users frequently use emoticons when communicating with friends or family, and emoticons are primarily used in a positive than negative context (Derks et al., 2008a).

However, there is still a need to conduct studies exploring the features of social media platforms that affect users' emotional expression and how they convey their emotions in CMC. Most of the literature in this chapter focuses on investigating Facebook, Twitter, and Instagram; however, Zalo, a Vietnamese social media platform, needs more attention from researchers. This study attempts to fill the aforementioned gap by comparing the features of Facebook and Zalo that influence users' emotional expression and how Facebook and Zalo users express their emotions. As motivated by the literature review, two following questions were formed:

RQ1: What features of social media platforms affect users' emotional self-expression?

RQ2: What strategies do people employ to express their positive and negative emotions on social media platforms?

METHODOLOGY

Participants

This study employed a convenience sampling strategy. Ten participants, aged 18 to 25 years (4 females & 6 males), were invited from the authors' Facebook friend list to participate in the study. Participants were Facebook and Zalo users who had used Facebook and Zalo for more than one year. All students installed Facebook and Zalo on their smartphones or computers. This criterion ensured that participants communicated on social networking sites regularly. The second inclusion criterion was that participants logged in to Facebook and Zalo at least five times weekly, ensuring that participants were aware of the differences in expressing emotions on Facebook and Zalo. It is important to note that participants only used Zalo to connect with their close friends and family relatives.

Instruments

The interview scheme was based on Hancock et al. (2007) and Prikhodko et al. (2020). It included two main parts. Based on Prikhodko et al. (2020), the first part collected data for the first research question related to factors influencing users' emotional expression on Facebook and Zalo. The second part, adapted from the work of Hancock et al. (2007), explored what strategies users employed to express their emotions in social media communication. It consists of verbal strategies (e.g., using affect terms or verbosity) and non-verbal strategies (e.g., using punctuation).

Data Collection

The study collected data through a series of semi-structured one-to-one interviews with ten selected students who volunteered to participate because they were interested in the research topic. The invitations were sent to participants via email. The first author, as the interviewer, informed them about the aims of the study and the confidentiality of their information. Therefore, participants are identified as P1 to P10 in the data report section.

Ten semi-structured one-to-one interviews were conducted via Zoom or Google Meet. The participants could suggest the platform (Zoom or Google Meet) which they found the most convenient. The prompt interview strategy was applied to collect data. In each session, the interviewer prompted the discussion. For instance, *"In this interview, we will discuss your experience communicating on social media."* In the first part of the interview, participants were asked to describe their experience of emotions on social networking sites such as Zalo and Facebook. To address the second research question, the researchers asked the participants to explain how they conveyed their feelings on Facebook (e.g., using more punctuation, emoticons, or verbosity). The participants could freely express their ideas relevant to the research topic because the interviewer used guiding questions like "What other factors affect your feelings on Facebook and Zalo?" and "Why do you think so?". During the interviews, the participants' responses were clarified and confirmed. All the interviews, which lasted 20-25 minutes each, were conducted in Vietnamese and recorded (participants informed) for further analysis (Hung et al., 2022).

DATA ANALYSIS

The recorded interviews were transcribed and classified into themes. To evaluate the data, we used the content-based method (Creswell & Creswell, 2018). Six stages were included in the data analysis process. The first stage is to organize the data, which includes putting all the recorded interviews in one folder and transcribing the information. In this step, we listened to the audio and transcribed the participants' responses for analysis. In the third stage, we began coding by fitting data to the preset codes from the literature review. The fourth stage is identifying the current study's categories through an iterative reading process. Following the fourth stage, we created a Word document with descriptive information on each participant's response. Finally, we interpreted the data by comparing the findings to previous research.

RESULTS

RQ1: What Features of Social Media Platforms Affect Users' Emotional Self-Expression?

Data analysis of social media features affecting users' emotional expression show emerging themes: online network population and density (e.g., the total number of Facebook and Zalo friends), visual properties (user interface design of Facebook and Zalo), content display (e.g., news, photos, status). Overall, the network size and density were mentioned the most frequently. Seven out of ten participants admitted that because of the large number of Facebook friends, they only shared positive emotions to avoid being judged by other users. P5 emphasized his need to portray himself as an optimistic person to Facebook friends. He said, "I always share motivational stories to express my feelings. I do not want people to see me as a pessimistic dude." The participants also acknowledged that sharing negative emotions may cause a wrong impression on their new acquaintances on Facebook. P4 said, "I recently accepted a friend request from my new co-workers, so I only update my status to share my happy moments so they can think I am a joyful person. Response by P7 shows her desire to convey herself as a respectful worker on Facebook, "Since I have connections with many colleagues, I usually share my career achievements on Facebook to show that I am proud of my jobs." P3 explained that she only shared negative events on Zalo because of the small network size. She said, "I feel it is all right if I express my anger on Zalo because my status is only viewed by a few people." P8 and P9 explained that they only share negative events to express their feelings on Zalo because they receive more social support on Zalo. P8 noted, "Only my best friends and relatives have connections with me on Zalo. I usually update my problems when I feel sad on Zalo. I know they will not judge me." Participants' answers show that they used Facebook and Zalo for different purposes. Facebook is more likely connected with online network size, while Zalo is related to online network density.

Regarding the visual properties of Facebook and Zalo, Four participants complained that Zalo has an unattractive interface which causes difficulty for them to express their emotions toward other users. P1 expressed her frustration when using Zalo to communicate with his friends. She said, "I feel like Zalo is only for sharing sadness or depression. It has only one reaction icon (the heart icon) for users to react to content updated by others." P2 explained that the mundane theme color of Zalo makes him feel that it is more appropriate to express negative emotions on Zalo, "There are only two main themes on Zalo (black & white). I feel that the dark theme color is suitable for sharing depression rather than update the positive emotion." In contrast, participants revealed that the attractive user interface design of Facebook facilitates their emotional self-expression. P10 explained, "Facebook allows me to react exactly what I feel towards my friends' content since they have five reaction emojis. I often use "Haha" and "Love" emojis to express my feelings." P6 also emphasizes that the attractiveness of Facebook visual properties. He noted, "The hashtag function of Facebook allows me to connect with more people. I usually share positive events with hashtags so that my post can gain more views or reactions." The results showed that the unattractive visual properties of social media platforms tied with users' negative emotional expression (and vice versa).

Some participants revealed that the content of Zalo and Facebook could make them feel both positive and negative. Results showed that sensitive content on Facebook and Zalo could negatively affect users' emotions. P4 said, "Sometimes I express disgust or shock when encountering gory content on Facebook." Also, P7 emphasizes his annoyance when encountering repetitive advertisements, "There was a time that I updated a status on both Facebook and Zalo to relieve my frustration toward the spam advertisements." P3 and P5 describe that some content on Facebook and Zalo make them feel empathy. P3 explained, "I often use the "love" and "heart" emoji to show my empathy towards sad stories on Facebook."

In general, network size and density, visual properties, and content display are three features of social media platforms that influence users' expressions of emotions. Network size and density are reported as the most noteworthy feature. Users' choices of sharing positive or negative emotions are based on the total number of online friends and close online friends. It is noteworthy that the results revealed that harmful content on social media platforms could trigger positive emotional self-expression.

RQ2: What Strategies Do Users Employ to Convey Emotions on Social Media Platforms?

Data showed two main categories of strategies employed by users to express their emotions: verbal strategies (e.g., using affect terms or verbosity) and non-verbal strategies (e.g., using punctuation or emoticons). Surprisingly, ten participants reported that they combined both strategies to express their emotions on social media platforms. Results revealed that the difference between verbal and nonverbal emotional expression strategies relied on whether the context was positive or negative.

Participants reported using more affective terms, punctuation, and mostly emoticons to add humor to conversations in positive contexts, such as chatting with friends. P6 described that she used many emoticons with the positive affective term, "I often use terms like ha-ha or LOL with emoticons like big laughing face ":D" or gentle smiling ":)" to express that I am having fun with my friends." Similarly, the response by P8 showed that he often used punctuation to communicate with his peers. He said, "I often put three exclamation marks or question marks at the end of each text to be hilarious when I chat with my close friend." P9 and P10 reported using the same emoticon (:P) when conversing with friends and family. P10 noted, "I often sent funny texts to my little brother with the:P emoticon, which means I tell a joke and tease him for fun."

As for negative contexts, such as arguments or expressions of disliking, results showed that emoticons are less likely to be used. At the same time, verbosity and punctuation are the preferred tools for participants to express their emotions. P1 explained, "If I were involved in an online argument, instead of using emoticons to show sarcasm or irony, I would respond more quickly with a longer number of words to express my anger." Also, P7 noted, "I tend to send longer messages to explain to someone that I disagree with them. I think more words mean that I express the feeling of disliking." P2 stated that he used the question marks to show confusion when receiving sensitive messages such as cursing. He said, "If my friend sent me a text with bad words, I immediately sent many question marks back to express my confusion."

Overall, participants used various ways to express their emotions, including using affective terms with emoticons. Results indicated that emoticons were used more frequently in the communication of emotion in positive contexts. Participants revealed that verbosity is used more often for expressing emotions in negative contexts.

DISCUSSION

The current study was guided by two research questions examining the features of Facebook and Zalo that impact Internet users and how they convey their emotions in CMC. Results were generally in line with the literature, as mentioned earlier. Regarding the first research question, the network density and size were reported as the most popular feature of Facebook and Zalo influencing users' emotional expression. Participants reported expressing only positive emotions, such as joy or pride, on Facebook. This finding ties in with the previous research by Bazarova (2012) and Lin et al. (2014), indicating that Internet users tend not to express negative emotions to create a favorable self-image when communicating with a broad audience. By contrast, Zalo is used for negative emotional expression because it only consists of participants' best friends and family relatives. This result echoes the work of Quan-Haase and Young (2010), explaining that negative events are more likely to be shared privately in intimate relationships.

Regarding the visual properties of Facebook and Zalo, reported data revealed that participants experienced more positive emotional expressions on Facebook. This finding is indirectly in line with the study of Sheldon and Bryant (2016), which also found that the attractive visual properties of social media can stimulate expressions of positive emotion. Also, participants stated that because of the poor visual properties, Zalo is mostly used for expressing negative emotions. To the authors' best knowledge, there is still a scarcity of studies investigating the effects of visual properties on social media users' emotions.

Moving onto the effects of social media content, the results showed that both positive and negative expressions of emotion could be triggered. Participants explained that they might express negative emotions when encountering sensitive or annoying content on Facebook and Zalo; these findings echo two previous studies (Kramer et al., 2014; Livingstone et al., 2014), emphasizing that the inappropriate content of social media platforms can provoke users' negative emotional self-expression. Interestingly, the study found that harmful content, such as stories with sad endings, could stimulate participants to express positive emotions, such as empathy. This interesting finding echoes the work of Pantti and Tikka (2014), indicating that users can express empathy towards disaster videos on YouTube.

Concerning the strategies that users employed to convey emotions, the obtained findings revealed that participants combined verbal strategies (e.g., using affect terms or verbosity) and non-verbal strategies (e.g., using punctuations or emoticons). In a positive context, participants used affect terms and emoticons to express emotions and add fun conversation interactions (Vu et al., 2021). This finding ties in with the study of Derks et al. (2008a), which also found that emoticons are used in positive contexts, such as family communication. Emoticons can also boost enjoyment in CMC (Thompson et al., 2016). Additionally, participants showed more frequent usage of verbosity in negative contexts. This result reflects the work of Hancock et al. (2007), concluding that verbosity is used for expressing negative emotions. Hancock et al. (2007) suggested that improving the interfaces of CMC can help humans express emotions more effectively.

CONCLUSION AND IMPLICATIONS

To the authors' best knowledge, this is one of the first studies in Vietnam investigating the features of Facebook and Zalo that influence users' expressions of emotion and how Internet users convey their emotions in CMC. Therefore, the findings in the current study provide fresh insights for future researchers to explore the relationship between emotions and CMC. Furthermore, companies or designers can use this study as a new reference to improve customers' experience using social media platforms.

The obtained data of this study provide several implications for social media companies such as Zalo. As indicated before, poor user interface design can trigger the negative experience of customers; social media platforms should modernize their visual properties to attract more customers. Also, social media platforms should provide appropriate functions based on users' needs to ensure that customers can communicate successfully on social media (Lin et al., 2014). Gruzd (2013) indicated that social media platforms, such as Twitter, should adjust their user interface to become more private so that people can share messages or statuses with a pre-defined group, such as friends or colleagues. Enhancing the private setting functionality may help users to communicate more freely on social networking sites (Gruzd, 2013). Another suggestion is that social networking sites should provide more policies to protect people from encountering sensitive content such as pornographic or gory videos. Livingstone et al. (2014) addressed the need to restrict children's access to certain content on social media so that they would not be exposed to harmful content. Finally, this study suggests that companies should be concerned about customers' emotions when promoting campaigns. According to Stieglitz and Dang-Xuan (2013), generating content that elicits favorable emotions can enhance the effectiveness of advertising.

Regardless of contribution, the current study has a few limitations. First, the sample size is relatively small to provide a variety of findings. Future research should investigate a larger sample size to enhance the generalizability of the results. Aside from the small sample size, the study is based mainly on participants' long-distance recollections of utilizing social media sites. Besides exploring participants' memories, future researchers should examine data from social media platforms, such as shared content or status. Finally, the study solely employed semi-structured interviews to acquire data. Future researchers should study the relationship between emotions and the Internet using more techniques, such as analyzing text messages.

REFERENCES

Adrianson, L. (2001). Gender and computer-mediated communication: Group processes in problem solving. *Computers in Human Behavior*, *17*(1), 71–94. doi:10.1016/S0747-5632(00)00033-9

Barasch, A., & Berger, J. (2014). Broadcasting and narrowcasting: How audience size affects what people share. *JMR, Journal of Marketing Research*, *51*(3), 286–299. doi:10.1509/jmr.13.0238

Bazarova, N. N. (2012). Public intimacy: Disclosure interpretation and social judgments on Facebook. *Journal of Communication*, *62*(5), 815–832. doi:10.1111/j.1460-2466.2012.01664.x

Bazarova, N. N., Taft, J. G., Choi, Y. H., & Cosley, D. (2013). Managing impressions and relationships on Facebook: Self-presentational and relational concerns revealed through the analysis of language style. *Journal of Language and Social Psychology*, *32*(2), 121–141. doi:10.1177/0261927X12456384

Biggest social media platforms 2022. (2022, July 26). Statista. https://www.statista.com/statistics/272014/global-social-net works-ranked-by-number-of-users/

Chaikin, A. L., & Derlega, V. J. (1974). Variables affecting the appropriateness of self-disclosure. *Journal of Consulting and Clinical Psychology*, *42*(4), 588–593. doi:10.1037/h0036614

Choi, M., & Toma, C. L. (2014). Social sharing through interpersonal media: Patterns and effects on emotional well-being. *Computers in Human Behavior*, *36*, 530–541. doi:10.1016/j.chb.2014.04.026

Creswell, J. W., & Creswell, J. D. (2018). *Research design: Qualitative, quantitative, and mixed methods approaches*. SAGE Publishing.

Derks, D., Bos, A. E., & von Grumbkow, J. (2008). Emoticons in computer-mediated communication: Social motives and social context. *Cyberpsychology & Behavior*, *11*(1), 99–101. doi:10.1089/cpb.2007.9926 PMID:18275321

Derks, D., Fischer, A. H., & Bos, A. E. R. (2008). The role of emotion in computer-mediated communication: A Review. *Computers in Human Behavior*, *24*(3), 766–785. doi:10.1016/j.chb.2007.04.004

Ellison, N. B., Steinfield, C., & Lampe, C. (2007). The benefits of Facebook "friends:" Social Capital and college students' use of online social network sites. *Journal of Computer-Mediated Communication*, *12*(4), 1143–1168. doi:10.1111/j.1083-6101.2007.00367.x

Fardouly, J., & Vartanian, L. R. (2015). Negative comparisons about one's appearance mediate the relationship between Facebook usage and body image concerns. *Body Image*, *12*, 82–88. doi:10.1016/j.bodyim.2014.10.004 PMID:25462886

Frost, R. L., & Rickwood, D. J. (2017). A systematic review of the mental health outcomes associated with Facebook use. *Computers in Human Behavior*, *76*, 576–600. doi:10.1016/j.chb.2017.08.001

Graciyal, D. G., & Viswam, D. (2021). Social media and emotional well-being: Pursuit of happiness or pleasure. *Asia Pacific Media Educator*, *31*(1), 99–115. doi:10.1177/1326365X211003737

Griffiths, M., & Light, B. (2008). Social networking and digital gaming media convergence: Classification and its consequences for appropriation. *Information Systems Frontiers*, *10*(4), 447–459. doi:10.100710796-008-9105-4

Hancock, J. T. (2004). Verbal Irony Use in Face-To-Face and Computer-Mediated Conversations. *Journal of Language and Social Psychology*, *23*(4), 447–463. doi:10.1177/0261927X04269587

Hancock, J. T., Landrigan, C., & Silver, C. (2007). Expressing emotion in text-based communication. *Proceedings of the SIGCHI Conference on Human Factors in Computing Systems*. ACM. 10.1145/1240624.1240764

Hey Tow, W. N., Dell, P., & Venable, J. (2010). Understanding information disclosure behaviour in Australian Facebook users. *Journal of Information Technology*, *25*(2), 126–136. doi:10.1057/jit.2010.18

Hung, B. P., Anh, D. P. T., & Purohit, P. (2022). Computer-mediated communication and second language education. In R. Sharma & D. Sharma (Eds.), *New trends and applications in Internet of things (IoT) and big data analytics* (pp. 109–122). Springer. doi:10.1007/978-3-030-99329-0_8

Hung, B. P., Khoa, B. T., & Hejsalembrahmi, M. (2022). Qualitative research in social sciences: Data collection, data analysis, and report writing. *International Journal of Public Sector Performance Management*, *9*(4), 10038439. doi:10.1504/IJPSPM.2022.10038439

Kallinikos, J., & Tempini, N. (2014). Patient data as medical facts: Social media practices as a foundation for medical knowledge creation. *Information Systems Research*, *25*(4), 817–833. doi:10.1287/isre.2014.0544

Kalpidou, M., Costin, D., & Morris, J. (2011). The relationship between Facebook and the well-being of undergraduate college students. *Cyberpsychology, Behavior, and Social Networking*, *14*(4), 183–189. doi:10.1089/cyber.2010.0061 PMID:21192765

Kapoor, K. K., Tamilmani, K., Rana, N. P., Patil, P., Dwivedi, Y. K., & Nerur, S. (2018). Advances in social media research: Past, present and future. *Information Systems Frontiers*, *20*(3), 531–558. doi:10.100710796-017-9810-y

Kramer, A. D., Guillory, J. E., & Hancock, J. T. (2014). Experimental evidence of massive-scale emotional contagion through social networks. *Proceedings of the National Academy of Sciences of the United States of America*, *111*(24), 8788–8790. doi:10.1073/pnas.1320040111 PMID:24889601

Kross, E., & Chandhok, S. (2020). How do online social networks influence people's emotional lives? In J. P. Forgas, W. D. Crano, & K. Fiedler (Eds.), Applications of social psychology: How social psychology can contribute to the solution of real-world problems (pp. 250–263). Routledge/Taylor & Francis Group. doi:10.4324/9780367816407-13

Kuss, D. J., & Griffiths, M. D. (2011). Online social networking and addiction a review of the psychological literature. *International Journal of Environmental Research and Public Health*, 8(9), 3528–3552. doi:10.3390/ijerph8093528 PMID:22016701

Leary, M. R., Allen, A. B., & Terry, M. L. (2011). Managing social images in naturalistic versus laboratory settings: Implications for understanding and studying self-presentation. *European Journal of Social Psychology*, 41(4), 411–421. doi:10.1002/ejsp.813

Lieberman, A., & Schroeder, J. (2020). Two social lives: How differences between online and offline interaction influence social outcomes. *Current Opinion in Psychology*, 31, 16–21. doi:10.1016/j.copsyc.2019.06.022 PMID:31386968

Lin, H., Tov, W., & Qiu, L. (2014). Emotional disclosure on social networking sites: The role of network structure and psychological needs. *Computers in Human Behavior*, 41, 342–350. doi:10.1016/j.chb.2014.09.045

Livingstone, S., Kirwil, L., Ponte, C., & Staksrud, E. (2014). In their own words: What bothers children online? *European Journal of Communication*, 29(3), 271–288. doi:10.1177/0267323114521045

Lo, S.-K. (2008). The nonverbal communication functions of emoticons in computer-mediated communication. *Cyberpsychology & Behavior*, 11(5), 595–597. doi:10.1089/cpb.2007.0132 PMID:18817486

Martin, K. A., Leary, M. R., & Rejeski, W. J. (2000). Self-presentational concerns in older adults: Implications for health and well-being. *Basic and Applied Social Psychology*, 22(3), 169–179. doi:10.1207/S15324834BASP2203_5

Matook, S., Cummings, J., & Bala, H. (2015). Are you feeling lonely? The impact of relationship characteristics and online social network features on loneliness. *Journal of Management Information Systems*, 31(4), 278–310. doi:10.1080/07421 222.2014.1001282

Moreno, M. A., Jelenchick, L. A., Egan, K. G., Cox, E., Young, H., Gannon, K. E., & Becker, T. (2011). Feeling bad on Facebook: Depression disclosures by college students on a social networking site. *Depression and Anxiety*, 28(6), 447–455. doi:10.1002/da.20805 PMID:21400639

Muñoz, C. L., & Towner, T. (2011). Back to the "wall": How to use Facebook in the college classroom. *First Monday*. doi:10.5210/fm.v16i12.3513

Myers, T. A., & Crowther, J. H. (2009). Social comparison as a predictor of body dissatisfaction: A meta-analytic review. *Journal of Abnormal Psychology, 118*(4), 683–698. doi:10.1037/a0016763 PMID:19899839

Naveed, N., Gottron, T., Kunegis, J., & Alhadi, A. C. (2011). Bad News Travel Fast. *Proceedings of the 3rd International Web Science Conference*. ACM. 10.1145/2527031.2527052

Nxumalo, L. K., & Chiweshe, N. (2019). Social enterprise digital marketing. *Strategic Marketing for Social Enterprises in Developing Nations*, 103-130. doi:10.4018/978-1-5225-7859-8.ch005

Pantti, M., & Tikka, M. (2014). Cosmopolitan Empathy and User Generated Disaster Appeal Videos on YouTube. In T. Benski & E. Fisher (Eds.), *Internet and Emotions* (pp. 178–192). Routledge.

Park, N., Lee, S., & Kim, J. H. (2012). Individuals' personal network characteristics and patterns of Facebook use: A Social Network approach. *Computers in Human Behavior, 28*(5), 1700–1707. doi:10.1016/j.chb.2012.04.009

Prikhodko, O. V., Cherdymova, E. I., Lopanova, E. V., Galchenko, N. A., Ikonnikov, A. I., Mechkovskaya, O. A., & Karamova, O. V. (2020). Ways of expressing emotions in social networks: Essential features, problems and features of manifestation in internet communication. *Online Journal of Communication and Media Technologies, 10*(2). doi:10.29333/ojcmt/7931

Quan-Haase, A., & Young, A. L. (2010). Uses and gratifications of social media: A comparison of Facebook and instant messaging. *Bulletin of Science, Technology & Society, 30*(5), 350–361. doi:10.1177/0270467610380009

Rice, R. E., & Love, G. (1987). Electronic Emotion: Socioemotional Content in a Computer-Mediated Communication Network. *Communication Research, 14*(1), 85–108. doi:10.1177/009365087014001005

Sagioglou, C., & Greitemeyer, T. (2014). Facebook's emotional consequences: Why Facebook causes a decrease in mood and why people still use it. *Computers in Human Behavior, 35*, 359–363. doi:10.1016/j.chb.2014.03.003

Satici, S. A., & Uysal, R. (2015). Well-being and problematic Facebook use. *Computers in Human Behavior, 49*, 185–190. doi:10.1016/j.chb.2015.03.005

Serrano-Puche, J. (2016). Internet and emotions: New trends in an emerging field of research. *Comunicar*, *24*(46), 19–26. doi:10.3916/C46-2016-02

Sheldon, P., & Bryant, K. (2016). Instagram: Motives for its use and relationship to narcissism and contextual age. *Computers in Human Behavior*, *58*, 89–97. doi:10.1016/j.chb.2015.12.059

Short, J., Williams, E., & Christie, B. (1976). *The social psychology of telecommunication*. Wiley.

Spagnoletti, P., Resca, A., & Sæbø, Ø. (2015). Design for social media engagement: Insights from elderly care assistance. *The Journal of Strategic Information Systems*, *24*(2), 128–145. doi:10.1016/j.jsis.2015.04.002

Sproull, L., & Kiesler, S. (1986). Reducing social context cues: Electronic Mail in organizational communication. *Management Science*, *32*(11), 1492–1512. doi:10.1287/mnsc.32.11.1492

Stieglitz, S., & Dang-Xuan, L. (2013). Emotions and information diffusion in social media—Sentiment of microblogs and sharing behavior. *Journal of Management Information Systems*, *29*(4), 217–248. doi:10.2753/MIS0742-1222290408

Thompson, D., Mackenzie, I. G., Leuthold, H., & Filik, R. (2016). Emotional responses to irony and emoticons in written language: Evidence from EDA and facial EMG. *Psychophysiology*, *53*(7), 1054–1062. doi:10.1111/psyp.12642 PMID:26989844

Tran, H. N., & Bui, H. P. (2022). Causes of and coping strategies for boredom in L2 classroom: A case in Vietnam. *Language Related Research*. https://lrr.modares.ac.ir/article-14-62169-fa.html (In Press).

Turel, O., & Serenko, A. (2012). The benefits and dangers of enjoyment with social networking websites. *European Journal of Information Systems*, *21*(5), 512–528. doi:10.1057/ejis.2012.1

Valenzuela, S., Park, N., & Kee, K. F. (2009). Is there social capital in a social network site? Facebook use and college students' life satisfaction, trust, and participation. *Journal of Computer-Mediated Communication*, *14*(4), 875–901. doi:10.1111/j.1083-6101.2009.01474.x

Valkenburg, P. M., Peter, J., & Schouten, A. P. (2006). Friend networking sites and their relationship to adolescents' well-being and social self-esteem. *Cyberpsychology & Behavior*, *9*(5), 584–590. doi:10.1089/cpb.2006.9.584 PMID:17034326

Vishwanath, A. (2014). Diffusion of deception in social media: Social contagion effects and its antecedents. *Information Systems Frontiers*, *17*(6), 1353–1367. doi:10.100710796-014-9509-2

Vu, N. N., Hung, B. P., Van, N. T. T., & Lien, N. T. H. (2021) Theoretical and Instructional Aspects of Using Multimedia Resources in Language Education: A Cognitive View. In: Kumar R., Sharma R., Pattnaik P.K. (eds) Multimedia Technologies in the Internet of Things Environment, Vol. 2. Studies in Big Data, Vol. 93 (pp. 165-194). Springer. doi:10.1007/978-981-16-3828-2_9

Walther, J. B., & D'Addario, K. P. (2001). The impacts of emoticons on message interpretation in computer-mediated communication. *Social Science Computer Review*, *19*(3), 324–347. doi:10.1177/089443930101900307

Waterloo, S. F., Baumgartner, S. E., Peter, J., & Valkenburg, P. M. (2018). Norms of online expressions of emotion: Comparing Facebook, Twitter, Instagram, and WhatsApp. *New Media & Society*, *20*(5), 1813–1831. doi:10.1177/1461444817707349 PMID:30581358

Yan, L., Peng, J., & Tan, Y. (2015). Network dynamics: How can we find patients like us? *Information Systems Research*, *26*(3), 496–512. doi:10.1287/isre.2015.0585

Yan, L., & Tan, Y. (2014). Feeling blue? Go online: An empirical study of social support among patients. *Information Systems Research*, *25*(4), 690–709. doi:10.1287/isre.2014.0538

ADDITIONAL READINGS

Baym, N. (2000). *Tune in, log on: Soaps, fandom, and online community*. SAGE Publications. doi:10.4135/9781452204710

Fiegerman, S. (2021, October 29). How twitter could still blow it and fade away into Social Network Oblivion. *Mashable*. https://mashable.com/article/how-twitter-could-fade-away

Hall, J. A. (2022). What we do in the Shadows: The consumption of mobile messaging by social media mobile apps in the twilight of the Social Networking Era. *Mobile Media & Communication*, *11*(1), 66–73. doi:10.1177/20501579221133610

Halliwell, J., & Wilkinson, S. (2021). Mobile phones, text messaging and social media. *Creative Methods for Human Geographers*, 259–272. doi:10.4135/9781529739152.n20

Herring, S. C., & Kapidzic, S. (2015). Teens, gender, and self-presentation in social media. International Encyclopedia of the Social & Behavioral Sciences, 146–152. doi:10.1016/B978-0-08-097086-8.64108-9

Hollenbaugh, E. E., & Ferris, A. L. (2014). Facebook self-disclosure: Examining the role of traits, social cohesion, and motives. *Computers in Human Behavior, 30*, 50–58. doi:10.1016/j.chb.2013.07.055

Junco, R. (2013). Comparing actual and self-reported measures of Facebook use. *Computers in Human Behavior, 29*(3), 626–631. doi:10.1016/j.chb.2012.11.007

Kim, J. (2017). How did the information flow in the #Alphago hashtag network? A social network analysis of the large-scale information network on Twitter. *Cyberpsychology, Behavior, and Social Networking, 20*(12), 746–752. doi:10.1089/cyber.2016.0572 PMID:29243963

Le, T. T., & Tran, B. T. (2022). Using Zalo application as a learning management system. *2022 The 8th International Conference on Frontiers of Educational Technologies (ICFET)*. 10.1145/3545862.3545863

Lenhart, A. (2022, August 10). *Teens, Social Media & Technology Overview 2015*. Pew Research Center: Internet, Science & Tech. https://www.pewresearch.org/internet/2015/04/09/teens-social -media-technology-2015/

Liu, I. L. B., Cheung, C. M. K., & Lee, M. K. O. (2010). *Understanding twitter usage: What drive people continue to tweet*. Hong Kong Baptist University. https://scholars.hkbu.edu.hk/en/publications/understanding-t witter-usage-what-drive-people-continue-to-tweet-11

Ruggiero, T. E. (2000). Uses and gratifications theory in the 21st Century. *Mass Communication & Society, 3*(1), 3–37. doi:10.1207/S15327825MCS0301_02

Schauer, P. (2015, June 28). *5 biggest differences between social media and Social Networking*. Social Media Today. https://www.socialmediatoday.com/social-business/peteschauer /2015-06-28/5-biggest-differences-between-social-media-and-s ocial

Social Media. (2015). *An overview*. Auditing Social Media. doi:10.1002/9781119202585.ch1

Vaterlaus, J. M., Barnett, K., Roche, C., & Young, J. A. (2016). "Snapchat is more personal": An exploratory study on Snapchat behaviors and young adult interpersonal relationships. *Computers in Human Behavior*, *62*, 594–601. doi:10.1016/j. chb.2016.04.029

KEY TERMS AND DEFINITIONS

Affect Terms: Affect terms refer to the feeling that a human has, such as anger, anxiety, depression, joy, or satisfaction.

Computer-Mediated Communication: Computer-mediated communication refers to the communication between people and digital devices, including personal computers, smartphones, smartwatches, and digital assistants (e.g., Alexa and Siri).

Emoticons: Emoticons are typographical symbols that replace facial expressions in computer-mediated communication. For example, the laughing face ":D," sad face ":(," smiling face ":)" are three standard emoticons.

User Interface Design: The user interface is the point of human-computer interaction and communication on a device, webpage, or app. This can include display screens, keyboards, a mouse, and the appearance of a desktop.

Verbosity: Verbosity is the act of performing wordiness in communication. This means using more words than are needed to convey a massage.

Chapter 11

Computer–Mediated Communication and the Business World

Ta Thi Nguyet Trang
International School, Thai Nguyen University, Vietnam

Pham Chien Thang
ⓘ https://orcid.org/0000-0002-5982-4173
Thai Nguyen University of Sciences, Vietnam

ABSTRACT

The importance of the Internet and online communication in business has significantly increased in recent years as more companies have been turning to digital tools to conduct their operations. This chapter presents how the Internet and online communication have impacted the business world, focusing on the B2B (business-to-business) sector. The chapter first will present the advantages and disadvantages of the Internet and online communication in the B2B sector, including the benefits of connecting and collaborating with other businesses in real time and the challenges of maintaining cybersecurity and building personal connections. It will then address the current state of the Internet and online communication in the business world, including updated facts and figures on its usage and impact. Also, strategies for businesses to effectively navigate the rapidly evolving landscape of the Internet and online communication, including best practices for utilizing these tools and overcoming challenges. By understanding and utilizing the internet and online communication, businesses can significantly enhance their operations and reach new levels of success.

DOI: 10.4018/978-1-6684-7034-3.ch011

INTRODUCTION

It isn't easy to imagine that the internet hasn't always been so integrated into our lives. Since the advent of the internet, many company communication duties have been carried out very differently, including seamless employee collaboration, an expanding number of communication channels, and the ability to instantly interact with clients anywhere in the globe. Because of how profoundly things have changed, it is nearly hard for people who were born after the internet's inception to fully comprehend life without it, and more developments are still to come (Apavaloaie, 2014). According to the Internet Society, researchers at the Defense Advanced Research Projects Agency (DARPA), under the direction of Lawrence G. Roberts and Leonard Kleinrock of MIT, developed the packet network known as ARPANET during the 1960s and 1970s. The internet evolved as a collection of various separate packet-switching networks utilizing ARPANET as the foundation. Over time, businesses and the armed services started communicating via ARPANET. Communities and scholars began utilizing the internet for email and research in 1985 (Crocker, 2021).

Before the internet, a businessman would start a store, publish advertisements in the local paper, join a networking group, and trust that the community would need what they had to give. Everything changed when the internet was invented. A company no longer depends on its local client because there is now a global market for its goods and services. The way that businesses engage with their personnel, find, and manage their competitors, and manage their customer base has all changed as a result of the internet (Liu et al., 2023). Small business owners may operate from anywhere thanks to the internet. For instance, a business owner on vacation can communicate with their workplace and exchange important documents with anyone, anywhere in the world, using live chat, messaging apps, or video meetings (Richmond et al., 2017). One example is the increasing importance of data security and protecting the company and customer information as the internet expands, and new communication technologies and channels emerge (Eddington et al., 2023).

Internal and external communication is essential for successful business operations. Contact between two or more entities and ensuring that each receives what they desire from the partnership are essential to a business's survival. Establishing the anticipated outcomes from both sides is essential to ensuring that the entities cooperate on the goals that have been set with the partnership (Matzler et al., 2015). The internet is a new method of communication, and in our opinion, various people use it and experience it differently. It is crucial to understand whether they believe that accurate information can be transmitted via the internet. Technology is heavily used in today's culture to facilitate communication and provide access to information (Hung & Khoa, 2021). There has never been a wider deployment of internet and online technology in history. Information and communication technologies have significantly influenced global economic growth over the past two decades. The use of online communication in businesses has considerably increased since the development of the internet. And this chapter is all about the importance of the internet and online communication in businesses (Rong et al., 2015).

CURRENT PERSPECTIVES ON THE ROLES OF THE INTERNET IN BUSINESS COMMUNICATION

Importance of Internet in Business World

The World Wide Web is one of the most potent and useful instruments for conducting business, so the importance of the internet in business cannot be ignored. There is no denying that the internet makes it easier for people to do their jobs in business world. The internet appears to be of utmost importance to those who engage in various business activities because it enables connection with suppliers, clients, etc. The pros of the internet are undeniable when it comes to saving clients' time, postage, mailing costs, etc. Employees may find and use the information on the internet, and it also satisfies business information needs (Combe, 2012). The most common approaches employed by businesses to gather specific data are based on what is known as business intelligence. In other terms, businesses now require more business intelligence considering the data that the internet makes available to different firms. Today, it is acknowledged that the internet is the key tool for a revolution in business. Everything in business can alter as a result, including how people interact, work, and make decisions. Most businesses use the World Wide Web to engage directly with customers for the first time (Choo et al., 2013). Others are strengthening their relationships with some of their trading partners using secure internet connections, and they're taking advantage of the internet's accessibility and reach to get quotes or auction off volatile supplies of products or services. Furthermore, it should be noted that the World Wide Web allows entrepreneurs to introduce their customers to new electronic marketplaces and other markets (Jagongo & Kinyua, 2013).

This study will delve deeper into the metaverse concept and its potential impact on the business world. The metaverse, a virtual world where people can interact and conduct business in a digital environment, is a novel theme that has gained increasing attention in recent years. Also, examine how the metaverse could revolutionize the way businesses communicate and conduct their operations, as well as the potential challenges and limitations of this technology. It also explores the role of the internet of things (IoT) in the business world, including how connected devices can streamline operations and gather valuable data for businesses. Information and Communications Technology (ICT) will also be discussed in the business world, including the various tools and platforms businesses can utilize to improve their communication and operations. In addition to incorporating these novel themes, we will also provide a more in-depth analysis of the impact of social media on the business world. We will examine the various types of social media platforms and their specific uses in the business world, as well as how businesses can effectively utilize social media to reach and engage with customers and clients. Overall, the aim is to provide a more comprehensive and cohesive understanding of the impact of the internet and online communication on the business world, including applying novel themes such as the metaverse, IoT, and ICT. By including a range of literature and applying illustrations to clarify concepts, we hope to provide a more coherent and clear analysis of this topic (Turban et al., 2018).

Concept of "Trust" on the Internet

Trust has often come up in earlier studies of commercial and online relationships. A relationship's foundation is trust, but trust typically exists only between people, not between businesses. To stay competitive, businesses and organizations have been essentially forced to adopt the use of the internet. The fact that consumer consciousness has started to rise is one of the key causes of this. Consumers today have access to far more information and are better equipped to compare costs, qualities, and offers than ever before, whether they are individuals or speaking on behalf of corporations. This has also given rise to a new mentality wherein a salesperson's own "know-how" is one of the things they must confront, refute, and persuade the customers of. Since the advent of the internet, it has grown to be the largest informational repository on the world, and it continues to grow daily. Stewart (2023) claims that one of the primary causes of this surge is the concurrent use of the internet by so many people across so many different frontiers. Every internet user can change it; when a modification is made, it is immediately available to all users, preventing the possibility of many minds working on the same development. People are better able to learn more about goods and services before contacting the company that offers them by being given the freedom to conduct independent research. They are also able to learn what other people think of goods or services through forums like the Flashback before trying them out for themselves (Bonneau et al., 2023). Customers may choose to use or skip a particular service or product based on attributes without fully understanding it. We believe that situations like these can happen and most likely are happening. They may cause customers to overlook products that may be better suited to their needs. Along with discussing the importance of effectively communicating trust in corporate relationships, it illustrates that organizations can use the internet for purposes other than only two-way direct contact and relationship building. This instance demonstrates how the internet can be used to obtain information from third parties.

Online Communication in The Business World

In business, there are many different forms of communication. The interaction between a company and its customers is the one that is most frequently linked to consumers. Business-to-consumer (B2C) communication is what this is. This includes all kinds of marketing, sales, and customer service. Customer-to-Company (C2B) and even Customer-to-Customer (C2C) communication are other methods of business communication (Paul et al., 2023). However, communication with other businesses is one type of communication that is very significant to a business which is called Business-to-business (B2B) communication (Kumar & Raheja, 2012).

The dynamic nature of communication makes it difficult for the scientific community to define it. Many definitions portray communication as a logical process that facilitates the transmission and comprehension of messages, whether they are verbal or non-verbal (Dwyer, 2012). Being frequently accompanied by branding and public relations, communication inside the businesses is crucial. Outlining steps for improved internal and external communication complete and supports the company plan. A company's performance speaks a thousand words. Without a proper manual, there is a chance that the company will engage in communication activities that will hinder its ability to accomplish its objectives. In order to ensure data security, boost speed and efficacy as well as provide for their clients more properly, not just huge businesses but even small local businesses of all shapes and masses operate and link using online mediums (Guffey & Loewy, 2014).

Online Business-To-Business Communication (B2B)

A "commercial transaction carried out by any two business organizations for profit" is referred to as B2B. The statement "intended to sell items or services to business, industry, or professionals rather than consumers" refers to B2B marketing. The entire B2B process must revolve around communication for a contemporary business. A company's entire B2B strategy would be unsuccessful without efficient communication channels. Business-to-business marketing will play a significant role in the sales strategy of a multinational corporation. Businesses that rely on them as customers greatly impact a developed B2B communication system (Murphy & Sashi, 2018). A purchase is a "personal buying decision" in business. The buyer considers whether they are purchasing the item for themselves, another consumer, or both. In business-to-business dealings, people are "consumers" but have the added duty of assessing and buying goods for a business they own or work for. With the internet, there is a chance to build relationships. This is a manner of defining the inter-organizational connections that can be cultivated via e-commerce methods. The fact that practically every type of business may be connected through e-commerce must also be emphasized. It is not just advantageous for businesses that conduct online business. Even a small local business selling goods to nearby residents might benefit from putting some effort and money into its online presence (Gupta, 2014).

Business-to-business (B2B) communication refers to the exchange of information and resources between businesses. This type of communication is critical for businesses as it plays a key role in building relationships, collaborating on projects, and conducting transactions with other businesses. There are several reasons why B2B communication is focused on in this study. Firstly, B2B communication is a vital component of business operations, as it allows businesses to connect with other companies and organizations that can provide the resources, services, or products they need to operate. This type of communication also plays a crucial role in developing strategic partnerships and collaborations, which can lead to increased efficiency, innovation, and growth for businesses. Secondly, the internet and online communication have had a major impact on the way businesses interact with each other, particularly in the B2B sector. The ability to connect and communicate with potential partners and clients in real-time has greatly enhanced the efficiency and effectiveness of B2B interactions (Adapa & Yarram, 2023). Online communication tools such as email, chat, and video conferencing have made it easier for businesses to communicate, collaborate, and exchange information. B2B communication is a key focus of this study due to its importance in business operations and the significant impact of the internet and online communication on how businesses interact in the B2B sector. In many sectors, a few dominant companies may control most purchases made in their sector.

In these circumstances, there are times when these big corporations' band together online to form an industry alliance. The firms are connected through an internet system that enables them to coordinate their purchasing strategies (Alnsour, 2018). A huge price decrease as a result of increased purchasing quantities is one advantage of doing this. Industry alliances are not administered by a third party, in contrast to online marketplaces, which lowers fees or commissions imposed by them. This technique also puts these businesses in close contact with one another, which is advantageous. This might facilitate the growth of the connection and future opportunities for increased collaboration. Businesses that would typically be competitors can work together for a similar goal due to this B2B communication technology. Private exchanges, sometimes known as "virtual private networks," can be thought of as a more sophisticated version of the long-established electronic data interchanges (EDI) which is one more benefit of B2B online communication. A business can develop stronger ties with its suppliers or clients through this network.

It facilitates the synchronization of numerous corporate operations and might even permit the automation of procedures like procurement (Soini & Eräranta, 2023).

REMAINING PROBLEMS AND THE FUTURE OF ONLINE COMMUNICATION

Advantages of Internet and Online Communication in Businesses

The internet has significantly impacted the way businesses operate. It has contributed to a more open, interconnected, diverse, and inclusive global economy. The internet provides numerous benefits for communication, interaction, and business development. A stable, high-speed internet connection is essential for smooth communication and efficient file transfer and can facilitate business growth. Many businesses and organizations are already utilizing high-speed internet to improve their operations and advance their plans. They heavily rely on this technology to increase production and achieve operational performance. With the aid of the internet, businesses see significant changes. They grow more adaptable, versatile, and flexible during the world's constant change (Combe, 2012). The internet is vital in enabling more efficient and convenient commercial communication. It offers a variety of tools to help manage daily tasks and conversations. Improved tools for enhancing internal and external corporate communication include chat services, integrated calendars with presence status, virtual phone networks, and emails. High-speed internet connectivity allows for seamless communication between local and remote staff, effortless task optimization, successful online video conferencing, and successful immediate message delivery. Many of the innovations that came about as a result of internet technology include VoIP phone systems, social networking sites, workgroup portals, and chat-bots. Any one of those facilitates efficient organizational communication and collaboration, which promotes business (Tiago & Verissimo, 2014).

No longer is it necessary for individuals to mail information or business documents to clients, coworkers, or business partners. You can share files over email more quickly and from any location if you have high-speed broadband. The information that peoples want to share should be accessible via email or other software programs, such as digital documents, digital records, or web-based records. People can access their corporate servers from any location and transfer essential data right away if they have a strong internet connection. As a result, business operations run more smoothly and quickly for the firm. However, accessing information now is far quicker than it was in the pre-internet period. To locate business information, documentation, or

217

services fast, search on Google, Bing, or in specialized directories for your industry. People have access to this data every single day. People can investigate competitors, find contacts, or just find company ideas quickly because of access to this limitless reservoir of information (Stewart, 2023). How you sell your goods and services has evolved due to digital and online marketing. Nowadays, every business, irrespective of size, needs to have a website. People can market your firm to a wider audience through online marketing. Among the modern tools for reaching the correct audiences are business websites, online networks, email marketing, and paid digital advertising (Tiago & Verissimo, 2014).

The internet's ability to boost sales and improve customer relationships is a significant advantage. It allows businesses to expand locally, nationally, or globally. The capacity to create a strong brand using tools that require a high-speed internet connection is just as important as the growth in online transactions. With carefully chosen content, tiny businesses can nonetheless attract huge audiences. If you want to interact more effectively, it's important to know the internet platforms where you may meet your potential clients and communicate openly about your products (Alnsour, 2018).

Disadvantages of Internet and Online Communication in Business

There is no doubt that the rapid growth of the internet has significantly transformed the way businesses operate. While it may seem that the internet is essential for conducting business, it is always advisable to consider the company's specific needs and weigh the pros and cons of using the internet as part of the business strategy. The internet has increased the market for a variety of business types. Customers now have easier access to rival products and services thanks to the internet, in addition to making it easier for them to purchase the products and services themselves. To stay at the front of customers' thoughts, businesses have turned to drastically lowering prices, which has also significantly reduced profit margins (Yurovskiy, 2014). If you run an online business, be ready to lose sales if customers can purchase the same goods elsewhere for less money. Nowadays, people communicate their grievances and thoughts about a company on social media. People tend to be more critical of a company than complimentary, which can harm its reputation. Many complaints have been spread widely online, forcing businesses to alter their tactics or products and leading to a loss of customers and revenue. Handle it carefully and in line with your company's requirements and business strategy to maximize the benefits and lessen the drawbacks. In addition, some of the most creative ideas come from face-to-face interactions. While the internet has effectively connected people, it cannot fully replace in-person interactions with clients and coworkers (Adapa & Yarram, 2023).

Impact of Social Media on Businesses

Recent years have seen a rise in social media's popularity and economic significance, enabling millions of active users to share data, information, and products and altering how businesses structure their networks of connected people. The interactive online and user-generated content is currently what keeps the internet alive. In the past ten years, thousands of social networks focused on fostering interpersonal connections have sprung up to give people new tools for collaboration and communication based on social aspects of technology use (Aral et al., 2013). Users are creating new communication techniques and adding content to media aggregators such as Amazon, Google, eBay, and Flickr. Online social networks like Facebook, Twitter, and LinkedIn have grown to unprecedented sizes. Social media sites like Facebook and Twitter allow people to become visible to a large audience and share their social networks, providing a technological platform for building and strengthening connections between users and businesses (Bria et al., 2013). The information-intensive environment in which businesses work is affected by the media's major position in modern society, but despite this awareness, institutional processes in a mediatized society are not well understood (Smits & Mogos, 2013). Additionally, not much scientific research clearly shows how changes in communication strategies and approaches support more significant changes in social institutions and how businesses design their interactions with stakeholder populations.

There are some variables that impact the use of the internet and online communication in the business world. These variables may include factors such as the type and frequency of online communication, the level of cybersecurity measures in place, and the presence of face-to-face interactions. The aim is to incorporate new variables and extend existing models in our analysis of the impact of the internet and online communication on the business world. These may include emerging technologies such as virtual and augmented reality and new forms of online communication such as messaging apps and social media platforms. It also examines the relationship between these variables, including how they may impact the effectiveness and efficiency of online communication for businesses (Soini & Eräranta, 2023). The issues and problems facing businesses regarding internet use and online communication. These issues may include cybersecurity concerns, difficulty building personal connections, and the potential for lack of face-to-face interactions. To address these problems and provide suggested recommendations and best practices for businesses to effectively navigate the rapidly evolving landscape of the internet and online communication.

These recommendations may include strategies for maintaining strong cybersecurity measures, utilizing social media to build personal connections with customers and clients, and incorporating tools such as video conferencing to facilitate face-to-face interactions in a digital environment. In terms of this chapter's contribution to the creation of new knowledge, our goal is to present a thorough and current examination of how the internet and online communication have affected business. We aim to offer insightful analysis for companies and scholars interested in this topic by looking at the most recent trends and technologies in this area, such as the metaverse and IoT. This study is valuable for businesses, researchers, and students interested in understanding the importance of the internet and online communication in the business world. It will also be relevant for those seeking to identify and address the challenges and opportunities presented by this technology and those interested in staying up to date on the latest trends and developments in this field (Donici et al., 2012).

CONCLUSION

The internet and online communication have significantly impacted how businesses interact with each other, particularly in the B2B sector. The ability to connect and communicate with potential partners and clients in real-time has greatly enhanced the efficiency and effectiveness of B2B interactions. Online communication tools such as email, chat, and video conferencing have made it easier for businesses to communicate, collaborate, and exchange information. However, the adoption of the internet and online communication in the B2B sector also comes with challenges. Cybersecurity concerns, such as data breaches and hacking, have become a major concern for businesses using online communication. In addition, the reliance on digital communication can lead to a lack of face-to-face interactions and personal connections, which can be important for building strong relationships in the B2B world.

Despite these challenges, the benefits of the internet and online communication in the B2B sector far outweigh the drawbacks. The ability to connect and collaborate with businesses worldwide has opened new opportunities and markets for B2B companies. The use of digital tools has also helped streamline operations and reduce costs for businesses. In conclusion, the internet and online communication have become an integral part of the B2B world, and understanding and utilizing these tools is essential for success in today's business landscape.

REFERENCES

Adapa, S., & Yarram, S. R. (2023). Communication of CSR practices and apparel industry in India—Perspectives of companies and consumers. In *Fashion Marketing in Emerging Economies* (Vol. I, pp. 137–161). Palgrave Macmillan. doi:10.1007/978-3-031-07326-7_6

Alnsour, M. (2018). Internet-Based Relationship Quality: A Model for Jordanian Business-to-Business Context. *Marketing and Management of Innovations*, *4*, 161–178. doi:10.21272/mmi.2018.4-15

Apăvăloaie, E. I. (2014). The impact of the internet on the business environment. *Procedia Economics and Finance*, *15*, 951–958. doi:10.1016/S2212-5671(14)00654-6

Aral, S., Dellarocas, C., & Godes, D. (2013). Introduction to the special issue—social media and business transformation: A framework for research. *Information Systems Research*, *24*(1), 3–13. doi:10.1287/isre.1120.0470

Bilton, J. (2014). Publishing in the digital age. In Stam, D., & Scott, A. (Eds.). (2014). Inside Magazine Publishing (1st ed.) (pp. 226-247). Routledge. doi:10.4324/9781315818528

Bonneau, C., Aroles, J., & Estagnasié, C. (2023). Romanticisation and monetisation of the digital nomad lifestyle: The role played by online narratives in shaping professional identity work. *Organization*, *30*(1), 65–88. doi:10.1177/13505084221131638

Bria, F. (2013). *Social media and their impact on organisations: building Firm Celebrity and organisational legitimacy through social media* [Doctoral dissertation, Imperial College London]. http://hdl.handle.net/10044/1/24944

Choo, C. W., Detlor, B., & Turnbull, D. (2013). *Web work: Information seeking and knowledge work on the World Wide Web* (Vol. 1). Springer Science & Business Media.

Crocker, S. D. (2021). Arpanet and Its Evolution—A Report Card. *IEEE Communications Magazine*, *59*(12), 118–124. doi:10.1109/MCOM.001.2100727

Dawe, K. (2015). Best practice in business-to-business email. *Journal of Direct, Data and Digital Marketing Practice*, *16*(4), 242–247. doi:10.1057/dddmp.2015.21

Demishkevich, M. (2015). *Small business use of internet marketing: Findings from case studies* [Doctoral dissertation, Walden University]. https://scholarworks.waldenu.edu/cgi/viewcontent.cgi?referer =&httpsredir=1&article=2339&context=dissertations

DONICI, A.N., Maha, A., Ignat, I. and MAHA, L.G. (2012). E-Commerce across United States of America: Amazon. com. *Economy Transdisciplinarity Cognition, 15*(1), 252–258.

Dwyer, J. (2012). *Communication for Business and the Professions: Strategie s and Skills.* Pearson Higher Education AU.

Eddington, S. M., Jarvis, C. M., & Buzzanell, P. M. (2023). Constituting affective identities: Understanding the communicative construction of identity in online men's rights spaces. *Organization, 30*(1), 116–139. doi:10.1177/13505084221137989

Guffey, M. E., & Loewy, D. (2014). *Business communication: Process and product.* Cengage Learning.

Gupta, A. (2014). E-Commerce: Role of E-Commerce in today's business. *International Journal of Computing and Corporate Research, 4*(1), 1–8.

Hendler, J. A. (2023). The future of the Web. In *The Internet and Philosophy of Science* (pp. 71–83). Routledge., doi:10.4324/9781003250470

Hung, B. P., & Khoa, B. T. (2021). Communication strategies for interaction in social networks: A multilingual perspective. In I. Priyadarshini & R. Sharma, Artificial Intelligence and Cybersecurity (pp. 195-208). Taylor & Francis.

Jagongo, A., & Kinyua, C. (2013). The social media and entrepreneurship growth. *International Journal of Humanities and Social Science, 3*(10), 213–227.

Kumar, V., & Raheja, G. (2012). Business to business (b2b) and business to consumer (b2c) management. *International Journal of Computers and Technology, 3*(3), 447–451.

Liu, Y., Alzahrani, I. R., Jaleel, R. A., & Al Sulaie, S. (2023). An efficient smart data mining framework based cloud internet of things for developing artificial intelligence of marketing information analysis. *Information Processing & Management, 60*(1), 103121. doi:10.1016/j.ipm.2022.103121

Matzler, K., Veider, V., & Kathan, W. (2015). Adapting to the sharing economy. Cambridge, MA, USA: Mit.

Miller, M. (2012). *B2B digital marketing: Using the web to market directly to businesses.* Que publishing.

Murphy, M., & Sashi, C. M. (2018). Communication, interactivity, and satisfaction in B2B relationships. *Industrial Marketing Management, 68*, 1–12. doi:10.1016/j. indmarman.2017.08.020

Nemat, R. (2011). Taking a look at different types of e-commerce. *World Applied Programming*, *1*(2), 100–104.

Paul, J., Alhassan, I., Binsaif, N., & Singh, P. (2023). Digital entrepreneurship research: A systematic review. *Journal of Business Research*, *156*, 113507. doi:10.1016/j.jbusres.2022.113507

Richmond, W., Rader, S., & Lanier, C. (2017). The "digital divide" for rural small businesses. *Journal of Research in Marketing and Entrepreneurship*, *19*(2), 94–104. doi:10.1108/JRME-02-2017-0006

Rong, K., Hu, G., Lin, Y., Shi, Y., & Guo, L. (2015). Understanding business ecosystem using a 6C framework in Internet-of-Things-based sectors. *International Journal of Production Economics*, *159*, 41–55. doi:10.1016/j.ijpe.2014.09.003

Smits, M. T., & Mogos, S. (2013). The impact of social media on business performance. *ECIS 2013 Completed Research*. 125. https://aisel.aisnet.org/ecis2013_cr/125

Soini, A., & Eräranta, K. (2023). Collaborative construction of the closet (in and out): The affordance of interactivity and gay and lesbian employees' identity work online. *Organization*, *30*(1), 21–41. doi:10.1177/13505084221115833

Stewart, B. E. (2023). The problem of the web: Can we prioritize both participatory practices and privacy. *Contemporary Educational Technology*, *15*(1), ep402. doi:10.30935/cedtech/12668

Tiago, M. T. P. M. B., & Veríssimo, J. M. C. (2014). Digital marketing and social media: Why bother? *Business Horizons*, *57*(6), 703–708. doi:10.1016/j.bushor.2014.07.002

Turban, E., Outland, J., King, D., Lee, J. K., Liang, T. P., & Turban, D. C. (2018). *Electronic commerce 2018: a managerial and social networks perspective.* Springer International Publishing., doi:10.1007/978-3-319-58715-8

Yurovskiy, V. (2014). Pros and cons of internet marketing. *Research Paper*, 1-12.

ADDITIONAL READINGS

Apăvăloaie, E. I. (2014). The impact of the internet on the business environment. *Procedia Economics and Finance*, *15*, 951–958. doi:10.1016/S2212-5671(14)00654-6

Bulearca, M., & Bulearca, S. (2010). Internet and interactive websites: Cornerstones of competitive advantage in the virtual economy. *Global Business and Management Research*, *1*(3 & 4), 44.

Chatterjee, D., & Sambamurthy, V. (1999). Business implications of web technology. *Electronic Markets*, *9*(1-2), 126–131. https://www.tandfonline.com/doi/abs/10.1080/101967899359355. doi:10.1080/101967899359355

Darics, E. (2015). Introduction: Business communication in the digital age—Fresh perspectives. In Digital business discourse (pp. 1-16). Palgrave Macmillan, London. doi:10.1057/9781137405579_1

Golden, W. (2005). Internet Communication in the Development of Business Relationships. *BLED 2005 Proceedings*. 29. https://aisel.aisnet.org/bled2005/29

Grubor, A., & Jakša, O. (2018). Internet marketing as a business necessity. *Interdisciplinary Description of Complex Systems: INDECS*, *16*(2), 265–274. doi:10.7906/indecs.16.2.6

Kabakchieva, T. (2021). Online Communications And Their Role In Business Sales. *Trakia Journal of Sciences*, *19*(1, Suppl.1), 107–116. doi:10.15547/tjs.2021.s.01.015

Kandampully, J. (2003). B2B relationships and networks in the Internet age. *Management Decision*, *41*(5), 443–451. doi:10.1108/00251740310479296

Kommers, P., Isaias, P., & Issa, K. (2014). *The evolution of the internet in the business sector: web 1.0 to web 3.0*. IGI Global.

Patrutiu-Baltes, L. (2016). The impact of digitalization on business communication. *SEA–Practical Application of Science, 4*(11), 319-325. https://www.ceeol.com/search/article-detail?id=740216

KEY TERMS AND DEFINITIONS

Business: Business refers to creating, producing, and exchanging goods and services to generate profit. The business study examines various aspects of business, including finance, marketing, and human resources.

Business-to-Business (B2B): Business-to-business (B2B) refers to commerce in which one business sells products or services to another. B2B companies may specialize in a particular industry or sector and operate in various business models.

Online Communication: Online communication refers to exchanging information and ideas through electronic communication technologies, such as the internet, social media platforms, and messaging apps.

Chapter 12
Electronic Word of Mouth (eWOM) in Consumer Communication

Ayushi Gupta
Indian Institute of Foreign Trade, New Delhi, India

ABSTRACT

Electronic word of mouth (eWOM) has garnered substantial interest from academic and market practitioners due to its considerable influence on consumer behavior. In the virtual world, consumer interactions have strengthened owing to the high use of digital technology. Although extant literature is available in this area, the corpus of academic literature is expanding due to fragmented published studies, which increases the complexity of the current research. In this study, the author attempts to integrate findings on the meaning of eWOM; theories used to study the area and its impact on influencing consumer behavior to synthesize existing literature. Finally, the paper explicates the scope for future research as identified from previous literature.

INTRODUCTION

The emerging growth in digital technologies is affecting marketing strategies in a significant manner (Rosario et al., 2022; Steinhoff & Palmatier, 2020). Organizations can reap the benefits of digital platforms by reaching out to their customers almost immediately and obtaining their rapid feedback through various digital mediums such as e-commerce, social commerce, online review sites and peer-to-peer networking sites (Donthu et al., 2021; Cheung & Thadani, 2012; Gupta & Harris, 2010). These

DOI: 10.4018/978-1-6684-7034-3.ch012

consumers' critiques and opinions about products and services, which they share through online mediums, are known as electronic word of mouth (eWOM) (Donthu et al., 2021). Current technologies have amplified the reach of word-of-mouth communications. With a wider reach compared to traditional WOM, previous research has shown that eWOM is a credible and influential source of information sharing and seeking among consumers (Cheung & Thadani, 2012).

Consumers use eWOM information to evaluate their purchase decisions (Godes & Mayzlin, 2004; Babic Rosario et al., 2016). Today's consumers are more informed since they look for unbiased reviews from anonymous people on online platforms before making any product-related decisions. The features of comments, likes, video-based eWOM and verified product/service reviewers have enhanced the understanding of potential customers about the marketer's offerings. Consumers feel more secure when scrutinising other consumers' experiences with using the product (Albayrak & Ceylan, 2021). It also tends to reduce their perceived likelihood of making impulsive purchases (Hennig- Thurau et al., 2004).

Gathering, presenting and encouraging consumers to post eWOM about products and services is becoming a priority for companies to exhilarate their sales (Duan et al.,2008; Xiong & Bharadwaj,2014). Previous studies have shown that eWOM impacts sales (Zhou & Duan, 2015; Hyrynsalmi et al., 2015; Sharma et al., 2011) and creates a rippling effect by generating more eWOM which induces higher sales (Ismagilova et al, 2017). Cedric Lacroix, managing director of Nestle, Japan, had claimed that eWOM created by their consumers when the company launched "Pink Kit Kat" made the product a viral sensation and overshadowed all types of paid advertising when the product became sold out within two days of its launch and was even sold through black markets (Indian Express, 2019). India alone has about 749 million active internet users, increasing to about 1.5 billion users by 2040 (Statista, 2021). The connected beauty consumer report by Google (2020) has postulated that nine out of ten Indian customers pass through digital touch points in their buying process journey. They critically analyze product reviews and seek eWOM recommendations for this purpose. Past research has noticed similar findings for countries such as America (Babic Rosario et al., 2019), China (Zhu et al., 2017) and other Asian countries (Mir & Rehman, 2013; Nusier, 2019; Hu & Kim, 2018). It has also been found that eWOM supports commercial advertising by reaching a larger set of audience in a faster and more cost-effective manner (Kim et al., 2017).

Previous studies explicate that applications of eWOM are not only limited to consumer behavior but also pertain to a company perspective, including quality control and management, revenue management and identification of service failures (Donthu et al., 2021; Serra Cantallops & Salvi, 2014). Companies have begun to use eWOM to promote their corporate social responsibility activities to consumers. A company that embraces social responsibility and communicates it to its customers might evoke more positive responses, leading to a positive consumer attitude (Donthu et al., 2021).

The domain of eWOM has been investigated from two perspectives: first, the market level research which studies the product sales making use of panel data from websites (Ismagilova et al., 2017; Baek et al., 2014) and second, the consumer level research wherein behavioral and social characteristics such as attitude, purchase intention, purchase decision, information adoption and influence have been explored (Kudeshia & Kumar, 2017; Erkan & Evans, 2016; Cheung et al., 2009; Hussain et al., 2017; Sokolova & Kefi, 2020). In this paper, we identify consumer-level research.

Although eWOM has emerged as a vast body of knowledge due to the increasing interest of marketing researchers in this domain, the extant literature is largely fragmented, which increases the difficulties in culminating the understanding from this area (Babic Rosario et al., 2020; Lamberton & Stephen, 2016). Moreover, technological advancements continue to enhance and change the manner and mode in which eWOM is disseminated amongst consumers, impacting their behavioral psychology in an ever-changing fashion (Wang et al., 2021).

Intending to synthesize the current literature on eWOM, the objectives of this study are twofold: (a) to encapsulate the existing literature on eWOM and (b) to assimilate the scope for future research. Therefore, this paper contributes to providing an overview of eWOM research and ascertaining the findings on the following themes occurring in eWOM research: Electronic Word of Mouth (eWOM) and consumer motivations to disseminate eWOM, Types of eWOM, Theoretical frameworks that have been used in literature to understand eWOM from the perspective of the consumers, Behavioral characteristics of users of eWOM information, Role of eWOM at various stages of the consumer buying process (as depicted in Figure1.) and Implications for marketers.

The review contributes to academic knowledge in a significant manner. Primarily, it corroborates previous research to showcase the state of research on eWOM. Moreover, it advances the future scope by suggesting the scope of research for future scholars.

The paper is organized as follows. First, the author defines eWOM, distinguishes it from User Generated Communication and identifies consumers' motivations to spread eWOM along with the types of eWOM. Second, the paper describes the most relevant theoretical frameworks occurring in eWOM literature. Third, the author elaborates upon the behavioral characteristics of eWOM users along with the attitudes towards eWOM information received. Fourthly, the role of eWOM at different stages of the consumer buying process. Lastly, the paper is concluded by discussing the implications for marketers and highlighting future research directions.

Figure 1. Illustrative representation of the present study

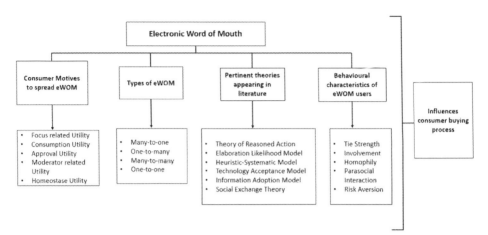

LITERATURE REVIEW

Electronic Word of Mouth (eWOM)

Digitalization has emerged as a vehicle for consumers to share their product-related thoughts via eWOM which was impossible through traditional WOM. Traditional WOM was limited to conversations with only the known ones (Brown et al., 2007; Steffes & Burgee, 2008; Lee & Youn, 2009), covering a sparse geographical area (Hennig-Thurau & Walsh, 2003). An increase in technology-mediated communication has paved the way for eWOM to act as a means of non-commercial advertising for customers (Cheung & Lee, 2012). It has also enabled customers to discuss their experiences with using the product of a brand (Chevalier & Mayzlin, 2006).

As postulated by Harrison-walker (2001), eWOM is "informal, in-person communication between a perceived non-commercial communicator and a receiver regarding a brand, a product, an organization or a service". Word of Mouth Marketing Association (2008) has explained word of mouth as "The act of consumers providing information to other consumers". It is any personal communication about the product that takes place between target buyers or near ones of the consumer (Kotler & Armstrong, 2006). Internet as a medium of communication differentiates eWOM from traditional word of mouth. Hennig-Thurau et al. (2004) have postulated eWOM as "any positive or negative statement made by potential, actual or former customers about a product or company, which is made available to a multitude of people and institutions via the Internet" (p.39).

eWOM can be initiated through various electronic mediums such as online discussion forums, blogs, product reviews websites and social media applications (Cheung & Thadani, 2012) which was not possible through traditional word-of-mouth. It is a type of electronic communication since word-of-mouth communication is transmitted through electronic mediums. eWOM has become an indispensable part of digital marketing since it is consumer generated, meaning that it can publicize a company's marketing activities without any financial resources invested thereby. Moreover, positive or negative information about the product can reach a wide geographical area without the company making any efforts. As a means of eWOM can be initiated through various, eWOM assists in image building (Donthu et al., 2021), obtaining product feedback and engaging in consumer interactions, creating brand awareness (Erkan & Evans, 2016).

Hu and Ha (2015) have classified eWOM into four categories based on its purpose and the platforms on which eWOM is exchanged: (1) Specialized eWOM is the consumer reviews that are posted on product rating websites and product comparison websites that do not sell products, e.g. Yelp; (2) Affiliated eWOM is through consumer reviews posted on e-commerce websites, e.g., Amazon and Flipkart; (3) Social eWOM is exchanged through product information on social networking sites, e.g., Face book, Twitter, YouTube, Instagram; (4) Miscellaneous eWOM is exchanged through blogs and discussions on social media networks.

These mediums enable "consumers to not only obtain information related to goods and services from the few people they know but also from vast, geographically dispersed groups of people who have experience with relevant products or services"(Cheung et al., 2008). eWOM can provide personal experiences about brand usage, which cannot be obtained from the company's dominated sources, which increases the utility of such information for the customers (Park and Lee, 2008). When such communication about a brand is positive, customers consider purchasing the brand. Previous customers who provide product-oriented information, ratings and suggestions act as both informers and advisors (Park et al., 2007). Since the sender of such information is independent, eWOM is considered more dependable (Brown et al., 2007) and effectual than conventional marketing tools (Trusov et al., 2009). In case of negative reviews, such information helps to reduce the consumer's risk (Hennig-Thurau et al., 2004). Table 1. represents the most prominent eWOM definitions appearing in literature.

With today's consumers being digitally aware of market offerings, it is imperative that word-of-mouth affects consumer behavior (Litvin et al., 2008). Firstly, it assists in consumers' attitude formation about existing and contemporary brands (Wolny et al., 2012). Secondly, the usefulness of such information in forming purchase intention and decisions induces consumers to base their final opinion about products through word of mouth (Chevalier & Mayzlin, 2006), thereby making it a crucial prerequisite for consumer buying process. Thirdly, word-of-mouth communications provides a platform for consumers to appreciate or vent their opinions, which directly assists brands understanding consumer reactions towards their products (Hennig-Thurau et al., 2004).

Table 1. Prominent definitions appearing in literature

S.No	Author(s)	Definition
1.	Harrison-Walker (2001)	eWOM is defined as "informal, person to person communication between a perceived non-commercial communicator and a receiver regarding a brand, a product, an organization or a service."
2.	Hennig-Thurau et al. (2004)	"Any positive or negative statement made by potential, actual or former customers about a product or company, which is made available to a multitude of people and institutions via the Internet."
3.	Litvin et al. (2008)	"eWOM can be defined as all informal communications directed at consumers through Internet-based technology related to the usage or characteristics of particular goods and services, or their sellers."
4.	Ismaagilova et al. (2017)	"The dynamic and ongoing information exchange process between potential, actual, or former customers regarding a product, service, brand or company, which is available to a multitude of people and institutions via the internet."
5.	Babic Rosario et al. (2020)	"eWOM is consumer-generated, consumption-related communication that employs digital tools and is directed primarily to other consumers."

Though eWOM and user-generated content (UGC) are used synonymously by most of the researchers, both the terms are divergent with regard to their explanation and extent (Daughtery et al., 2008; Thao & Shurong, 2020). Bruns (2016) has defined UGC as, "User-generated content (UGC), sometimes also referred to as user-created content (UCC), is a generic term that encompasses a wide range of media and creative content types that were created or at least substantially co-created by "users"- that is, by contributors working outside of conventional professional environments." (p.1).

As observed by Smith et al. (2012), the distinction lies in the fact that users themselves primarily create UGC by applying their innovative omnific skills, for example, through (video creation and blog posting) which might not be consumption related to eWOM, which by its definition itself is consumption related and determines the opinions and attitudes towards products and services. The second distinction relates to the compensation involved. While the company does not compensate creators of UGC for promoting the product, eWOM is free from such boundation and may include benefits to the communicator of information in the form of cash backs, discounts or rewards.

Why Consumers Spread eWOM

Hennig-Thurau et al. (2004) have comprehensively studied customer stimulations behind spreading eWOM. The first has been described as Focus Related Utility wherein the reason for spreading eWOM is altruism towards both fellow consumers and the company, social identity and recognition (McWilliam, 2000) and exerting power over the company, particularly in case of negative reviews. Through Consumption Utility, customers giving product opinions encourage other customers to write reviews and provide useful feedbacks that can be helpful to a larger audience (Balasubramanian & Mahajan,2001, p.125). Approval Utility provides a sense of satisfaction to product reviewers when other users agree or approve of their reviews (Balasubramanian & Mahajan,2001, p.126).

Moderator Related Utility states that eWOM acts as a moderator between customer and company, which helps customers to express their complaints and dissatisfaction related to the product (Harrison-Walker,2001). The Homeostase Utility is based on the Balance Theory, which states that an individual wants to reinstate equilibrium after their balanced state has been displaced. Customers will express positive feelings when their satisfaction with a product or service creates inner tension to express their joy or experience (Dichter, 1966). Similarly, customers will also express their negative experiences to vent unpleasant feelings and reach their equilibrium (Sundaram et al., 1998).

Types of eWOM

Applying the information systems approach to the marketing-consumer environment (Xia et al.2009), Weisfeld-Spolter et al. (2013) have conceptualized eWOM into four definite categories: many-to-one, one-to-many, many-to-many and one-to-one. The first category, many-to-one, showcases the opinions and preferences of many consumers who have already used the product (Weisfeld-Spolter et al., 2013). Product reviews on e-commerce websites are present in the form of ratings and usage experience (Lin & Huang, 2006; Xia et al., 2009), which is a significant source of information for potential customers (Petrescu et al., 2016). Previous studies have posited that such reviews induce strong purchase intentions and play an important role in deciding the brand's reputation and the time spent on the seller's website (Mudambi & Schuff, 2010). Anderson (2018) has found that consumers are sceptical about making purchases if they do not find product ratings or reviews on shopping websites. Though positive reviews encourage customers to consider purchasing, negative reviews reduce the probability of making wrong choices (Hennig-Thurau et al., 2004).

The second category is of one-to-many, where a market maven (Feick & Price,1987), who can be either an opinion leader or an influencer, creates awareness about new product offerings through eWOM (Kotler & Zaltman,1976). Social media platforms are a crucial source of eWOM dissemination where independent influencers share their product opinions (Yang et al.,2011). The eWOM shared by social media influencers has higher credibility (O'Reilly & Marx, 2011; Han et al., 2009; Latane,1981). The parasocial interaction generated in the mind of eWOM user (Kim et al.,2015; Sokolova & Kefi, 2020), positively influences purchase intention (Hwang & Zhang,2018) and impacts brand perceptions (Lee & Watkins, 2016).

The third category, many-to-many, includes eWOM transmitted through discussion forums (Bickart & Schindler, 2001), where consumers constantly communicate about the product (Weisfeld-Spolter et al., 2013; Hennig-Thurau et al., 2004). Here, the website's reputation, source credibility, information usefulness and social orientation significantly influence purchase intention and usage attitudes (Chih et al., 2013. Brown and Reingen (1987) have postulated that users with strong ties on discussion forums are more likely to participate in eWOM on such platforms than those with weak ties. Online discussion forums exhibit reciprocatory behaviors since consumers receive eWOM information and social-emotional behaviors (Chan et al., 2010; Brown et al., 2007).

The last category, one-to-one, is usually a personal and non-transparent form of eWOM (Weisfeld-Spolter et al., 2013). The perceived interactivity (McMillan, 2002) users have with each other in terms of cognitive, affective and behavioral dimensions (Burgoon et al., 2000) helps both users in simulating their social presence and product experience (Kiousis, 2002).

THEORETICAL BACKGROUND

Theoretical background aims to illustrate a comprehensive foundation for understanding concepts and helps researchers critically evaluate the area under research. Further, it helps in understanding the reasons for the existence of the research problem. Although eWOM extends the scope of traditional WOM, a literature review indicates that various theories have been used to study eWOM communication from the perspective of individual and group behaviors. But previous studies have found that the Theory of Reasoned Action, Elaboration Likelihood Model, Heuristic System Model, Technology Acceptance Model, Information Adoption Model, Information Acceptance model and Social Exchange Theory are particularly important and have been widely used in predicting the relationship of eWOM with purchase intention.

We have therefore reviewed the above-stated theories and attempt to provide a concise understanding and comparative insights on the same.

Theory of Reasoned Action

The Theory of Reasoned Action propounded that human behavior depends upon intentions to execute behaviors (Ajzen, 1991; Park, 2000). Such determination to perform behaviors depends on attitudes and subjective norms (Fishbein & Ajzen, 1977; Ajzen & Fishbein, 1980). The more the acceptability for attitudes and subjective standards, the stronger the propensity to engage in the behavior (Fishbein & Ajzen, 1977). Attitudes are derived as a consequence of perceived outcomes related to behaviors and their importance to the user.

While subjective norms depend upon existing beliefs about an individual's viewpoints and motivation to adhere to them (Ajzen & Fishbein,1980). Past marketing research has used the TRA model to understand purchase intentions due to eWOM (Mir & Rehman, 2013; Cheung & Thadani, 2012; Yuksel, 2016). A study undertaken by Ngah et al. (2018) has found that purchase intention had a strong positive relationship with attitude and subjective norm component of TRA. Further, it was concluded that these components displayed a positive relationship with destination visit intentions. Wolny et al. (2013) postulated that both attitudes and subjective norms influence eWOM engagement when dispersed through social media. The theoretical model of this theory is represented in Figure 2.

Figure 2. Theory of Reasoned Action

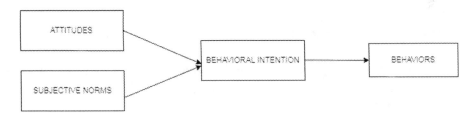

Elaboration Likelihood Model

Petty and Cacioppo (1986) have stated that a user of information undergoes two routes before being persuaded by such information: central and peripheral routes. A user passes through the central route when they are highly self-motivated to pay attention to the essence and quality of the message (Sussman & Siegal, 2003). The peripheral route occurs when the user indirectly encounters the information without any motivation to process it (Petty & Cacioppo, 1986; Cheung et al., 2008).

Regarding eWOM communications, users pass through the central route when they are highly attentive towards information processing and aware of the importance of information. While, in the peripheral route, users engage in non-content-related information due to the bandwagon effect regarding the demand for products (Park & Lee, 2008) or the convincing identity of the speaker (Burnstein et al., 1973). Several studies have used this model to predict consumer purchase intentions. Li et al. (2021) found that eWOM acts as a peripheral cue and positively impacts patient follow-up intention with their doctors. Fan (2012) found that the level of involvement, that is, the route through which the user receives information, has a significant impact on credibility, acceptance and intentions to purchase for e-commerce websites. The theory has been conceptualized in Figure 3.

Figure 3. Elaboration Likelihood Model

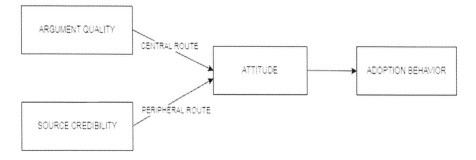

Heuristic – Systematic Model

The IS literature has applied dual-process models to comprehend the effects of eWOM on user behaviors (Chaiken, 1980). Although many scholars have approved using the Elaboration Likelihood Model to probe into eWOM research, the seminal works of Gupta and Harris(2010), Zhang and Watts (2014) and Cheung and Thadani(2012) have recommended the use of the heuristic systematic model overelaboration likelihood model grounding on two reasons: first, this model demonstrates a wider range of information processing activities(Chaiken,1989) and second, this model is more theoretically extensive as compared with elaboration likelihood model(Zhang et al.,2014). Moreover, Petty and Cacioppo (1986) posited that the central and peripheral information processing routes are mutually exclusive. But Chaiken (1980) examined and concluded that heuristic and systematic information processing can coexist.

The heuristic–systematic model elucidates three motivations of users to process information, namely, Accuracy, Defense and Impression motivation (Chaiken, 1989). Todorov et al. (2002) have postulated that "people consider all relevant pieces of information, elaborate on these pieces of information, and form a judgment based on these elaborations". In the systematic route, information seekers apply their cognitive abilities in scrutinizing the arguments put forward and their validity (Chaiken, 1980). Tam and Yo (2005) have referred to heuristic information processing as a situation in which information seekers apply little time and effort in information processing and mainly depend on attainable informational cues to form their conclusions (Chaiken,1980). Consumers having low motivation for information processing are likely to follow the heuristic cue (Gupta & Harris, 2010). eWOM information users are exposed to heuristic processes before evaluating the arguments of systematic processes. Therefore, systematic information processing can unveil opinions in detail that are actually lost in heuristic processing (Hlee et al., 2018).

Technology Acceptance Model

The technology acceptance model proposed by Davis (1989) is considered an important model in academic literature to describe how users accept the information they receive through eWOM communications (Venkatesh & Davis,2000). Past researchers have extensively used TAM to understand how users behave while accepting a new technology (Venkatesh et al, 2003). The model mainly studies two constructs: ' perceived usefulness' and 'perceived ease of use' in understanding acceptance-related behaviors (Davis,1889). The theory is particularly important to understand user's intentions to communicate on social media platforms (Pinho & Soares,2011; Chen & Berger, 2016) and has showcased how consumer engagement moderated the impact of eWOM on buying intentions (Lee et al.,2008). Nevertheless, this model has been subject to criticism from researchers due to its inability to explain user's point of view and their intentions of using products (Ayeh, 2015). Figure 4 provides a conceptual understanding of TAM.

Information Adoption Model

IAM was developed by synthesizing Technology Acceptance Model (Davis, 1989) and Elaboration Likelihood Model (Petty & Cacioppo, 1986). IAM extends the TAM model and ELM model and adopts information usefulness as a mediating variable between information adoption, message quality and source quality. Though users might pass through any of these two routes, they will usually reach the same conclusion: information adoption (Sussman & Siegel, 2003). The components of this model: information quality, information credibility, information usefulness and

Figure 4. Technology Acceptance Model

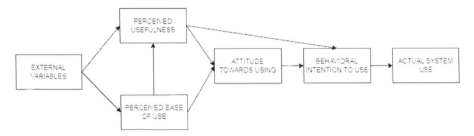

information adoption, have been used by researchers to study the impacts on buying interests and buying decisions (Erkan & Evans, 2016; Bueno & Gallego,2021). It has also been used to study how users understand and react to computer-mediated communication (CMC) and interpret how receivers grasp eWOM (Sussman & Siegal, 2003; Cheung et al., 2009).

Park et al. (2007) have explained that the quality of reviews significantly impacts consumers' purchase intentions. Wathen and Burkell (2002) have advocated that information credibility is essential to persuasion. Further, a message coming from a highly credible source has more trustworthiness than compared with a message coming from a moderately credible source (Sternthal et al., 1978). Consumers using eWOM information show a higher tendency towards purchase intentions (Cheung et al., 2009). Perceived usefulness can reduce the waste of time and improve access to only that information which is useful and relevant for the consumer (Bouhlel et al., 2010). Lu et al. (2010) have explained that trust is an important factor consumers look for before adopting eWOM. Information is more likely to be adopted when supported with trust and credibility (Dabholkar & Sheng, 2012). Since consumers in online markets cannot touch, test or try the products, their purchase judgment depends upon product information available online. It has been found that eWOM-based information about a product has a better impact on consumers (Brinol et al., 2007). Figure 5 represents the theoretical model of IAM.

Social Exchange Theory

Social exchange theory has been used substantially in marketing to understand the relational exchange communication process between two parties (Lambe et al., 2001). According to Homans' (1958) thesis, people seek to maximize their welfare and minimize expenses to maximize their profits. The advantages of information sharing, such reciprocity and recognition, boost and improve eWOM and aid in establishing relationships and benefits for both parties. Moreover, the willingness

Figure 5. Information Adoption Model

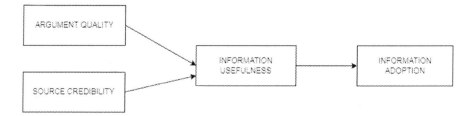

to share eWOM increases on digital platforms as compared to traditional WOM when the mutual gain for information users is of interest.

Seminal works of Hennig Thurau et al. (2004) and Cheung and Lee (2012) have explained that altruism, social recognition and feeling of belongingness positively influence the spread of eWOM in online communities. Toth et al. (2022) applied SET in their study and concluded that relationship intimacy could occur even in online B2B eWOM, evidencing that in-person contact is not necessary for developing mutual interests and bonding. Bakshi et al. (2021) postulated that venting out negative feelings, altruism, recognition amongst online peers and receiving economic rewards positively influence posting online travel review intentions. Table 2 represents previous studies that have utilized the above-discussed theories to study eWOM's relation with consumer behavior.

Table 2. Author's compilation

S.No	Theory used	Studies applying the theory
1.	**Theory of Reasoned Action**	Ngah et al. (2018); Orzan et al. (2013); Bulut & Karabulut (2018); Khwaja & Zaman (2020); Yusuf & Busalim(2018); Mir & Rehman(2013);Wolny & Mueller(2013)
2.	**Elaboration Likelihood Model**	Leong et al. (2019); Hussain et al. (2017); Luo et al. (2014); Fan & Miao (2012); Li et al. (2021); Cao et al. (2017)
3.	**Heuristic System Model**	Zhang et al. (2014); Hlee et al. (2018); Xiao et al. (2018); Lee & Hong (2021); Abedin et al. (2021); Yang et al. (2017)
4.	**Technology Acceptance Model**	Lee et al. (2021); Shaker et al (2021); Bilal et al. (2020); Liao et al. (2022); Pinho & Soares (2011); Chen & Berger (2016); Ayeh (2015); Lee et al. (2008)
5.	**Information Adoption Model**	Erkan and Evans (2016); Bueno & Gallego (2021); Arora & Lata (2020); Chen et al. (2014); Song et al. (2021); Sardar et al. (2021); Shen (2021)
6.	**Social Exchange Theory**	Toth et al. (2022); Bakshi et al. (2021); Chang & Hsiao (2013); Phan et al. (2020); Mohammed & Al-Swidi (2020); Hayes & King (2014)

An overview of the above-discussed theories helps in concluding that the foundation of eWOM is mainly embedded in information systems and communication research. Early theories, for example, TRA and ELM, have evolved to accommodate the role of technology in spreading eWOM and understanding how it affects consumer behaviors. The theories mentioned above have a multidisciplinary perspective enabling researchers to expand their theoretical lens towards newer areas in order to investigate eWOM and its relation with consumer behaviors through a holistic approach. The next section of this paper discusses the behaviors of eWOM users, which actually aids in understanding eWOM itself.

BEHAVIORAL CHARACTERISTICS OF eWOM INFORMATION USERS

Previous works emphasize on how eWOM marketing focus and efforts are shifting towards consumer behavior. Moreover, it is being further investigated how prospective consumers and users spreading eWOM can influence various consumer behaviors (Donthu et al., 2021). Henceforth, we discuss certain behaviors of existing and prospective consumers in relation to eWOM in the following sections.

Tie Strength

Tie strength determines the intensity of the relationship that exists between individuals (ME Abd-Elaziz, 2015). Tie strength with family members and friends is usually more than that with unknown individuals (Hennig-Thurau et al., 2004; Godes & Mayzlin, 2004). However, the tie strength is greater with anonymous customers who give unbiased product reviews on social media platforms (Erkan & Evans, 2018; Filieri et al., 2018). This has been attributed to a multitude of product information and opinions that cannot be shared by family and friends alone (Pan, 2014; Levy & Gvili, 2015). Social media users have strong ties with information providers when product information is provided by influencers who become market mavens (Fan & Sun, 2012).

Involvement

Involvement is defined as robustness and closeness with emotional cues related to a product or service. Two types of users are known based on involvement. First are the high-involvement users who participate in the communication process by liking, commenting, rating, or sharing eWOM with other consumers (Sohaib et al., 2018). Others are the low-involvement consumers who show low involvement in the information exchange process. They might not completely rely on eWOM and depend upon their own judgment skills to evaluate the coming information (Ngarmwongnoi et al., 2020).

Homophily

Individuals prefer to interact with those who share similar values, beliefs, social status and intellectual levels. This attitude is called homophily (Prisbell & Andersen, 1980). On YouTube also, subscribers identify with those influencers with whom they share common characteristics. Consumers see an opinion leader and a trustworthy advisor in such providers of product information. Homophily increases the influence and confidence towards eWOM (Fan & Miao, 2012). Sokolova and Kefi (2020) found that homophily is an important characteristic of eWOM that strongly influences persuasion towards the believability of information.

Parasocial Interaction

Parasocial interaction is a psychological connection encountered by the receiver of information, specifically in online communication (Horton & Wohl, 1956). Such an association is self-manifested by the receiver, whereas the sender of information is uninformed about such bonding. The interaction is usually one-directional since it is not always possible for influencers to respond back to every subscriber. A report by Google (2016) posited that 40% of YouTube users affirmed that they confide in video creators more than their own friends. Moreover, 70% of teenagers have claimed they identify with YouTubers more than celebrities. Past studies have found that Parasocial interaction on eWOM sharing platforms positively influences brand perception and purchase intention (Lee & Watkins, 2016; Hwang & Zhang, 2018).

Risk Aversion

The main reason for seeking eWOM information is to reduce the likelihood of economic risk of making actual purchases (Hennig-Thurau et al., 2004). Previous studies have found that consumers actively seek eWOM when they perceive a higher risk. This is seen in high-involvement goods such as cars and jewellery than for low-involvement goods like groceries (Sakkthivel, 2010). Due to this risk aversion, consumers prefer receiving information only from trustworthy sources (Akdim, 2021). On social media platforms like Instagram and YouTube, consumers can verify the trustworthiness of the user and their content through comments of other consumers and the number of likes the video receives, demonstrating whether the information provided by the video creator is actually true or not.

ROLE OF eWOM IN THE CONSUMER BUYING PROCESS

Understanding the current consumer purchase process is becoming increasingly critical for marketers in order to understand customer wants and provide value to them. (Pelau, 2011), which is technology driven. Companies need to focus on Omni-channel marketing strategies due to digitalization. This has been attributed to the numerous digital contact points that a buyer experiences during the purchasing process. The survey also discovered that eWOM information influences nine out of ten Indian customers throughout the product purchasing process.

Since the outcomes associated with purchases are unknown until the actual purchase is made, every choice made by a consumer involves risk (Taylor, 1974). Consumers first find information about the product to reduce this risk before purchasing them. They prefer product information from other consumers who have already used the product by understanding their experiences (Brown et al., 2007). They believe that word-of-mouth information is more reliable than company-generated marketing (Trusov et al., 2009). Kotler's model (2009) is an extensively used model for understanding the buying process of consumers. It comprises of need recognition, information search, evaluation of alternatives, purchase decision and post-purchase phase. It has been found that the duration of the buying process will vary with different types of goods. It will be longer for higher-value products like cars and will have high customer involvement. While the process will be shorter for low-cost goods such as clothes and will have lower customer involvement (Sakkthivel, 2010).

The discussion starts with the need recognition stage. The ongoing elevation in digitalization has made the consumer buying process more personalized and the latest trends play a significant role in purchasing decisions. In today's time, external stimuli to need recognition have become more dominant than internal stimuli. Social media creates a bandwagon effect on consumers through eWOM, which induces the needs of consumers (Park & Lee, 2008; Burnstein et al., 1973). Word of mouth works as a buzz marketing when market mavens give such reviews. The second stage, which is information search, is well explained through Elaboration Likelihood Model (ELM) (Petty & Cacioppo, 1986; Zaichkowsky, 1985). This is a dual process theory of persuasion in communication which assumes that a consumer can encounter product information through the central or peripheral routes. Consumer passing through the central route apply their cognition, are attentive to the information they receive, and evaluate the pros and cons of eWOM reviews (Bhattacherjee & Sanford, 2006).

A consumer passes through the central route for high-involvement goods like vehicles and electronics. While consumers passing through the peripheral route are less attentive towards the information received and apply their affective domain in evaluating eWOM. Here, consumers encounter the information unintentionally and, therefore, do not show any impulse to evaluate it (Petty & Cacioppo, 1986).

Previous studies have found that the peripheral route occurs for low-involvement goods (Sakkthivel, 2010) or the latest trends in fashion, cosmetics and personal care products. eWOM provides wider product opinions than that gathered from family and friends (Erkan & Evans, 2016). During the assessment of alternatives stage, it has been observed that eWOM opinion is crucial for customers and has the power to affect their behavior (Solomon, 2009). It also assists in determining the best product for various groups of consumers (Google, 2020).

The purchase decision is the fourth stage in the consumer buying process wherein the actual purchase takes place. Here, consumers intend to purchase since they have already evaluated the alternatives and have complete information about what value the product brings. Finally, consumers can have various reasons to share eWOM, which is the last stage of the consumer buying stage: post-purchase behavior. Ngarmwongnoi et al. (2020) found that consumers having a negative experience with a product are more active and participative in sharing their feelings with others than those having positive experiences. Hennig-Thurau et al. (2004) attributed this consumer behavior to Homeostats Utility, which states that consumers express their negative feelings with others to regain their lost balance.

IMPLICATIONS & LIMITATIONS

The primary objective of this study was to comprehensively understand the extant literature on eWOM communication since the body of knowledge on the area is vast and fragmented. The author concentrated on the individual-level analysis in the present study and reviewed existing studies from various theoretical perspectives occurring in eWOM literature. The study shows how eWOM has evolved regarding its meaning, theories and consumer behaviors. Although the author synthesized some of the most prominent theories to study eWOM literature, the author strongly believes that future studies can extend and expand these theories to report new findings or present existing findings in an enhanced manner through conducting a systematic literature review. Further, it is urged to investigate eWOM from the perspective of other disciplines to understand different facets of social communication that influence eWOM; for instance, since companies are already considering incorporating AI in eWOM communication, it would be interesting to study if AI-powered voice assistants assisting in product recommendations influence consumer behavior. Also, since culture is an essential aspect that shapes human behaviors (Donthu et al., 2021), it would be interesting to study if the country of origin relates to the acceptance and credibility of eWOM. Further, word of mouth possesses complexities for marketing practitioners due to its availability on multiple digital mediums. Managers should insulate themselves against bad eWOM that spreads quickly and harms company image, while ensuring that positive eWOM reaches potential consumers before rival's marketing communication.

Since eWOM has an extant presence in marketing literature, covering the entire theoretical and behavioral aspects was out of the scope of this study. Further, we studied individual-level studies and did not encompass market-level studies in this paper. However, it was noted that market-level studies employed a completely distinct perspective in understanding eWOM. It is strongly advised to study eWOM from the perspectives of various industries, developing economies, and technology.

CONCLUSION

To conclude, this paper brings forth an outline of the current status of literature in literature in eWOM domain. Furthermore, we highlighted the scope for future research in light of evolving nature of the domain. Firstly, we examined eWOM in the context of marketing and differentiated it from user-generated communication to clarify two phrases that appeared synonymous but are fundamentally distinct. Secondly, frequently occurring eWOM theories were elaborated to portray the existing state of research. Thirdly, we illustrated eWOM in consumer and buying-related behaviors. Finally, we postulate that this study will induce further studies to investigate various theories linked to studying eWOM regarding consumer behavior in light of the discussed future scope. The main limitation of this study was that a narrative review was conducted to corroborate the present state of research on eWOM. Future researchers may expand this study's scope through a systematic literature review. Conducting platform-specific or industry-specific literature reviews can contribute to academic knowledge. Further, this study investigated the eWOM phenomenon from a marketing perspective through the lens of consumers. Subsequent studies can extend the scope through companies' perspectives, particularly focusing on eWOM's role identification of service failures and risk reduction.

REFERENCES

Abedin, E., Mendoza, A., & Karunasekera, S. (2021). Exploring the Moderating Role of Readers' Perspective in Evaluations of Online Consumer Reviews. *Journal of Theoretical and Applied Electronic Commerce Research*, *16*(7), 3406–3424. doi:10.3390/jtaer16070184

Ajzen, I. (1991). The theory of planned behavior. *Organizational Behavior and Human Decision Processes*, *50*(2), 179–211. doi:10.1016/0749-5978(91)90020-T

Ajzen, I., & Fishbein, M. (1980). *Understanding attitudes and predicting social behavior*. Prentice-Hall.

Albayrak, M., & Ceylan, C. (2021). Effect of eWOM on purchase intention: Meta-analysis. *Data Technologies and Applications, 55*(5), 810–840. doi:10.1108/DTA-03-2020-0068

Arora, N., & Lata, S. (2020). YouTube channels influence on destination visit intentions: An empirical analysis on the base of information adoption model. *Journal of Indian Business Research, 12*(1), 23–42. doi:10.1108/JIBR-09-2019-0269

Babić Rosario, A., de Valck, K., & Sotgiu, F. (2020). Conceptualizing the electronic word-of-mouth process: What we know and need to know about eWOM creation, exposure, and evaluation. *Journal of the Academy of Marketing Science, 48*(3), 422–448. doi:10.100711747-019-00706-1

Babić Rosario, A., Sotgiu, F., De Valck, K., & Bijmolt, T. H. A. (2016). The Effect of Electronic Word of Mouth on Sales: A Meta-Analytic Review of Platform, Product, and Metric Factors. *JMR, Journal of Marketing Research, 53*(3), 297–318. doi:10.1509/jmr.14.0380

Baek, H., Ahn, J., & Oh, S. (2014). Impact of Tweets on Box Office Revenue: Focusing on When Tweets are Written. *ETRI Journal, 36*(4), 581–590. doi:10.4218/etrij.14.0113.0732

Bakshi, S., Gupta, D. R., & Gupta, A. (2021). Online travel review posting intentions: A social exchange theory perspective. *Leisure/Loisir, 45*(4), 603–633. doi:10.1080/14927713.2021.1924076

Balasubramanian, S., & Mahajan, V. (2001). The Economic Leverage of the Virtual Community. *International Journal of Electronic Commerce, 5*(3), 103–138. doi:10.1080/10864415.2001.11044212

Bhattacherjee, ASanford., C. (2006). Influence Processes for Information Technology Acceptance: An Elaboration Likelihood Model. *Management Information Systems Quarterly, 30*(4), 805. doi:10.2307/25148755

Bickart, B., & Schindler, R. M. (2001). Internet forums as influential sources of consumer information. *Journal of Interactive Marketing, 15*(3), 31–40. doi:10.1002/dir.1014

Bilal, M., Jianqiu, Z., Akram, U., Tanveer, Y., Sardar, T., & Rasool, H. (2020). Understanding the effects of Internet usage behavior on eWOM. [IJISSS]. *International Journal of Information Systems in the Service Sector, 12*(3), 93–113. doi:10.4018/IJISSS.2020070106

Brown, J., Broderick, A. J., & Lee, N. (2007). Word of mouth communication within online communities: Conceptualizing the online social network. *Journal of Interactive Marketing, 21*(3), 2–20. doi:10.1002/dir.20082

Brown, J. J., & Reingen, P. H. (1987). Social Ties and Word-of-Mouth Referral Behavior. *The Journal of Consumer Research, 14*(3), 350. doi:10.1086/209118

Bueno, S., & Gallego, M. D. (2021). eWOM in C2C Platforms: Combining IAM and Customer Satisfaction to Examine the Impact on Purchase Intention. *Journal of Theoretical and Applied Electronic Commerce Research, 16*(5), 1612–1630. doi:10.3390/jtaer16050091

Bulut, Z. A., & Karabulut, A. N. (2018). Examining the role of two aspects of eWOM in online repurchase intention: An integrated trust-loyalty perspective. *Journal of Consumer Behaviour, 17*(4), 407–417. doi:10.1002/cb.1721

Burgoon, J. K., Bonito, J. A., Bengtsson, B., Ramirez, A. Jr, Dunbar, N. E., & Miczo, N. (1999). Testing the Interactivity Model: Communication Processes, Partner Assessments, and the Quality of Collaborative Work. *Journal of Management Information Systems, 16*(3), 33–56. doi:10.1080/07421222.1999.11518255

Cao, X., Liu, Y., Zhu, Z., Hu, J., & Chen, X. (2017). Online selection of a physician by patients: Empirical study from elaboration likelihood perspective. *Computers in Human Behavior, 73*, 403–412. doi:10.1016/j.chb.2017.03.060

Chaiken, S. (1980). Heuristic versus systematic information processing and the use of source versus message cues in persuasion. *Journal of Personality and Social Psychology, 39*(5), 752–766. doi:10.1037/0022-3514.39.5.752

Chaiken, S. (1989). Heuristic and systematic information processing within and beyond the persuasion context. *Unintended thought*, 212-252.

Chan, K. W., & Li, S. Y. (2010). Understanding consumer-to-consumer interactions in virtual communities: The salience of reciprocity. *Journal of Business Research, 63*(9–10), 1033–1040. doi:10.1016/j.jbusres.2008.08.009

Chang, T.-S., & Hsiao, W.-H. (2013). Factors Influencing Intentions to Use Social Recommender Systems: A Social Exchange Perspective. *Cyberpsychology, Behavior, and Social Networking, 16*(5), 357–363. doi:10.1089/cyber.2012.0278 PMID:23374171

Chen, C.-W. (2014). UNDERSTANDING THE EFFECTS OF EWOM ON COSMETIC CONSUMER BEHAVIORAL INTENTION. *International Journal of Electronic Commerce Studies, 5*(1), 97–102. doi:10.7903/ijecs.1030

Cheung, C. M. K., Lee, M. K. O., & Rabjohn, N. (2008). The impact of electronic word-of-mouth: The adoption of online opinions in online customer communities. *Internet Research, 18*(3), 229–247. doi:10.1108/10662240810883290

Cheung, C. M. K., Lee, M. K. O., & Thadani, D. R. (2009). The Impact of Positive Electronic Word-of-Mouth on Consumer Online Purchasing Decision. In M. D. Lytras, E. Damiani, J. M. Carroll, R. D. Tennyson, D. Avison, A. Naeve, A. Dale, P. Lefrere, F. Tan, J. Sipior, & G. Vossen (Eds.), *Visioning and Engineering the Knowledge Society. A Web Science Perspective* (Vol. 5736, pp. 501–510). Springer Berlin Heidelberg., doi:10.1007/978-3-642-04754-1_51

Cheung, C. M. K., & Thadani, D. R. (2012). The impact of electronic word-of-mouth communication: A literature analysis and integrative model. *Decision Support Systems, 54*(1), 461–470. doi:10.1016/j.dss.2012.06.008

Chevalier, J. A., & Mayzlin, D. (2006). The Effect of Word of Mouth on Sales: Online Book Reviews. *JMR, Journal of Marketing Research, 43*(3), 345–354. doi:10.1509/jmkr.43.3.345

Chih, W.-H., Wang, K.-Y., Hsu, L.-C., & Huang, S.-C. (2013). Investigating Electronic Word-of-Mouth Effects on Online Discussion Forums: The Role of Perceived Positive Electronic Word-of-Mouth Review Credibility. *Cyberpsychology, Behavior, and Social Networking, 16*(9), 658–668. doi:10.1089/cyber.2012.0364 PMID:23895466

Daugherty, T., Eastin, M. S., & Bright, L. (2008). Exploring Consumer Motivations for Creating User-Generated Content. *Journal of Interactive Advertising, 8*(2), 16–25. doi:10.1080/15252019.2008.10722139

Dichter, E. (1966). How Word-of-Mouth Advertising Works. *Harvard Business Review, 44*(November–December), 147–166.

Donthu, N., Kumar, S., Pandey, N., Pandey, N., & Mishra, A. (2021). Mapping the electronic word-of-mouth (eWOM) research: A systematic review and bibliometric analysis. *Journal of Business Research, 135*, 758–773. doi:10.1016/j.jbusres.2021.07.015

Duan, W., Gu, B., & Whinston, A. (2008). The dynamics of online word-of-mouth and product sales—An empirical investigation of the movie industry. *Journal of Retailing, 84*(2), 233–242. doi:10.1016/j.jretai.2008.04.005

Eisenhart, M. (1991). *Conceptual frameworks for research circa 1991: Ideas from a cultural anthropologist; implications for mathematics education researchers.* Paper presented at the Thirteenth Annual Meeting North American Paper of the International Group for the Psychology of Mathematics Education, Blacksburg, Virginia, USA.

Fan, Y. W., & Miao, Y. F. (2012). Effect of electronic word-of-mouth on consumer purchase intention: The perspective of gender differences. *International Journal of Electronic Business Management, 10*(3), 175.

Feick, L. F., & Price, L. L. (1987). The Market Maven: A Diffuser of Marketplace Information. *Journal of Marketing, 51*(1), 83–97. doi:10.1177/002224298705100107

Filieri, R., McLeay, F., Tsui, B., & Lin, Z. (2018). Consumer perceptions of information helpfulness and determinants of purchase intention in online consumer reviews of services. *Information & Management, 55*(8), 956–970. doi:10.1016/j.im.2018.04.010

Fishbein, M., Ajzen, I., & Belief, A. (1975). *Intention and Behavior: An introduction to theory and research.* University of Massachusetts.

Godes, D., & Mayzlin, D. (2004). Using Online Conversations to Study Word-of-Mouth Communication. *Marketing Science, 23*(4), 545–560. doi:10.1287/mksc.1040.0071

Gupta, P., & Harris, J. (2010). How e-WOM recommendations influence product consideration and quality of choice: A motivation to process information perspective. *Journal of Business Research, 63*(9–10), 1041–1049. doi:10.1016/j.jbusres.2009.01.015

Harrison-Walker, L. J. (2001). The Measurement of Word-of-Mouth Communication and an Investigation of Service Quality and Customer Commitment As Potential Antecedents. *Journal of Service Research, 4*(1), 60–75. doi:10.1177/109467050141006

Hayes, J. L., & King, K. W. (2014). The Social Exchange of Viral Ads: Referral and Coreferral of Ads Among College Students. *Journal of Interactive Advertising, 14*(2), 98–109. doi:10.1080/15252019.2014.942473

Hennig-Thurau, T., Gwinner, K. P., Walsh, G., & Gremler, D. D. (2004). Electronic word-of-mouth via consumer-opinion platforms: What motivates consumers to articulate themselves on the Internet? *Journal of Interactive Marketing, 18*(1), 38–52. doi:10.1002/dir.10073

Hennig-Thurau, T., Walsh, G., & Walsh, G. (2003). Electronic Word-of-Mouth: Motives for and Consequences of Reading Customer Articulations on the Internet. *International Journal of Electronic Commerce, 8*(2), 51–74. doi:10.1080/108644 15.2003.11044293

Hlee, S., Lee, H., & Koo, C. (2018). Hospitality and Tourism Online Review Research: A Systematic Analysis and Heuristic-Systematic Model. *Sustainability, 10*(4), 1141. doi:10.3390u10041141

Homans, G. C. (1958). Social Behavior as Exchange. *American Journal of Sociology, 63*(6), 597–606. doi:10.1086/222355

Hu, X., & Ha, L. (2015). Which form of word-of-mouth is more important to online shoppers? A comparative study of WOM use between general population and college students. *Journal of Communication and Media Research, 7*(2), 15–35.

Hu, Y., & Kim, H. J. (2018). Positive and negative eWOM motivations and hotel customers' eWOM behavior: Does personality matter? *International Journal of Hospitality Management, 75*, 27–37. doi:10.1016/j.ijhm.2018.03.004

Hussain, S., Ahmed, W., Jafar, R. M. S., Rabnawaz, A., & Jianzhou, Y. (2017). EWOM source credibility, perceived risk and food product customer's information adoption. *Computers in Human Behavior, 66*, 96–102. doi:10.1016/j.chb.2016.09.034

Hwang, K., & Zhang, Q. (2018). Influence of parasocial relationship between digital celebrities and their followers on followers' purchase and electronic word-of-mouth intentions, and persuasion knowledge. *Computers in Human Behavior, 87*, 155–173. doi:10.1016/j.chb.2018.05.029

Hyrynsalmi, S., Seppänen, M., Aarikka-Stenroos, L., Suominen, A., Järveläinen, J., & Harkke, V. (2015). Busting Myths of Electronic Word of Mouth: The Relationship between Customer Ratings and the Sales of Mobile Applications. *Journal of Theoretical and Applied Electronic Commerce Research, 10*(2), 1–18. doi:10.4067/ S0718-18762015000200002

Ismagilova, E., Dwivedi, Y. K., Slade, E., & Williams, M. D. (2017). Electronic Word of Mouth (eWOM) in the Marketing Context: A State of the Art Analysis and Future Directions (SpringerBriefs in Business) (1st ed. 2017 ed.). Springer.

Khwaja, M. G., & Zaman, U. (2020). Configuring the Evolving Role of eWOM on the Consumers Information Adoption. *Journal of Open Innovation, 6*(4), 125. doi:10.3390/joitmc6040125

Kim, K., Cheong, Y., & Kim, H. (2017). User-generated product reviews on the internet: The drivers and outcomes of the perceived usefulness of product reviews. *International Journal of Advertising, 36*(2), 227–245. doi:10.1080/02650487.201 5.1096100

Kiousis, S. (2002). Interactivity: A concept explication. *New Media & Society, 4*(3), 355–383. doi:10.1177/146144480200400303

Kotler, P. & Armstrong, G. (2006). Principles of Marketing 11th Edition. Prentice Hall.

Kotler, P., & Zaltman, G. (1976). Targeting prospects for a new product. *Journal of Advertising Research, 16*, 7–20.

Kudeshia, C., & Kumar, A. (2017). Social eWOM: Does it affect the brand attitude and purchase intention of brands? *Management Research Review, 40*(3), 310–330. doi:10.1108/MRR-07-2015-0161

Lambe, C. J., Wittmann, C. M., & Spekman, R. E. (2001). Social Exchange Theory and Research on Business-to-Business Relational Exchange. *Journal of Business-To-Business Marketing, 8*(3), 1–36. doi:10.1300/J033v08n03_01

Lamberton, C., & Stephen, A. T. (2016). A Thematic Exploration of Digital, Social Media, and Mobile Marketing: Research Evolution from 2000 to 2015 and an Agenda for Future Inquiry. *Journal of Marketing, 80*(6), 146–172. doi:10.1509/jm.15.0415

Lee, H., Min, J., & Yuan, J. (2021). The influence of eWOM on intentions for booking luxury hotels by Generation Y. *Journal of Vacation Marketing, 27*(3), 237–251. doi:10.1177/1356766720987872

Lee, J., & Hong, I. B. (2021). The Influence of Situational Constraints on Consumers' Evaluation and Use of Online Reviews: A Heuristic-Systematic Model Perspective. *Journal of Theoretical and Applied Electronic Commerce Research, 16*(5), 1517–1536. doi:10.3390/jtaer16050085

Lee, J., Park, D. H., & Han, I. (2008). The effect of negative online consumer reviews on product attitude: An information processing view. *Electronic Commerce Research and Applications, 7*(3), 341–352. doi:10.1016/j.elerap.2007.05.004

Lee, M., & Youn, S. (2009). Electronic word of mouth (eWOM): How eWOM platforms influence consumer product judgement. *International Journal of Advertising, 28*(3), 473–499. doi:10.2501/S0265048709200709

Leong, C. M., Loi, A. M. W., & Woon, S. (2021). The influence of social media eWOM information on purchase intention. *Journal of Marketing Analytics.* doi:10.1057/s41270-021-00132-9

Leong, L.-Y., Hew, T.-S., Ooi, K.-B., & Lin, B. (2019). Do Electronic Word-of-Mouth and Elaboration Likelihood Model Influence Hotel Booking? *Journal of Computer Information Systems, 59*(2), 146–160. doi:10.1080/08874417.2017.1320953

Levy, S., & Gvili, Y. (2015). How Credible is E-Word of Mouth Across Digital-Marketing Channels?: The Roles of Social Capital, Information Richness, and Interactivity. *Journal of Advertising Research, 55*(1), 95–109. doi:10.2501/JAR-55-1-095-109

Liao, Y.-K., Wu, W.-Y., Le, T. Q., & Phung, T. T. T. (2022). The Integration of the Technology Acceptance Model and Value-Based Adoption Model to Study the Adoption of E-Learning: The Moderating Role of e-WOM. *Sustainability, 14*(2), 815. doi:10.3390u14020815

Litvin, S. W., Goldsmith, R. E., & Pan, B. (2008). Electronic word-of-mouth in hospitality and tourism management. *Tourism Management, 29*(3), 458–468. doi:10.1016/j.tourman.2007.05.011

Luo, C., Wu, J., Shi, Y., & Xu, Y. (2014). The effects of individualism–collectivism cultural orientation on eWOM information. *International Journal of Information Management, 34*(4), 446–456. doi:10.1016/j.ijinfomgt.2014.04.001

McMillan, S. J. (2002). Exploring Models of Interactivity from Multiple Research Traditions: Users, Documents, and Systems. In L. A. Lievrouw & S. Livingstone (Eds.), *Handbook of New Media: Social Shaping and Consequences of ICTs* (pp. 163–182). Sage. doi:10.4135/9781848608245.n13

McWilliam, G. (2000). Building Stronger Brands Through Online Communities. *Sloan Management Review, 41*(Spring), 43–54.

Mir, I. A., & Rehman, K. U. (2013). Factors Affecting Consumer Attitudes and Intentions toward User-Generated Product Content on YouTube. *Management & Marketing Challenges for Knowledge Society, 8*(4), 637–654.

Mohammed, A., & Al-Swidi, A. (2021). The mediating role of affective commitment between corporate social responsibility and eWOM in the hospitality industry. *Journal of Sustainable Tourism, 29*(4), 570–594. doi:10.1080/09669582.2020.1818086

Nestle boosts Ruby chocolate range after driving viral sensation. (2019). Indianexpress.Com. https://indianexpress.com/article/business/nestle-boosts-ruby-chocolate-kitkat-viral-sensation-5634042/

Ngah, A. H., Halim, M. R., & Aziz, N. A. (2018). The influence of electronic word of mouth on theory of *reasoned* action and the visit intention to the world monument fund site.

Ngarmwongnoi, C., Oliveira, J. S., AbedRabbo, M., & Mousavi, S. (2020). The implications of eWOM adoption on the customer journey. *Journal of Consumer Marketing, 37*(7), 749–759. doi:10.1108/JCM-10-2019-3450

Number of internet users in India from 2010 to 2020, with estimates until 2040. (2022). Statista. https://www.statista.com/statistics/255146/number-of-internet-users-in-india/

Nuseir, M. T. (2019). The impact of electronic word of mouth (e-WOM) on the online purchase intention of consumers in the Islamic countries – a case of (UAE). *Journal of Islamic Marketing, 10*(3), 759–767. doi:10.1108/JIMA-03-2018-0059

Orzan, G., Iconaru, C., Popescu, I. C., Orzan, M., & Macovei, O. I. (2013). PLS-based SEM analysis of apparel online buying behavior. *The importance of eWOM. Industria Textila, 64*(6), 362–367.

Park, D.-H., & Lee, J. (2008). EWOM overload and its effect on consumer behavioral intention depending on consumer involvement. *Electronic Commerce Research and Applications, 7*(4), 386–398. doi:10.1016/j.elerap.2007.11.004

Park, D.-H., Lee, J., & Han, I. (2007). The Effect of On-Line Consumer Reviews on Consumer Purchasing Intention: The Moderating Role of Involvement. *International Journal of Electronic Commerce, 11*(4), 125–148. doi:10.2753/JEC1086-4415110405

Park, H. S. (2000). Relationships among attitudes and subjective norms: Testing the theory of reasoned action across cultures. *Communication Studies, 51*(2), 162–175. doi:10.1080/10510970009388516

Petrescu, M., O'Leary, K., Goldring, D., & Ben Mrad, S. (2016). Incentivized reviews: Promising the moon for a few stars. *Journal of Retailing and Consumer Services, 41*, 288–295. doi:10.1016/j.jretconser.2017.04.005

Phan, Q. P. T., Pham, N. T., & Nguyen, L. H. L. (2020). How to Drive Brand Engagement and eWOM Intention in Social Commerce: A Competitive Strategy for the Emerging Market. *Journal of Competitiveness, 12*(3), 136–155. doi:10.7441/joc.2020.03.08

Rosário, A. T., Raimundo, R. G., & Cruz, R. (2022). The Impact of Digital Technologies on Marketing and Communication in the Tourism Industry: In I. R. Management Association (Ed.), Research Anthology on Business Continuity and Navigating Times of Crisis (pp. 748–760). IGI Global. doi:10.4018/978-1-6684-4503-7.ch037

Sardar, A., Manzoor, A., Shaikh, K. A., & Ali, L. (2021). An Empirical Examination of the Impact of eWOM Information on Young Consumers' Online Purchase Intention: Mediating Role of eWOM Information Adoption. *SAGE Open, 11*(4), 215824402110525. doi:10.1177/21582440211052547

Serra Cantallops, A., & Salvi, F. (2014). New consumer behavior: A review of research on eWOM and hotels. *International Journal of Hospitality Management, 36*, 41–51. doi:10.1016/j.ijhm.2013.08.007

Shaker, A. K., Mostafa, R. H. A., & Elseidi, R. I. (2021). Predicting intention to follow online restaurant community advice: A trust-integrated technology acceptance model. *European Journal of Management and Business Economics.* doi:10.1108/EJMBE-01-2021-0036

S Sharma, R., Morales-Arroyo, M., & Pandey, T. (2011). The Emergence of Electronic Word-of-Mouth as a Marketing Channel for the Digital Marketplace. *Journal of Information, Information Technology, and Organizations (Years 1-3), 6*, 041–061. doi:10.28945/1695

Shen, Z. (2021). A persuasive eWOM model for increasing consumer engagement on social media: Evidence from Irish fashion micro-influencers. *Journal of Research in Interactive Marketing, 15*(2), 181–199. doi:10.1108/JRIM-10-2019-0161

Smith, A. N., Fischer, E., & Yongjian, C. (2012). How Does Brand-related User-generated Content Differ across YouTube, Facebook, and Twitter? *Journal of Interactive Marketing, 26*(2), 102–113. doi:10.1016/j.intmar.2012.01.002

Sokolova, K., & Kefi, H. (2020). Instagram and YouTube bloggers promote it, why should I buy? How credibility and parasocial interaction influence purchase intentions. *Journal of Retailing and Consumer Services, 53*, 101742. doi:10.1016/j.jretconser.2019.01.011

Song, B. L., Liew, C. Y., Sia, J. Y., & Gopal, K. (2021). Electronic word-of-mouth in travel social networking sites and young consumers' purchase intentions: An extended information adoption model. *Young Consumers, 22*(4), 521–538. doi:10.1108/YC-03-2021-1288

Steffes, E. M., & Burgee, L. E. (2009). Social ties and online word of mouth. *Internet Research, 19*(1), 42–59. doi:10.1108/10662240910927812

Steinhoff, L., & Palmatier, R. W. (2021). Commentary: Opportunities and challenges of technology in relationship marketing. *Australasian Marketing Journal, 29*(2), 111–117. doi:10.1016/j.ausmj.2020.07.003

Sundaram, D. S., Mitra, K., & Webster, C. (1998). Word of-Mouth Communications: A Motivational Analysis. *Advances in Consumer Research. Association for Consumer Research (U. S.), 25*, 527–531.

Tam, K. Y., & Ho, S. Y. (2005). Web Personalization as a Persuasion Strategy: An Elaboration Likelihood Model Perspective. *Information Systems Research, 16*(3), 271–291. doi:10.1287/isre.1050.0058

Thao, N., & Shurong, T. (2020). Is It Possible for "Electronic Word-of-Mouth" and "User-Generated Content" to be Used Interchangeably? *Journal of Marketing and Consumer Research.* doi:10.7176/JMCR/65-04

Todorov, A., Chaiken, S., & Henderson, M. D. (2002). The heuristic-systematic model of social information processing. *The persuasion handbook: Developments in theory and practice, 23*, 195-211.

Tóth, Z., Mrad, M., Itani, O. S., Luo, J., & Liu, M. J. (2022). B2B eWOM on Alibaba: Signaling through online reviews in platform-based social exchange. *Industrial Marketing Management, 104*, 226–240. doi:10.1016/j.indmarman.2022.04.019

Trusov, M., Bucklin, R. E., & Pauwels, K. (2009). Effects of Word-of-Mouth versus Traditional Marketing: Findings from an Internet Social Networking Site. *Journal of Marketing, 73*(5), 90–102. doi:10.1509/jmkg.73.5.90

Wang, L., Wang, Z., Wang, X., & Zhao, Y. (2022). Assessing word-of-mouth reputation of influencers on B2C live streaming platforms: The role of the characteristics of information source. *Asia Pacific Journal of Marketing and Logistics, 34*(7), 1544–1570. doi:10.1108/APJML-03-2021-0197

Weisfeld-Spolter, S., Sussan, F., & Gould, S. (2014). An integrative approach to eWOM and marketing communications. *Corporate Communications, 19*(3), 260–274. doi:10.1108/CCIJ-03-2013-0015

Wolny, J., & Mueller, C. (2013). Analysis of fashion consumers' motives to engage in electronic word-of-mouth communication through social media platforms. *Journal of Marketing Management, 29*(5–6), 562–583. doi:10.1080/0267257X.2013.778324

Xia, M., Huang, Y., Duan, W., & Whinston, A. (2009). Ballot box communication in online communities. *Communications of the ACM*, *52*(9), 138–142. doi:10.1145/1562164.1562199

Xiao, M., Wang, R., & Chan-Olmsted, S. (2018a). Factors affecting YouTube influencer marketing credibility: A heuristic-systematic model. *Journal of Media Business Studies*, *15*(3), 188–213. doi:10.1080/16522354.2018.1501146

Xiao, M., Wang, R., & Chan-Olmsted, S. (2018b). Factors affecting YouTube influencer marketing credibility: A heuristic-systematic model. *Journal of Media Business Studies*, *15*(3), 188–213. doi:10.1080/16522354.2018.1501146

Xiong, G., & Bharadwaj, S. (2014). Prerelease Buzz Evolution Patterns and New Product Performance. *Marketing Science*, *33*(3), 401–421. doi:10.1287/mksc.2013.0828

Yang, S.-B., Shin, S.-H., Joun, Y., & Koo, C. (2017). Exploring the comparative importance of online hotel reviews' heuristic attributes in review helpfulness: A conjoint analysis approach. *Journal of Travel & Tourism Marketing*, *34*(7), 963–985. doi:10.1080/10548408.2016.1251872

Yüksel, H. F. (2016). Factors affecting purchase intention in YouTube videos. *The Journal of Knowledge Economy & Knowledge Management*, *11*(2), 33–47.

Yusuf, A. S., Che Hussin, A. R., & Busalim, A. H. (2018). Influence of e-WOM engagement on consumer purchase intention in social commerce. *Journal of Services Marketing*, *32*(4), 493–504. doi:10.1108/JSM-01-2017-0031

Zaichkowsky, J. L. (1985). Measuring the Involvement Construct. *The Journal of Consumer Research*, *12*(3), 341. doi:10.1086/208520

Zhang, K. Z. K., Zhao, S. J., Cheung, C. M. K., & Lee, M. K. O. (2014). Examining the influence of online reviews on consumers' decision-making: A heuristic–systematic model. *Decision Support Systems*, *67*, 78–89. doi:10.1016/j.dss.2014.08.005

Zhou, W., & Duan, W. (2013). How Does the Distribution of Word-of-Mouth Across Websites Affect Online Retail Sales? SSRN *Electronic Journal*. doi:10.2139/ssrn.2396072

Zhu, D. H., Ye, Z. Q., & Chang, Y. P. (2017). Understanding the textual content of online customer reviews in B2C websites: A cross-cultural comparison between the U.S. and China. *Computers in Human Behavior*, *76*, 483–493. doi:10.1016/j.chb.2017.07.045

ADDITIONAL READINGS

Belhadi, A., Kamble, S., Benkhati, I., Gupta, S., & Mangla, S. K. (2023). Does strategic management of digital technologies influence electronic word-of-mouth (eWOM) and customer loyalty? Empirical insights from B2B platform economy. *Journal of Business Research, 156*, 113548. doi:10.1016/j.jbusres.2022.113548

Hsieh, J. K., Hsieh, Y. C., & Tang, Y. C. (2012). Exploring the disseminating behaviors of eWOM marketing: Persuasion in online video. *Electronic Commerce Research, 12*(2), 201–224. doi:10.100710660-012-9091-y

Ismagilova, E., Rana, N. P., Slade, E. L., & Dwivedi, Y. K. (2020). A meta-analysis of the factors affecting eWOM providing behaviour. *European Journal of Marketing, 55*(4), 1067–1102. doi:10.1108/EJM-07-2018-0472

Mishra, A., & Satish, S. M. (2016). eWOM: Extant Research Review and Future Research Avenues. *Vikalpa, 41*(3), 222–233. doi:10.1177/0256090916650952

Mishra, A., Shukla, A., & Sharma, S. K. (2022). Psychological determinants of users' adoption and word-of-mouth recommendations of smart voice assistants. *International Journal of Information Management, 67*, 102413. doi:10.1016/j.ijinfomgt.2021.102413

Mukhopadhyay, S., Pandey, R., & Rishi, B. (2022). Electronic word of mouth (eWOM) research – a comparative bibliometric analysis and future research insight. *Journal of Hospitality and Tourism Insights*. doi:10.1108/JHTI-07-2021-0174

Pashchenko, Y., Rahman, M. F., Hossain, M. S., Uddin, M. K., & Islam, T. (2022). Emotional and the normative aspects of customers' reviews. *Journal of Retailing and Consumer Services, 68*, 103011. doi:10.1016/j.jretconser.2022.103011

Serra Cantallops, A., & Salvi, F. (2014). New consumer behavior: A review of research on eWOM and hotels. *International Journal of Hospitality Management, 36*, 41–51. doi:10.1016/j.ijhm.2013.08.007

Shen, Z. (2021). A persuasive eWOM model for increasing consumer engagement on social media: Evidence from Irish fashion micro-influencers. *Journal of Research in Interactive Marketing, 15*(2), 181–199. doi:10.1108/JRIM-10-2019-0161

Verma, S., & Yadav, N. (2021). Past, Present, and Future of Electronic Word of Mouth (EWOM). *Journal of Interactive Marketing, 53*, 111–128. doi:10.1016/j.intmar.2020.07.001

KEY TERMS AND DEFINITIONS

Behavior: The manner in which an individual reacts in a particular situation.

Communicator: An individual who communicates, shares or disseminates information amongst people.

Consumer: An individual who purchases a good or service.

Credible: Someone who can be relied upon.

Electronic Word of Mouth: The word of mouth that is generated through digital mediums.

Market Maven: An individual who is highly credible since he or she shares the most appropriate and reliable information.

Purchase Intention: The chance that a potential consumer will purchase the product.

Theory: An explanation of knowledge that can be supported with empirical evidence.

Chapter 13
Customer Satisfaction With a Named Entity Recognition (NER) Store-Based Management System Using Computer-Mediated Communication

Le Thi Hong Vo
University of Economics Ho Chi Minh City, Vietnam

Thien Tuan Hang

Mobile World Investment Corporation, Vietnam

Ayman Youssef Nassif
University of Portsmouth, UK

ABSTRACT

With the rise of the popularity of e-commerce, it is evident that the service retail industries aim to reduce inventory and increase sales and profit margins. To achieve this, it is of paramount importance to establish excellent and effective interaction between customers and customer support. When a customer orders a product online, it is essential that the store demonstrates whether the products are in stock and the nearest stores to where the customers are. Currently, the needs of the customers are unlikely to be effectively met. Hence, the stores are unlikely to provide desirable products to customers even with high inventory. This paper investigates this issue at a typical and popular retail store in Vietnam. The authors present an investigation of this issue through two main stages. Corpus analysis for a set of collected text

DOI: 10.4018/978-1-6684-7034-3.ch013

messages posted on the stores' websites for customer support was first carried out to explore the lexical patterns that indicate the customers' needs. This analysis revealed the frequency of customers' requests for the stores' locations where they can buy the goods and/or whether they are in stock. In the second stage of the investigation, the valuable findings from the corpus analysis were used for data extraction based on Named Entity Recognition (NER) software. The NER recognizes entities, including locations and names.

INTRODUCTION

The increase of sales and profit margins is a major objective of the retail service industries. Good customer services, i.e., a good interaction between customers and customer-support, is considered the most important role in achieving the goal. E-commerce has been becoming popular and widespread. Customers, therefore, contact stores mainly through websites and smartphone Apps (Ek et al 2011). However, there have been difficulties when a customer orders a product online. The problems concern the store searching for the products to establish whether the products are in stock and establish the nearest stores to where the customers are. The process often takes a while when adopting current computer-based systems. In many cases, the needs of customers are not met.

There is an evident concern in the service retail industries on how to reduce the inventory to increase sales and profit margins (Sirimanna & Gunawardana 2020). In order to achieve this goal, the ideal scenario relates to a good interaction between customers and customer-support. The context of this research investigation involved popular retail stores where the importance of this interaction is clearly highlighted.

As e-commerce becomes increasingly popular and widespread, customers have been considered online shoppers who contact stores mainly through websites and smartphone Apps. However, when a customer orders a product online, it takes a while for the store to search for the products to establish whether the products are in stock and install the nearest stores to where the customers are. Such activity, when adopting current computer-based systems, would take time and in many cases, customers' needs are not met. As a result, the stores are unlikely to provide desirable products to customers even with high inventory. The increasing customers' complaints can be found in the customer support chat box in the company's website. The current research investigates this issue at a typical and popular retail store in Vietnam and how Named Entity Recognition (NER) software supports quicker and more efficient buying and selling processes. This research supports the utilization of text-messages (Thurlow & Poff, 2011) as means for good customer services in service sectors.

LITERATURE REVIEW

Computer-mediated communication (CMC) is the communication medium individuals use to establish relational and social meaning, making use of technological advances. There are many styles used for communications online, such as emails, blogs, messaging apps, online conferencing, and emails. The term CMC refers to all these communications online or digital platforms used for human interactions (Altohalmi 2020; Sheblom 2020).

The question related to how CMC impacts different aspects of human life, either individual communications or organised group communications, for pleasure business or scholarly activities, has been explored by researchers (December, 1996). CMC was evidenced to facilitate effective learning experience in an environment where constructive alignment learning philosophy can be implemented. This is with particular reference to designing learning experiences for learners of second languages (Romiszowski & Mason 2004). CMC enables users to enhance their interactions and exchange of ideas and opinions with other people of similar interests and learning goals with similar interest in such conversations.

CMC tools can be synchronous or asynchronous. Synchronous CMC mode allows users to have real time interaction and discussions, such as real time meeting rooms, and video conferencing. This is similar to spoken interaction (Sykes 2005).

Asynchronous CMC does not happen in real-time. However, it is based on recorded or published information, which users access in their own time and not in real time. The time lag between receiving the CMC and contributing an answer allows the user to contemplate their answer. Users are also able to reread or re-watch the recorded communication anytime they wish (Altohami 2020, Newhagen 1996).

According to Berry (2004), the main advantages of asynchronous and synchronous communication media include flexibility over time and distance combined, more active and equal team member participation, and the ability of users to reflect or collect data before responding. The following section discusses how text messages are considered one of the most popular CMC modes that reflects the mentioned advantages.

Text Messages

It is evident that the use of text messaging, referred to as SMS and texting, has gained huge popularity in the last three decades (Altohami 2020). Texting has become appealing because it is cheap, personal, and unobtrusive. There are some issues related to a clear understanding of evolving texting messages which could result in misunderstandings (Thurlow & Poff 2011). Despite these issues, text messages are effective means for social interactions such as greetings, exchanging ideas, as well as providing help and support (Crystal 2008). Text messaging is also used to organise real-time synchronous interactions with families, friends, and colleagues. The instant messaging (IM) technology allows users to chat over the internet in real-time using text (Larson 2011).

Sirimanna and Gunawardana (2020) observed the development of technology, noting that instant messaging (IM) apps can now allow users to transmit texts and images and files. Such files are usually sent through the chat app as hyperlinks, voice, or video. Instant messaging is now prolific in the workplace as it allows effective and quick messages communications.

Cho et al. (2005) indicate that IM can support better work performance of employees based on the quality of communication and trust though it could sometimes interrupt work. IM, therefore, is effective in enhancing other CMC tools. A more effective and comprehensive work communication environment can be created for the employees.

It is evident that previous studies related to CMC were concerned with the effectiveness of such medium and the user's satisfaction level (Giri & Kumar, 2010; Mukahi et al., 2003).

There seems to be a dearth of information related to previous research examining the impact of CMC on customer satisfaction of how Named-Entity-Recognisation (NER) is applied. The following sections will outline previous research efforts investigating NER.

Previous Research into NER

There seems to be few published research efforts concerning NER in constrained environments. Jiang et al. (2010) investigated Hidden Markov Models (HMM), an SMS database of 1000 messages. Named Entities related to events and activities were extracted from Chinese SMSes. Such findings were specifically aimed at the use of handsets.

The research carried out by Polifroni et al. (2010) adopted logistic regression to recognize named entities, such as time, date, and location, from spoken and typed messages. This research was carried out in a controlled laboratory setting where the SMSes and spoken words of the participants were transcribed to establish a database. The research effort was focused on automatic speech recognition for mobile phones.

Hård af Segerstad (2002) provided the analysis of linguistics of Swedish test messages characteristics and usage of SMS in Sweden. The research findings showed interesting aspects related to the use of abbreviations borrowed from other languages and not known in the Swedish language.

Ek et al. (2011) investigated the implementation of the NER application for text messages in Swedish on a mobile flatform. The research supports the idea that the extraction of entities including names, locations, dates, time and telephone number could be used by other applications on the phones.

In Vietnam, there is little research on NER. Pham (2020) discusses VLSP 2016 as a Vietnamese training set established by Association for Vietnamese Language and Speech Processing-vlsp.org.vn). It provides two data sets: Named Entity Recognition and Sentiment analysis. Four named-entity categories were adopted in VLSP 2016. These are similar to those considered in the shared task of CoNLL 2003. They are person (PER), organization (ORG), location (LOC), and miscellaneous entities (MISC).

There are two approaches to machine learning. One approach is a conventional machine-learning methodology, and the other is a deep-learning one. Such models and methodologies can be applied to NER systems in VLSP2016, such as Hidden Markov Models, Support Vector Machines, Conditional Random Fields (CRFs), Maximum-Entropy-Markov Models (MEMMs) or recurrent neural network (RNN) with LSTM units (Li et al. 2020, Pham 2019), PhoBERT (Nguyen and Nguyen 2020).

Among the above approach, BERT (the Bidirectional Encoder Representations from Transformers) is considered as one of the most common pre-trained language models (Devlin et al. 2019). BERT has become popular as an open source like Spacy and Open NLP. Huge improvements for a variety of NLP tasks have been achieved through BERT, which is now well-supported and prolifically used. The currently available BERT tools have been mainly applied for English language communications. However, the architecture of such BERT tools could be retrained, or existing pre-trained multilingual BERT-based models can be employed (Devlin et al. 2019; Lample & Conneau 2019).

In this study, the authors use BERT to name entities. NER software in this research recognizes entities, including locations and names. This NER supports quicker and more efficient processes of buying and selling goods at a typical and popular retail store in Vietnam.

METHODOLOGY

The study aims to establish good and effective interaction between customers and customer support through analysing text messages. The two research questions are:

(1) To what extent does Named Entity Recognition (NER) software support more efficient processes to increase sales and profit margins in the retail industry?
(2) How satisfied are the customers with the support of the NER?

To answer the two research questions, the study includes two main stages and a pilot evaluation. Stage 1 relates to corpus analysis for collected text messages to explore the lexical patterns that indicate the customers' needs. In the second stage of the investigation, the valuable findings from the corpus analysis were used for data extraction based on Named Entity Recognition (NER) software including locations and names. Finally, a pilot evaluation of customers' satisfaction of the revised app is carried out based on the data collected via customers' comments on the website.

Stage 1

In the first stage of the research, the authors collected messages exchanges (Altohami 2020) between customers and customer support on the company's website including chat box support, comments on the website, and chat applications, such as Zalo, Messengers. Messages, could convey short conversations, mainly on customers' asking whether products they wanted to buy were in stock, the addresses of the stores where they could collect the products, or providing their own addresses for deliveries. The messages also related to customers' complaints about products, and supporters' replies identifying the addresses of the buyers before transferring the issues to departments that have responsibilities.

These messages were written in Vietnamese, some of which were common 'borrowed' English such as 'thanks', 'inbox', 'chat'. These English words were often seen among young customers. There were about 3500 text messages collected. The data collection took into consideration any ethical and confidentiality issues. The researchers applied the company's policy of keeping customers' information confidential. The data covered aspects of customer-customer support as it was collected from all the means of communication to customers offered by the company. We annotated the dataset with the categories of name entities: date, time, product, location and address.

Text messages, posted on the stores' websites between customers and customer support were collected from December 2021 to February 2022. This is when the needs to buy products increase due to the Tet holiday (the biggest holiday in Vietnam). This period was chosen as the demand for effective customer support services increases and affects the retailing process.

We divide our system into two components, as shown in Figure 1. The first component consisted of real-life text-based communication core, which was collected from three sources: mobile chat application, chatbox support on the website, and comments on websites. This component of the system was utilised to generate the datasets for the implication of the second stage. Secondly, with the collected raw dataset, some issues concerning spelling mistakes, redundancy, or punctuation

Figure 1. The two stages of the research

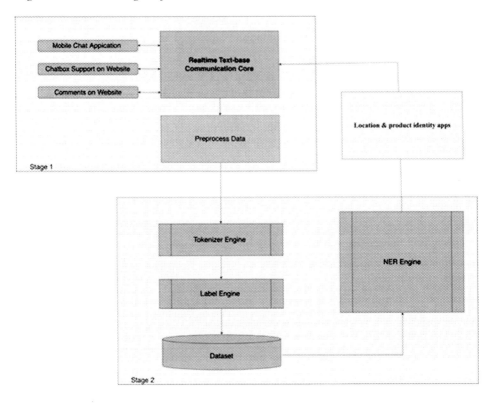

must be addressed before proceeding with the data. In details, the issues that need to sort out include:

The first issue concern Unicode application. Vietnamese applies two Unicode types, including two different byte (UTF-8). This causes difficulties in proceeding the text. For example:

b'Ti\xe1\xba\xbfng Vi\xe1\xbb\x87t'b'Tie\xcc\x82\xcc\x81ng Vie\xcc\xa3\xcc\x82t'

The difference in showing the tone (e.g. 'oề' and 'oề') which could lead to the difference in the meaning. The solution for this is the use of Normalization Form C (a defined normalization of Unicode strings which make it possible to determine whether two Unicode strings are equivalent to each other-Wiki Media) to standardize one typy of Unicode.

The second issue relates to the error of using space. This is solved by space deletion. The third issue concern the error of writing the name in lower case which could convey different meanings. For example, 'hồ Xuân Hương' means 'the lake named Xuân Hương' instead of 'Hồ' as surname. The solution for this issue is to set lower case for all texts.

Stage 2

In the second stage of the research, the text messages corpus collected in stage 1 is used as input. With 3500 conversations, there are about 15000 common words in Vietnamese, including addresses, customers' names, names of products, and materials. The recognition procedure was carried out using Workpiece.

Wordpiece is a tokenizer method which is used to split a sentence into parts including words in this study. In comparison with old versions, e.g., BPE PhoBERT (Nguyen and Nguyen 2020). The common issue is that some of these words cannot be found in dictionaries, which are called UNK words. Wordpiece helps solve the problem by splitting words with high frequency into subcategories (clusters, vowels, and consonants) (Figure 2). This method helps to have UNK words as subwords which can be easier to predict the meanings.

Figure 2. Wordpiece spliting sentences into parts

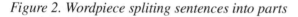

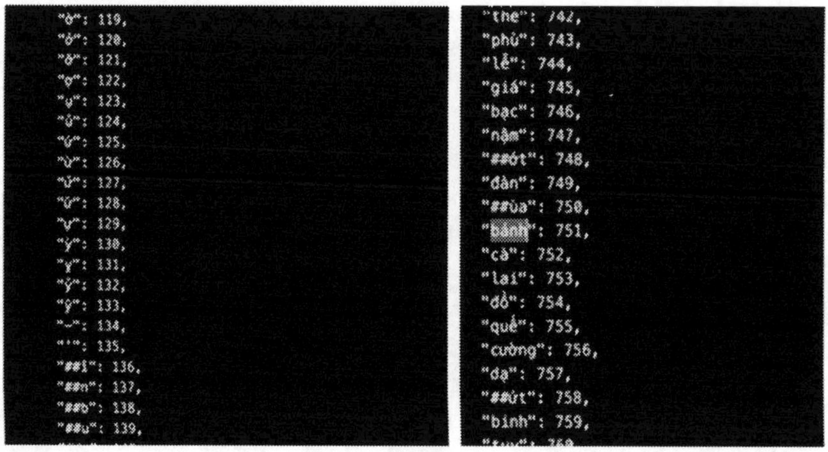

For instance, 'cty Thế Giới Di Động' is splited to 'cty', 'thế', 'giới', 'di', 'động'. If 'cty' (means 'công ty'-company) cannot be found in dictionaries because it is written in shorthand technique, Wordpiece would help to have two subwords, 'c' and 'ty'. The meaning can be revealed from the two tokens.

The other example concerns the lack of space from customers' texting 'hiện gà này ở bhx chợ phú thọ trảngdàibiênhòa đn có không ạ', Wordpiece will split all the strings into words 'hiện, gà, này, ở, bhx, chợ, phú, thọ, trảng,##dài,##biên,##hòa, đ,#n, có, không, ạ' to proceed for the meanings.

Table 1. Label classification

	Label	Description	Examples
1	B-LOC	Begin Location	Chung [B-LOC] (The intial of 'apartment building': chung cư)
2	I-LOC	Inside Location	Cư [I-LOC] (The ending of 'apartment building': chung cư)
3	B-ADD	Begin Address	Alexander [B-ADD]
4	I-ADD	Inside Address	de [I-ADD] Rhodes[I-ADD]
5	B-PROD	Begin Product	Nước [B-PROD] (The intial of 'fish sauce':nước mắm)
6	I-PROD	Inside Product	Mắm [I-PROD] (The ending of 'fish sauce': nước mắm)
0	O	Words which do not belong to the above groups	Chai [O] bottle

After splitting words, the statistics is carried out to find out the frequency of the 15.000 words. This is to build the list of words that can merge to each other, as shown in Figure 2. above.

Label engine (Lafferty et al. 2008) is the next step in this stage. In this step, the list of words provided from the step of tokenizer engine and is given labels as it is shown in Table 1 below:

There are 3 groups of label classification: Location, Address, Product. Each group includes 2 subgroups: B- (begin), I-(inside) (Pham 2018). The last group 0 includes words which do not belong to the first six groups. This follows a common tagging format for tagging tokens in computational linguistics presented by Ramshaw and Marcus (1995). The B-prefix before a tag indicates that the tag is the beginning of a block. The tag is shown inside a chunk with I- prefix before a tag. An O tag indicates that a token belongs to no block.

The third step is the step of mapping tokens to labels. For example, "Một chai nước mắm đến chung cư đường Alexander de Rhodes" is encoded as below (Table 2)

Table 2. Mapping tokens to labels

Tokens	Một	chai	nước	mắm	đến	Chung	Cư	đường	Alexander	De	Rhodes
IDs	11	13	15	16	8	56	34	17	88	98	332
Labels	O	O	B-PROD	I-PROD	O	B-LOC	I-LOC	O	B-ADD	I-ADD	I-ADD
IDs	0	0	5	6	0	1	2	0	3	4	4

The combination of each couple of token ID and label ID is call an offset. A number of offsets form an encoded sentence. For example, these offsets from the table x.x [(11,0),(13,0), (15,5),(16,6),(8,0),(56,1),(34,2),(17,0),(88,3),(98,4),(332, 4)] makes the sentence "Một chai nước mắm đến chung cư đường Alexander de Rhodes". A dataset contains such encoded sentences. In this study, 12250 sentences from 3500 conversations are proceeded through this process.

In the final step, the dataset is input to the NER engine. Tokens are recognized to classify according to the labels such as B-LOC, I-LOC, B-ADD, I-ADD, B-PRO, I-PRO, O. For example, the encoded input of "Hồ Xuân Hương" will be classified "Hồ" as B-ADD, "Xuân" as I-ADD, "Hương" as I-ADD. Then, the merging process will happen to form the entity of address "Hồ Xuân Hương". The similar processes are applied to identify the entities of location and product.

A Pilot Evaluation on Customer Satisfaction

A pilot experiment was carried out for a preliminary evaluation of customer satisfaction. The proposed chat facility within the NER Application was used for a pilot investigation for a duration of two weeks. Around 600 conversations in the chatbox were collected during this time. The messages were mainly related to customers asking for the products they wanted to buy and the nearest stores to where the customers were. These messages were reviewed to establish the overall satisfaction of the customers who chatted with customer support.

In order for the feedback of the customers on using the revised Application, an online feedback form was designed and attached to the website so that the customers can fill it in after the conversation exchanges with customer support. The form contains five main questions which attempted to quantify the customer's self-assessment of their use of the revised chatting app. These questions were designed to determine how satisfied the customers were with the customer support service. 387 out of 600 customers completed the online feedback form. Participants answered the questions on Likert's scale that describes their choice from 1 (very bad) to 5 (very good). One open-ended question (question 6) was used for aspects of triangulation of the answers and further comments.

These questions in the feedback form are as follows:

1. How much do you like the chatting app?
2. How fast do you have the answers for the products you want to buy?
3. How fast do you receive the answers for the nearest stores?
4. How much are the suggestions on the website appropriate to your needs?
5. What scale do you choose to rate the level of service satisfaction? (from 1 very bad – 5 very good)

6. Which Application version do you prefer? The old version or the new version?

Table 3. The frequency of common entities

	Details	Frequency	Labels	3500 for total
Sample date	Friday	79%	DATE	2765
Sample time	5.PM - 10.PM		TIME	
Sample telephone no	+092.xxxxx		PHONE	
Product	Bò Úc, rau, sản phẩm 4kFarm, sữa Vinamilk (Australian beef, vegetable, 4kFarm, Vinamilk)	96%	PRODUCT	3360
Location	Gần trường học, Vinhome, Nhà văn hoá thanh niên. (near schools, Vinhome, behind Youth Cultural House)	89%	LOCATION	3115
Address	Phường Tỉnh quận huyện thành phố (Ward, District, City)	83%	ADDRESS	2905
First name & Titles	Cô Lê, Chú Ba, Anh Bảy, chị Hai chủ nhà, chủ quán kế bên nhà (house owner, the shop owner who lives next door)	72%	NAME	2520

RESULTS AND DISCUSSION

In this section, the data collected from corpus analysis and NER software (Larson 2011, Pham 2018) will be explored to find out how NER facilitates customer services. In the first part, locations refer to customers' addresses, as provided by them, to establish the nearest store to them which can supply the ordered products. The locations database, therefore, also refers to the stores' addresses, thus linking the customers to the nearest locations of stores. Names database refers to the products and the names of customers. In the second part of the section, the customers' satisfaction with the support of the NER will be discussed.

\NER s Software Supports More Efficient Processes to Increase Sales and Profit Margins in the Retail Industry

The Common Entities Reflecting Customer's Needs

The entities of 'products' and 'locations' appear in high frequency in the text messages, 96% and 89% respectively. Table. 3 indicates that the customers who

Table 4. Types of products according to date and time

Date	Time	Product
Monday	10AM - 1PM, 6PM - 10PM	Thịt, cá, trứng, sữa, rau tươi, trái cây (Meat, fish, eggs, vegetables, fruits)
Tuesday	10AM - 1PM, 6PM - 10PM	Thịt, cá, trứng, sữa, rau tươi, trái cây (Meat, fish, eggs, vegetables, fruits)
Wednesday	11AM - 1PM, 6PM - 10PM	Thịt, cá, trứng, sữa, rau tươi, trái cây (Meat, fish, eggs, vegetables, fruits)
Thursday	11AM - 1PM, 6PM - 11PM	Thịt, cá, trứng, sữa, rau tươi, trái cây (Meat, fish, eggs, vegetables, fruits)
Friday	6 PM - 12 PM	Hải sản, thịt, rau củ (Seafood, meat, vegetables)
Weekend	9AM - 11 PM6	Bánh kẹo, nước ngọt, trái cây, đồ dùng một lần (Sweets, soft drink, fruits, disposable products)
Holiday	9AM - 11 PM6	Bánh kẹo, nước ngọt, trái cây, đồ dùng một lần (Sweets, soft drink, fruits, disposable products, disposable products) Thịt & hải sản (Meat & sea food)

contact to buy products often enquires about where they can find the products they want. Customer support, in return, ask the customers for their addresses to serve and meet their needs.

The sample data indicated that 89% of customers provided their location while 83% provided their address as well. Customers provide their addresses to be supported regarding the stores where they can find the products they need. This shows that these entities appear in almost all customers to customer-support conversations. Interestingly, these entities appear together with high frequency.

Date and Time Related to Customers' Purchases and Needs

Table 4. shows that customers often enquired about ordering food around 10 a.m-1p.m and 6 p.m-11p.m in weekdays and 9 a.m to 11 p.m at weekends and holidays. The required Products, mainly fish, meat, eggs, milk, vegetables and fruits, were quite varied in weekdays.

Evidently, the customers enquiries are varied according to date and time. For example, at weekends, entertainment related food such as beer, soft drinks, cakes, sea food as well as disposable products have higher frequency of orders, particularly from the suburbs of Ho Chi Minh City (HCMC).

Table 5. Products – location and location-products

Residential areas ● **High population areas** e.g. Bình Thạnh, Phú Nhuận, Gò Vấp, Tân Bình, Tân Phú districts. ● **Apartment buildings** e.g. Vinhome Center Park, Vinhome Grand Park ● **Dormitories** e.g. University Dormitory in Thu Duc)		Office buildings & Industrial Parks ● District 1, 3 ● High Tech Park ● Quang Trung Software City, district 12 ● Tan Tao Industrial Park, Tân Bình District Vinh Loc Industrial Park, Binh Chanh district	
Item	**%**	**Item**	**%**
Đồ vệ sinh cá nhân (Toiletries) Nước rửa tay (Hand sanitizer) Khẩu trang (Face Mask)	43%	Đồ dùng 1 lần (Disposable items) Nước rửa tay và khăn giấy (Hand sanitizer and tissue)	83%
Đồ khô (instant noodle) - Gạo và ngũ cốc (Rice & cereal) - Đồ chay (Vegan food) - Gia vị (Flavour) - Sweet food (Bánh kẹo)	85%	Thức ăn nấu sẵn (Ready-to-eat food) Snack & Sandwich Bánh kẹo (Sweets) Mì ly, xúc xích (Instant noodle and sausage)	89%
Đồ uống (Beverages) Rượu, bia, nước ngọt (wine, beer, soft drink) Nước khoáng (Mineral water) Nước yến (bird net) Trà và cà phê (Coffee & tea) Nước uống sô cô la (Chocolate drink)	71%	Đồ uống (Beverages): Rượu, bia, nước ngọt (wine, beer, soft drink) Nước khoáng (Mineral water) Trà và cà phê (Coffee & tea)	81%
Thịt, cá, trứng, rau, Trái cây (Meat, fish, eggs, vegetables, fruits)	92%	Trái cây (fruits)	65%
Sữa, phomai, yaourt, ngũ cốc (Milk, cheese, yogurt, cereal)	82%	Sữa, phomai, yaourt, ngũ cốc (Milk, cheese, yogurt, cereal)	85%
Đồ đông lạnh (frozen food)	38%	Đồ đóng hộp (canned food)	23%

Products Location Relationship Related to Customers' Needs

The analysis of the collection of the written texts (Altohami 2020) with Wordpiece showed that common products such as meat, fish, eggs, vegetables, fruits, milk, cheese, yogurt, cereal are similarly required by customers from residential areas, office buildings and industrial parks (Table 5). This demonstrates the high needs for these products.

However, customers who live in areas where there are more apartment buildings (e.g. Vinhome Centre Park in District 2, Vinhome Grand Park in district 9, Tan Binh and Tan Phu where the population is high) requires toiletries more. Ready-to-eat food, disposable products are required more in areas where there are more office buildings and industrial parks. It is worth noting that the requirement of all products is more from the suburbs such as Thu Duc rather than from the centre of the city (district 1,3). This reflects the recent trend of expansion of the residential areas from the center to the suburbs of HCMC.

Based on such information related to the required products and locations, which was established with the use of NER (Li et al. 2020), effective suggestions to match customers to products and/or location could be provided. This is crucial to provide good customer service responding to customer needs.

The Use of Lexical Features for Good Interaction Between Customers and Customer-Support

In this section a number of lexical features of the communication data collected for this research, such as, titles prefixing people's names, will be explored (Table 6).

Titles Prefixes

In Vietnamese, it is important to show politeness by using the titles. 'Anh' for male customers and 'chị', 'em' for female customer are most commonly used as they are for the majority of customers. In the case that the customers are known to be older people, 'cô' 'chú'. 'bác' are used.

Shorthand Techniques

Common shorthand techniques which are used by customers are: Tks = thanks, ac = anh chị, e = em, k/ko = không(can't), dc = được(can), hsd = hạn sử dụng(expiry), while NSX (Ngày sản xuất (Manufatured date), HSD (Hạn sử dụng (expire date), SL (Số lượng, số lô (quantity, Lot number) are often used by customer support.

Common Emojis and Emoticons

Customers usually expressed their positive feelings or negative feelings by using emojis and emoticons, mainly for happy (:-)😊) or sad (:-(☹).

Greetings and Questions

Some questions that are often seen in the collected text messages are: "Ac có đang inbox ko?", "Allo ac" (Are you inbox?, Allo?). In many cases, these questions are considered as greetings.

It results in the high frequency of the replies such as 'Dạ có ở cửa hàng Nguyen Dinh Chieu, gần chỗ chị" (Yes, we sell them in Nguyen Dinh Chieu near your place) than 'Xin lỗi, hiện chúng tôi hết hàng' ('Sorry, we are out of it) or 'Anh/chị chờ

Table 6. Lexical features

Lexical features	Customers	Customer support
Greetings	Chào ac	Chào anh/chị
Titles prefixes	Male: Anh	Male: Anh, chú
	Female: Chị	Female: Chị, cô
	Both: Em, tôi	Both: Em, quý khách, bác
Shorthand techniques	Tks = thanks ac = anh chị e = em k/ko = không dc= được hsd = ha …	Cảm ơn (Thanks) NSX (Ngày sản xuất- Manufactured date) HSD (Hạn sử dụng – Expiry) SL (Số lượng, số lô- Quantity, Lot number) cty (Công Ty - Company), tnhh (Ltd)
Hedges/ Hegdes	Vâng (Yeah) ạ dạ ơi	Vâng (Yeah) ạ dạ
Emoticons	Happy: happy:-) or 'hehe'	Happy: happy:-) or 'hehe'
	Unhappy: sad:(, confused =/, cool B-)	Sorry feelings when the needs of customers are not met yet :((:-((

một chút ạ' (Please can you wait for a while') from customer supports. The more the replies were 'Yes, (có) than 'Không có' ('No'), the more happy emoticons were shown in customers' messages.

In summary, there are two common questions the customers made: (1) whether the products they need are in stock (products-location) and (2) providing their addresses to request the nearest stores' locations where they can buy the products (location- products).

As for products-location, the results show that, as expected, customers require products close their locations. Generally, all customers enquires were related to, not just the products but, as expected, to the nearest store to them. The findings reveal that with the support of NER engine (Nguyen and Nguyen 2020), when the needs of customers increased, the retail system of the company was able to meet such needs effectively as evident by the high frequency of the replies of 'Yes, (có) than 'Không có' ('No') and high frequency of more happy emoticons (happy:-)) in the customers' messages.

As for location-products, if the customer's desired store is out of stock of the product, the proposed NER system could provide appropriate alternative stores where the product is in stock. This resulted in significant reduction of complaints from the customers in relation to clarity of product availability in different locations. NER has supported the process of searching for the addresses and give it to the delivers to make sure products can be delivered to the customers quickly.

In addition, the research data included evidence of the current trend of expansion of residential areas from the center to the suburbs of HCMC. Hence, the proposed NER system will provide appropriate arrangements for products and delivery to meet customers' needs in the wider suburbs of HCMC.

4.2 The Customers' Satisfaction with the Support of the Ner

In this section, the results from a pilot evaluation of customers' satisfaction are discussed. As it is shown in table 7, the majority of customers found that the proposed App is supportive in helping them to buy the products they want.

The percentage of customers who chose "5" or "4" (A great deal/Good) to the question related to "how fast they have the answers" was 60%. Similarly, the percentage of customers who chose "5" or "4" (A great deal/Good) to the question related to "how much the suggestions on the web are appropriate to your needs" was 80%. The responses of "How much do you like the app" showed that the majority of the customers, total of 86%, responded with either 5 or 4 (a great deal) similar to the chatting app.

As for the last question related to the customers' preference of the version of the app (old or new version), 90% of the customers said they preferred the revised app. The reasons for their answers were as follows:

"I prefer the new version because I received quick answers. It makes me feel I am cared for."

"I prefer the new version because it is very friendly and natural. I don't feel like I am talking with a machine."

"I often receive the right products to buy and can find the right shops easier."

CONCLUSION AND RECOMEMDATION

The research showed that customers had strong positive responses to the proposed App. This suggests that NER system can provide more effective practices for interaction between customers and customer-support customer-support. Therefore, the design and implementation of such apps is recommended. This leads to the following conclusion.

The Named Entity Recognition (NER) system, which is proposed by the authors in this paper, was found to increase the efficiency of the retail process in the age of increased popularity and wide-spread of e-commerce. The effectiveness

Table 7. The customers' feedback on using the revised app

	A little/ Bad		Average	A great deal/Good	
	1	**2**	**3**	**4**	**5**
How much do you like the chatting app?	4 (1%)	50 (12.9%)	100 (25.8%)	153 (39.5%)	180 (46.5%)
How fast do you have the answers for the products you want to buy?	35 (9%)	45 (11.6%)	50 (12.9%)	100 (25.8%)	157 (40.6%)
How fast do you receive the answers for the nearest shops?	45 (11.6%)	50 (12.9%)	50 (12.9%)	70 (18.1%)	172 (44.4%)
How much are the suggestions on the website appropriate to your need?	10 (2.6%)	32 (8.2%)	88 (22.7%)	120 (31%)	130 (33.5%)
What scale do you choose to rate the level of service satisfaction?	4 (1%)	56 (14.5%)	63 (16.3%)	98 (25.3%)	166 (42.9%)

was evident in both of the investigated aspects of the process, namely: location-products and products-location. The NER system provided a platform for quick and effective customer interface to find the products of interest as well as the nearest store location where the products are in stock. The user interface of NER provides the first option of *location-products* with output statements related to whether the product is available in stock at the location of the customer. In the situation when the product is not available at the customer's location, NER data engine switches to the second option of *products-location*. This option integrates the database to list all possible store where the product is available in stock.

The authors are now seeking collaborations from various e-commerce retail sectors in various geographical locations to expand this research with a view to refine a targeted database linking customers to their preferences and needs. This aspect of further research will employ tools form Artificial Intelligence (AI).

LIMITATIONS

The paper was limited in several ways. Firstly, the data was collected mainly through the company website. Though it is a major source to communicate with the customers at the company, the authors would like to explore the research issues in other means of communication, e.g. email. However, this kind of information is unable to access because it is internal company documents.

Secondly, the data is limited in one company. In order for the better validity and generalization of the findings, other company data should be incooperated in the development of the research in the future.

This research is partly funded by University of Economics Ho Chi Minh City (UEH), Vietnam.

REFERENCES

Altohami, M. A. (2020). Text messages: A computer-mediated discourse analysis. *International Journal of Advanced Computer Science and Applications, 11*(7), 79–87. doi:10.14569/IJACSA.2020.0110711

Berry, G. R. (2004). Lessons from the Online Teaching Experience. *Journal of the Academy of Business Education, 5,* 88–97.

Cho, H.-K., Trier, M., & Kim, E. (2005). The use of instant messaging in working relationship development: A case study. *Journal of Computer-Mediated Communication, 10*(4), 00. http://jcmc.indiana.edu/vol10/issue4/cho.html. doi:10.1111/j.1083-6101.2005.tb00280.x

Crystal, D. (2008). *Txtng: The gr8 db8.* Oxford University Press.

December, J. (1996). Units of analysis for Internet communication. *Journal of Computer-Mediated Communication, 1*(4), 0. Advance online publication. doi:10.1111/j.1083-6101.1996.tb00173.x

Devlin, J., Chang, M.-W., Lee, K., & Toutanova, K. (2019). Bert: Pre-training of deep bidirectional transformers for language understanding. In *Proceedings of NAACL,* (pp. 4171-4186). ACL.

Ek, T., Kirkegaard, C., Jonsson, H., & Nugues, P. (2011). Named Entity Recognition for Short Text Messages. *Social and Behavioral Sciences, 27,* 178–187. doi:10.1016/j.sbspro.2011.10.596

Giri, V. N., & Kumar, P. (2010). Assessing the Impact of Organizational Communication on Job Satisfaction and Job Performance. *Psychological Studies, 55*(2), 137–143. doi:10.100712646-010-0013-6

Hård af Segerstad, Y. (2002). *Use and adaptation of written language to the conditions of computer-mediated communication.* [Doctoral thesis, Göteborg University].

Jiang, H., Wang, X., & Tian, J. (2010). Second-order HMM for event extraction from short message. In *Proceedings of NLDB,* (pp. 149–156). Springer. 10.1007/978-3-642-13881-2_15

Kingma, D. P., & Ba, J. 2014. Adam: A method for stochastic optimization. *ICLR Conference Proceedings,* (pp. 1-15). Scientific Research Publishing.

Lafferty, J., McCallum, A., & Pereira, F. (2008). *Conditional random fields: Probabilistic models for segmenting and labelling sequence data.* University of Pennsylvania. https://repository.upenn.edu/cgi/viewcontent.cgi?article=116 2&context=cis_papers

Lample, G., & Conneau, A. (2019). Cross-lingual language model pretraining. Proceedings of NeurIPS, (pp. 7059–7069).

Larson, G. W. (2011). *Instant Messaging.* Encyclopedia Britannica. https://www.britannica.com/topic/instant-messaging

Li, P. H., Fu, T. J., & Ma, W. Y. (2020). Why attention? Analyze BiLSTM deficiency and its remedies in the case of NER. *Proceedings of the AAAI Conference on Artificial Intelligence*, (pp. 8236—8244). PKP Publishing. 10.1609/aaai.v34i05.6338

Mukahi, T., Nakamura, M., & Not, R. D. (2003). *An Empirical Study on Impacts of Computer-Mediated Communication Management on Job Satisfaction.* 7th Pacific Asia Conference on Information Systems, Adelaid, South Australia.

Newhagen, J. E., & Rafaeli, S. (1996). Why communication researchers should study the internet: A dialogue. *Journal of Communication*, *46*(1), 4–13. doi:10.1111/j.1460-2466.1996.tb01458.x

Nguyen, D. Q., & Nguyen, A. T. (2022). PhoBERT: Pre-trained language models for Vietnamese. *Findings of the Association for Computational Linguistics: EMNLP, 2020*, 1037–1042. doi:10.18653/v1/2020.findings-emnlp.92

Pham, P. Q. M. (2018). *A feature-rich Vietnamese Named-Entity Recognition Model.* Cornell University. doi:10.48550/arXiv.1803.04375

Polifroni, J., Kiss, I., & Adler, M. (2010). Bootstrapping named entity extraction for the creation of mobile services. *Proceedings of the Seventh International Conference on Language Resources and Evaluation (LREC'10)*, (pp. 1515-1520). European Language Resources Association (ELRA).

Ramshaw, L. A., & Marcus, M. P. (1995). Text chunking using transformation-based learning. *Computer Languages*, *11*, 82–94. https://aclanthology.org/W95-0107/

Romiszowski, A., & Mason, R. (2004). Computer-mediated communication. In A. Romiszowski & R. Mason, Handbook of research on educational communications and technology (pp. 391-431). Lawrence Erlbaum Associates.

Sirimanna, U. I., & Gunawardana, T. S. L. W. (2020). Impact of computer mediated communication systems on job satisfaction: Employees in the transmission division of ceylon electricity board (pp. 379-402). *The 9th International Conference on Management and Economics.* Sri Lanka.

Sykes, J. M. (2005). Synchronous CMC and pragmatic development: Effects of oral and written chat. *CALICO Journal, 22*(3), 399–431. doi:10.1558/cj.v22i3.399-431

Thurlow, C., & Poff, M. (2011). Text messaging. C. Susan, D. S. Herring, & V. Tuija Virtanen (Eds.), Handbook of the Pragmatics of CMC. Mouton de Gruyter

ADDITIONAL READINGS

Chiu, J., & Nichols, E. (2016). Named entity recognition with bidirectional LSTM-CNNs. *Transactions of the Association for Computational Linguistics, 4*, 357–370. doi:10.1162/tacl_a_00104

Cui, Y., Che, W., Liu, T., Qin, B., Yang, Z., Wang, S., & Hu, G. (2019). Pre-Training with Whole Word Masking for Chinese BERT. *arXiv:1906.08101.*

De Vries, W., Van Cranenburgh, A., Bisazza, A., Caselli, T., Van Noord, G., & Nissim, M. (2019). BERTje: A Dutch BERT Model. *arXiv:*1912.09582.

Huang, Z., Xu, W., & Yu, K. (2015). Bidirectional LSTM-CRF models for sequence tagging. *arXiv:1508.01991.*

Lin, B. Y., Xu, F., Luo, Z., & Zhu, K. (2017). Multi-channel BiLSTM-CRF model for emerging named entity recognition in social media. In *Proceedings of the 3rd Workshop on Noisy User-generated Text.* ACL. 10.18653/v1/W17-4421

Ma, X., & Hovy, E. (2016). End-to-end sequence labeling via bi-directional LSTM-CNNs-CRF. *In Proceedings of the 54th Annual Meeting of the Association for Computational Linguistics (*Volume 1*).* ACL. 10.18653/v1/P16-1101

Nguyen, H., Ngo, H., Vu, L., Chan, V., & Nguyen, H. (2019). VLSP Shared Task: Named Entity Recognition. *Journal of Computer Science and Cybernetics, 34*(4), 283–294. doi:10.15625/1813-9663/34/4/13161

Nguyen, K. A., & Dong, N., and Nguyen Cam-Tu. (2019). Attentive Neural Network for Named Entity Recognition in Vietnamese. In *Proceedings of RIVF.* ACL.

Ratinov, L., & Roth, D. 2009. Design challenges and misconceptions in named entity recognition. In *Proceedings of the Thirteenth Conference on Computational Natural Language Learning (CoNLL-2009)*. ACM. 10.3115/1596374.1596399

Wu, S., & Dredze, M. (2019). Beto, bentz, becas: The surprising cross-lingual effectiveness of BERT. In *Proceedings of EMNLP-IJCNLP*, (pages 833–844). ACL. 10.18653/v1/D19-1077

KEY TERMS AND EXPLANATIONS

Named-Entity Recognition: A subtask of information extraction that seeks to locate and classify named entities mentioned in unstructed text into pre-defined categories such as person names, organizations, locations, time expressions, quantities, monetary values, percentages.

Customer Satisfaction: A term frequently used in marketing. It is a measure of how products and services supplied by a company meet or surpass customer expectation.

Text Message: Real-time text transmission over the Internet.

Compilation of References

Abedin, E., Mendoza, A., & Karunasekera, S. (2021). Exploring the Moderating Role of Readers' Perspective in Evaluations of Online Consumer Reviews. *Journal of Theoretical and Applied Electronic Commerce Research, 16*(7), 3406–3424. doi:10.3390/jtaer16070184

Abramenka, V. (2015). *Students' motivations and barriers to online education.* [Masters Thesis, Grand Valley State University]. https://scholarworks.gvsu.edu/theses/776

Adapa, S., & Yarram, S. R. (2023). Communication of CSR practices and apparel industry in India—Perspectives of companies and consumers. In *Fashion Marketing in Emerging Economies* (Vol. I, pp. 137–161). Palgrave Macmillan. doi:10.1007/978-3-031-07326-7_6

Adrianson, L. (2001). Gender and computer-mediated communication: Group processes in problem solving. *Computers in Human Behavior, 17*(1), 71–94. doi:10.1016/S0747-5632(00)00033-9

Aflatoony, L., Wakkary, R., & Neustaedter, C. (2018). Becoming a design thinker: Assessing the learning process of students in a secondary level design thinking course. *International Journal of Art & Design Education, 37*(3), 438–453. doi:10.1111/jade.12139

Ahmed, A. K. (2013). Teacher-centered versus learner-centered teaching style. *Journal of Global Business Management, 9*(1), 22.

Ajzen, I. (1991). The theory of planned behavior. *Organizational Behavior and Human Decision Processes, 50*(2), 179–211. doi:10.1016/0749-5978(91)90020-T

Ajzen, I., & Fishbein, M. (1980). *Understanding attitudes and predicting social behavior.* Prentice-Hall.

Al Kurdi, B., Alshurideh, M., Salloum, S. A., Obeidat, Z. M., & Al-Dweeri, R. M. (2020). An empirical investigation into examination of factors influencing university students' behavior towards ELearning acceptance using SEM approach. *Int. J. Interact. Mob. Technol., 14*(2), 19–41. doi:10.3991/ijim.v14i02.11115

Albayrak, M., & Ceylan, C. (2021). Effect of eWOM on purchase intention: Meta-analysis. *Data Technologies and Applications, 55*(5), 810–840. doi:10.1108/DTA-03-2020-0068

Alnsour, M. (2018). Internet-Based Relationship Quality: A Model for Jordanian Business-to-Business Context. *Marketing and Management of Innovations*, *4*, 161–178. doi:10.21272/mmi.2018.4-15

Alshamrani, M. (2019). *An investigation of the advantages and disadvantages of online education* [Doctoral dissertation, Auckland University of Technology].

Altohami, M. A. (2020). Text messages: A computer-mediated discourse analysis. *International Journal of Advanced Computer Science and Applications*, *11*(7), 79–87. doi:10.14569/IJACSA.2020.0110711

Alzahrani, L., & Seth, K. P. (2021). Factors influencing students' satisfaction with continuous use of learning management systems during the COVID-19 pandemic: An empirical study. *Education and Information Technologies*, *26*(6), 6787–6805. doi:10.100710639-021-10492-5 PMID:33841029

Al-Zu'be, A. F. M. (2013). The difference between the learner-centred approach and the teacher-centred approach in teaching English as a foreign language. *Education Research International*, *2*(2), 24–31.

Apăvăloaie, E. I. (2014). The impact of the internet on the business environment. *Procedia Economics and Finance*, *15*, 951–958. doi:10.1016/S2212-5671(14)00654-6

Aral, S., Dellarocas, C., & Godes, D. (2013). Introduction to the special issue—social media and business transformation: A framework for research. *Information Systems Research*, *24*(1), 3–13. doi:10.1287/isre.1120.0470

Arkorful, V., & Abaidoo, N. (2015). The role of e-learning, advantages and disadvantages of its adoption in higher education. *International journal of instructional technology and distance learning, 12*(1), 29-42.

Arora, N., & Lata, S. (2020). YouTube channels influence on destination visit intentions: An empirical analysis on the base of information adoption model. *Journal of Indian Business Research*, *12*(1), 23–42. doi:10.1108/JIBR-09-2019-0269

Aryadoust, V. (2019). An Integrated Cognitive Theory of Comprehension. *International Journal of Listening*, *33*(2), 71–100. doi:10.1080/10904018.2017.1397519

Babić Rosario, A., de Valck, K., & Sotgiu, F. (2020). Conceptualizing the electronic word-of-mouth process: What we know and need to know about eWOM creation, exposure, and evaluation. *Journal of the Academy of Marketing Science*, *48*(3), 422–448. doi:10.100711747-019-00706-1

Babić Rosario, A., Sotgiu, F., De Valck, K., & Bijmolt, T. H. A. (2016). The Effect of Electronic Word of Mouth on Sales: A Meta-Analytic Review of Platform, Product, and Metric Factors. *JMR, Journal of Marketing Research*, *53*(3), 297–318. doi:10.1509/jmr.14.0380

Baek, H., Ahn, J., & Oh, S. (2014). Impact of Tweets on Box Office Revenue: Focusing on When Tweets are Written. *ETRI Journal*, *36*(4), 581–590. doi:10.4218/etrij.14.0113.0732

Baharom, S. S. (2013). Designing Mobile Learning Activities. In *The Malaysian He Context: A Social Constructivist Approach* (p. 395). Salford Business School University of Salford.

Bahns, J., & Eldaw, M. (1993). Should we teach EFL students collocations? *System, 21*(1), 101–114. doi:10.1016/0346-251X(93)90010-E

Bailey, K. M., Curtis, A., & Nunan, D. (2001). *Pursuing Professional Development: The Self as Course*. Heinle ELT.

Baker, C. (1992). *Attitudes and language*. Multilingual Matters.

Bakshi, S., Gupta, D. R., & Gupta, A. (2021). Online travel review posting intentions: A social exchange theory perspective. *Leisure/Loisir, 45*(4), 603–633. doi:10.1080/14927713.2021.1924076

Balasubramanian, S., & Mahajan, V. (2001). The Economic Leverage of the Virtual Community. *International Journal of Electronic Commerce, 5*(3), 103–138. doi:10.1080/10864415.2001.1 1044212

Bannink, A., & van Dam, J. (2021). Teaching via Zoom: Emergent Discourse Practices and Complex Footings in the Online/Offline Classroom Interface. *Languages, 6*(3), 148. doi:10.3390/languages6030148

Bano, N., Arshad, F., Khan, S., & Aqeel Safdar, C. (2015). Case based learning and traditional teaching strategies: Where lies the future? *Pakistan Armed Forces Medical Journal, 65*(1).

Barasch, A., & Berger, J. (2014). Broadcasting and narrowcasting: How audience size affects what people share. *JMR, Journal of Marketing Research, 51*(3), 286–299. doi:10.1509/jmr.13.0238

Bazarova, N. N. (2012). Public intimacy: Disclosure interpretation and social judgments on Facebook. *Journal of Communication, 62*(5), 815–832. doi:10.1111/j.1460-2466.2012.01664.x

Bazarova, N. N., Taft, J. G., Choi, Y. H., & Cosley, D. (2013). Managing impressions and relationships on Facebook: Self-presentational and relational concerns revealed through the analysis of language style. *Journal of Language and Social Psychology, 32*(2), 121–141. doi:10.1177/0261927X12456384

Benson, M., Benson, E., & Ilson, R. (1997). *The BBI dictionary of English word combinations* (2nd ed.). John Benjamins. doi:10.1075/z.bbi1(2nd)

Benta, D., Bologa, G., Dzitac, S., & Dzitac, I. (2015). University level learning and teaching via e-learning platforms. *Procedia Computer Science, 55*, 1366–1373. doi:10.1016/j.procs.2015.07.123

Berry, G. R. (2004). Lessons from the Online Teaching Experience. *Journal of the Academy of Business Education, 5*, 88–97.

Bertea, P. (2009). Measuring students' attitude toward E-learning: A case study. *The 5th International Scientific Conference: eLearning and Software for Education, Bucharest.*

Bhattacherjee, A Sanford., C. (2006). Influence Processes for Information Technology Acceptance: An Elaboration Likelihood Model. *Management Information Systems Quarterly*, *30*(4), 805. doi:10.2307/25148755

Bickart, B., & Schindler, R. M. (2001). Internet forums as influential sources of consumer information. *Journal of Interactive Marketing*, *15*(3), 31–40. doi:10.1002/dir.1014

Biggest social media platforms 2022. (2022, July 26). Statista. https://www.statista.com/statistics/272014/global-social-net works-ranked-by-number-of-users/

Bilal, M., Jianqiu, Z., Akram, U., Tanveer, Y., Sardar, T., & Rasool, H. (2020). Understanding the effects of Internet usage behavior on eWOM. [IJISSS]. *International Journal of Information Systems in the Service Sector*, *12*(3), 93–113. doi:10.4018/IJISSS.2020070106

Bilton, J. (2014). Publishing in the digital age. In Stam, D., & Scott, A. (Eds.). (2014). Inside Magazine Publishing (1st ed.) (pp. 226-247). Routledge. doi:10.4324/9781315818528

Boca, G. D. (2021). Factors influencing students' behavior and attitude towards online education during COVID-19. Sustainability, 13(13), 7469. <jrn> Bui, H. P. (2022). Students' and teachers' perceptions of effective ESP teaching. [PubMed]. *Heliyon*, *10628*(9), e10628. doi:10.1016/j.heliyon.2022.e10628

Bonneau, C., Aroles, J., & Estagnasié, C. (2023). Romanticisation and monetisation of the digital nomad lifestyle: The role played by online narratives in shaping professional identity work. *Organization*, *30*(1), 65–88. doi:10.1177/13505084221131638

Boon, A. (2007). Building bridges: Instant messenger cooperative development. *Language Teaching*, *31*(12), 9–13. https://jalt-publications.org/files/pdf/the_language_teacher/12_2007tlt.pdf#page=10

Boon, A. (2011). Developing Instant Messenger Cooperative Development. *Bulletin of Toyo Gakuen University*, *19*, 109–120. doi:10.24547/00000239

Boon, A. (2019). Facilitating reflective practice via Instant Messenger Cooperative Development. *Indonesian Journal of English Language Teaching*, *14*(1), 35–54. doi:10.25170/ijelt.v14i1.1417

Borg, S. (2016). Researching Language Teacher Education. In B. Paltridge & A. Phakiti (Eds.), *Research Methods in Applied Linguistics - a practical course* (pp. 541–555). Bloomsbury.

Bosch, T. E. (2009). Using online social networking for teaching and learning: Facebook use at the University of Cape Town. *Communication*, *35*(2), 185–200. doi:10.1080/02500160903250648

Boyd, D. M., & Ellison, N. B. (2007). Social network sites: Definition, history, and scholarship. *Journal of Computer-Mediated Communication*, *13*(1), 210–230. https://bit.ly/2uo0IAl. doi:10.1111/j.1083-6101.2007.00393.x

Bradley, J. P., Cabell, C., Cole, D. R., Kennedy, D. H., & Poje, J. (2018). From which point do we begin?: On combining the multiliteral and multiperspectival. *Stem Journal, 19*(2), 65–93. doi:10.16875tem.2018.19.2.65

Bria, F. (2013). *Social media and their impact on organisations: building Firm Celebrity and organisational legitimacy through social media* [Doctoral dissertation, Imperial College London]. http://hdl.handle.net/10044/1/24944

Brown, J. J., & Reingen, P. H. (1987). Social Ties and Word-of-Mouth Referral Behavior. *The Journal of Consumer Research, 14*(3), 350. doi:10.1086/209118

Brown, J., Broderick, A. J., & Lee, N. (2007). Word of mouth communication within online communities: Conceptualizing the online social network. *Journal of Interactive Marketing, 21*(3), 2–20. doi:10.1002/dir.20082

Bueno, S., & Gallego, M. D. (2021). eWOM in C2C Platforms: Combining IAM and Customer Satisfaction to Examine the Impact on Purchase Intention. *Journal of Theoretical and Applied Electronic Commerce Research, 16*(5), 1612–1630. doi:10.3390/jtaer16050091

Bui, H. P., & Anh, D. P. T. (2022). Computer-mediated communication and second language education. In R. Sharma & D. Sharma (Eds.), *New trends and applications in Internet of things (IoT) and big data analytics* (pp. 109–122). Springer.

Bui, H. P., & Nguyen, L. T. (2022). Scaffolding language learning in the online classroom. In R. Sharma & D. Sharma (Eds.), *New trends and applications in Internet of things (IoT) and big data analytics* (pp. 45–60). Springer., doi:10.1007/978-3-030-99329-0_8

Bui, L. T. (2021). The role of collocations in the English teaching and learning. *International Journal of TESOL & Education, 1*(2), 99–109. https://i-jte.org/index.php/journal/article/view/26

Bulut, Z. A., & Karabulut, A. N. (2018). Examining the role of two aspects of eWOM in online repurchase intention: An integrated trust-loyalty perspective. *Journal of Consumer Behaviour, 17*(4), 407–417. doi:10.1002/cb.1721

Burgoon, J. K., Bonito, J. A., Bengtsson, B., Ramirez, A. Jr, Dunbar, N. E., & Miczo, N. (1999). Testing the Interactivity Model: Communication Processes, Partner Assessments, and the Quality of Collaborative Work. *Journal of Management Information Systems, 16*(3), 33–56. doi:10.1080/07421222.1999.11518255

Butler, K. C. (2012). A model of successful adaptation to online learning for college-bound native American high school students. *Multicultural Education & Technology Journal, 6*(2), 60–76. doi:10.1108/17504971211236245

Cakrawati, L. M. (2017). Students' perceptions on the use of online learning platforms in EFL classroom. *English Language Teaching and Technology Journal, 1*(1), 22–30. doi:10.17509/elt%20tech.v1i1.9428

Cammarano, A., Michelino, F., & Caputo, M. (2019). Open innovation practices for knowledge acquisition and their effects on innovation output. *Technology Analysis and Strategic Management*, *31*(11), 1297–1313. doi:10.1080/09537325.2019.1606420

Camus, M., Hurt, N. E., Larson, L. R., & Prevost, L. (2016). Facebook as an online teaching tool: Effects on student participation, learning, and overall course performance. *College Teaching*, *64*(2), 84–94. doi:10.1080/87567555.2015.1099093

Cao, X., Liu, Y., Zhu, Z., Hu, J., & Chen, X. (2017). Online selection of a physician by patients: Empirical study from elaboration likelihood perspective. *Computers in Human Behavior*, *73*, 403–412. doi:10.1016/j.chb.2017.03.060

Carter, R., & McCarthy, M. (1988). *Vocabulary and language teaching*. Longman.

Chaiken, S. (1989). Heuristic and systematic information processing within and beyond the persuasion context. *Unintended thought*, 212-252.

Chaiken, S. (1980). Heuristic versus systematic information processing and the use of source versus message cues in persuasion. *Journal of Personality and Social Psychology*, *39*(5), 752–766. doi:10.1037/0022-3514.39.5.752

Chaikin, A. L., & Derlega, V. J. (1974). Variables affecting the appropriateness of self-disclosure. *Journal of Consulting and Clinical Psychology*, *42*(4), 588–593. doi:10.1037/h0036614

Chakraborty, M., & Muyia, N. F. (2014). Strengthening student engagement: What do students want in online courses? *European Journal of Training and Development*, *38*(9), 782–802. doi:10.1108/EJTD-11-2013-0123

Chang, L. (2011). Integrating collocation instruction into a writing class: A case study of Taiwanese EFL students' writing production. *Journal of Kao - Tech University - Humanities and Social Sciences*, 281 - 304.

Chang, T.-S., & Hsiao, W.-H. (2013). Factors Influencing Intentions to Use Social Recommender Systems: A Social Exchange Perspective. *Cyberpsychology, Behavior, and Social Networking*, *16*(5), 357–363. doi:10.1089/cyber.2012.0278 PMID:23374171

Chan, K. W., & Li, S. Y. (2010). Understanding consumer-to-consumer interactions in virtual communities: The salience of reciprocity. *Journal of Business Research*, *63*(9–10), 1033–1040. doi:10.1016/j.jbusres.2008.08.009

Channel, J. (1981). Applying semantic theory to vocabulary teaching. *ELT Journal*, 115 - 122. doi:10.1093/elt/XXXV.2.115

Chen, C.-W. (2014). UNDERSTANDING THE EFFECTS OF EWOM ON COSMETIC CONSUMER BEHAVIORAL INTENTION. *International Journal of Electronic Commerce Studies*, *5*(1), 97–102. doi:10.7903/ijecs.1030

Cheok, M. L., & Wong, S. L. (2015). Predictors of e-learning satisfaction in teaching and learning for school teachers: A literature review. *International Journal of Instruction*, *8*(1), 75–90. doi:10.12973/iji.2015.816a

Cheung, C. M. K., Lee, M. K. O., & Rabjohn, N. (2008). The impact of electronic word-of-mouth: The adoption of online opinions in online customer communities. *Internet Research*, *18*(3), 229–247. doi:10.1108/10662240810883290

Cheung, C. M. K., Lee, M. K. O., & Thadani, D. R. (2009). The Impact of Positive Electronic Word-of-Mouth on Consumer Online Purchasing Decision. In M. D. Lytras, E. Damiani, J. M. Carroll, R. D. Tennyson, D. Avison, A. Naeve, A. Dale, P. Lefrere, F. Tan, J. Sipior, & G. Vossen (Eds.), *Visioning and Engineering the Knowledge Society. A Web Science Perspective* (Vol. 5736, pp. 501–510). Springer Berlin Heidelberg., doi:10.1007/978-3-642-04754-1_51

Cheung, C. M. K., & Thadani, D. R. (2012). The impact of electronic word-of-mouth communication: A literature analysis and integrative model. *Decision Support Systems*, *54*(1), 461–470. doi:10.1016/j.dss.2012.06.008

Chevalier, J. A., & Mayzlin, D. (2006). The Effect of Word of Mouth on Sales: Online Book Reviews. *JMR, Journal of Marketing Research*, *43*(3), 345–354. doi:10.1509/jmkr.43.3.345

Chih, W.-H., Wang, K.-Y., Hsu, L.-C., & Huang, S.-C. (2013). Investigating Electronic Word-of-Mouth Effects on Online Discussion Forums: The Role of Perceived Positive Electronic Word-of-Mouth Review Credibility. *Cyberpsychology, Behavior, and Social Networking*, *16*(9), 658–668. doi:10.1089/cyber.2012.0364 PMID:23895466

Cho, H.-K., Trier, M., & Kim, E. (2005). The use of instant messaging in working relationship development: A case study. *Journal of Computer-Mediated Communication*, *10*(4), 00. http://jcmc.indiana.edu/vol10/issue4/cho.html. doi:10.1111/j.1083-6101.2005.tb00280.x

Choi, M., & Toma, C. L. (2014). Social sharing through interpersonal media: Patterns and effects on emotional well-being. *Computers in Human Behavior*, *36*, 530–541. doi:10.1016/j.chb.2014.04.026

Choo, C. W., Detlor, B., & Turnbull, D. (2013). *Web work: Information seeking and knowledge work on the World Wide Web* (Vol. 1). Springer Science & Business Media.

Chugh, R., & Ruhi, U. (2018). Social media in higher education: A literature review of Facebook. *Education and Information Technologies*, *23*(2), 605–616. doi:10.100710639-017-9621-2

Cochrane, T. D. (2010). Exploring mobile learning success factors. *ALT-J: Research in Learning Technology*, *18*(2), 133–148. doi:10.1080/09687769.2010.494718

Cohen, L., Manion, L., & Morrison, K. (2018). *Research Methods in Education* (8th ed.). Routledge.

Conference on English Education (CEE). (2005) *Beliefs about technology and the preparation of English teachers*. NCTE. http://www.ncte.org/cee/positions/beliefsontechnology

Cook, G. (2010). Sweet talking: Food, language, and democracy. *Language Teaching, 43*(2), 168–181. doi:10.1017/S0261444809990140

Cowie, N. (2002). CD by email. In Edge. J (Ed), Continuing Cooperative Development (pp. 225–229). University of Michigan Press.

Creswell, J. W. (2014). *Research design: Qualitative, quantitative, and mixed methods approaches* (4th ed.). Sage.

Crocker, S. D. (2021). Arpanet and Its Evolution—A Report Card. *IEEE Communications Magazine, 59*(12), 118–124. doi:10.1109/MCOM.001.2100727

Croft, N., Dalton, A., & Grant, M. (2010). Overcoming isolation in distance learning: Building a learning community through time and space. *The Journal for Education in the Built Environment, 5*(1), 27–64. doi:10.11120/jebe.2010.05010027

Crowther, D., & Lancaster, G. (2008). *Research methods: A concise introduction to research in Management and Business consultancy*. Butterworth-Heinemann.

Crystal, D. (2008). *Txtng: The gr8 db8*. Oxford University Press.

Daniels, K., Elliott, C., Finley, S., & Chapman, C. (2019). Learning and Teaching in Higher Education. In *Learning and Teaching in Higher Education*. Edward Elgar Publishing. doi:10.4337/9781788975087

Daniels, L. M., Tze, V. M. C., & Goetz, T. (2015). Examining boredom: Different causes for different coping profiles. *Learning and Individual Differences, 37*, 255–261. doi:10.1016/j.lindif.2014.11.004

Daugherty, T., Eastin, M. S., & Bright, L. (2008). Exploring Consumer Motivations for Creating User-Generated Content. *Journal of Interactive Advertising, 8*(2), 16–25. doi:10.1080/152520 19.2008.10722139

Davis, F. D. (1993). User acceptance of information technology: System characteristics, user perceptions and behavioral impacts. *International Journal of Man-Machine Studies, 38*(3), 475–487. doi:10.1006/imms.1993.1022

Dawe, K. (2015). Best practice in business-to-business email. *Journal of Direct, Data and Digital Marketing Practice, 16*(4), 242–247. doi:10.1057/dddmp.2015.21

December, J. (1996). Units of analysis for Internet communication. *Journal of Computer-Mediated Communication, 1*(4), 0. Advance online publication. doi:10.1111/j.1083-6101.1996.tb00173.x

Demishkevich, M. (2015). *Small business use of internet marketing: Findings from case studies* [Doctoral dissertation, Walden University]. https://scholarworks.waldenu.edu/cgi/viewcontent.cgi?referer =&httpsredir=1&article=2339&context=dissertations

Derakhshan, A., Fathi, J., Pawlak, M., & Kruk, M. (2022). Classroom social climate, growth language mindset, and student engagement: The mediating role of boredom in learning English as a foreign language. *Journal of Multilingual and Multicultural Development*, 1–19. doi:10.1 080/01434632.2022.2099407

Derakhshan, A., Kruk, M., Mehdizadeh, M., & Pawlak, M. (2021). Boredom in online classes in the Iranian EFL context: Sources and solutions. *System*, *101*(February), 102556. doi:10.1016/j. system.2021.102556

Derks, D., Bos, A. E., & von Grumbkow, J. (2008). Emoticons in computer-mediated communication: Social motives and social context. *Cyberpsychology & Behavior*, *11*(1), 99–101. doi:10.1089/cpb.2007.9926 PMID:18275321

Derks, D., Fischer, A. H., & Bos, A. E. R. (2008). The role of emotion in computer-mediated communication: A Review. *Computers in Human Behavior*, *24*(3), 766–785. doi:10.1016/j. chb.2007.04.004

Devlin, J., Chang, M.-W., Lee, K., & Toutanova, K. (2019). Bert: Pre-training of deep bidirectional transformers for language understanding. In *Proceedings of NAACL*, (pp. 4171-4186). ACL.

Dichter, E. (1966). How Word-of-Mouth Advertising Works. *Harvard Business Review*, *44*(November–December), 147–166.

Do, K. (2022). *Video-mediated Cooperative Development in Vietnam: a case study* [Unpublished Master Thesis, University of Warwick].

Doğan, D., & Gulbahar, Y. (2018). Using Facebook as social learning environment. *Informatics in Education*, *17*(2), 207–228. doi:10.15388/infedu.2018.11

DONICI, A.N., Maha, A., Ignat, I. and MAHA, L.G. (2012). E-Commerce across United States of America: Amazon. com. *Economy Transdisciplinarity Cognition*, *15*(1), 252–258.

Donthu, N., Kumar, S., Pandey, N., Pandey, N., & Mishra, A. (2021). Mapping the electronic word-of-mouth (eWOM) research: A systematic review and bibliometric analysis. *Journal of Business Research*, *135*, 758–773. doi:10.1016/j.jbusres.2021.07.015

Dooly, M., & Sadler, R. (2013). Filling in the gaps: Linking theory and practice through telecollaboration in teacher education. *ReCALL*, *25*(1), 4–29. doi:10.1017/S0958344012000237

Dörnyei, Z. (2005). *The psychology of the language learner: Individual differences in second language acquisition*. Lawrence Erlbaum.

Duan, W., Gu, B., & Whinston, A. (2008). The dynamics of online word-of-mouth and product sales—An empirical investigation of the movie industry. *Journal of Retailing*, *84*(2), 233–242. doi:10.1016/j.jretai.2008.04.005

Duc-Long, L., Thien-Vu, G., & Dieu-Khuon, H. (2021). The impact of the COVID-19 pandemic on online learning in higher education: A Vietnamese case. *European Journal of Educational Research*, *10*(4), 1683–1695. doi:10.12973/eu-jer.10.4.1683

Dudeney, G., Hockly, N., & Pegrum, M. (2013). *Digital literacies.* Pearson Education.

Dumford, A. D., & Miller, A. L. (2018). Online learning in higher education: Exploring advantages and disadvantages for engagement. *Journal of Computing in Higher Education, 30*(3), 452–465. doi:10.100712528-018-9179-z

Duncum, P. (2011). Youth on YouTube: Prosumers in a peer-to-peer participatory culture. *The International Journal of Arts Education, 9*(2), 24–39.

Duong, T. M., Tran, T. Q., & Nguyen, T. T. P. (2021). Non-English majored students' use of English vocabulary learning strategies with technology-enhanced language learning tools. [AJUE]. *Asian Journal of University Education, 17*(4), 455–463. doi:10.24191/ajue.v17i4.16252

Dwyer, J. (2012). *Communication for Business and the Professions: Strategie s and Skills.* Pearson Higher Education AU.

Eagly, A., & Chaiken, S. (1998). *Attitude structure and function. Handbook of social psychology.* McGrow Company.

Early, M., Kendrick, M., & Potts, D. (2015). Multimodality: Out from the margins of English language teaching. *TESOL Quarterly, 49*(3), 447–460. doi:10.1002/tesq.246

Eddington, S. M., Jarvis, C. M., & Buzzanell, P. M. (2023). Constituting affective identities: Understanding the communicative construction of identity in online men's rights spaces. *Organization, 30*(1), 116–139. doi:10.1177/13505084221137989

Edge, J. (1992a). *Cooperative Development.* Longman Group UK Limited.

Edge, J. (1992b). Co-operative Development. *ELT Journal, 46*(1), 62–70. doi:10.1093/elt/46.1.62

Edge, J. (2002). *Continuing Cooperative Development - A Discourse Framework for Individuals and Colleagues.* The University of Michigan Press. doi:10.3998/mpub.8915

Edge, J. (2006). Computer-mediated cooperative development: Non-judgemental discourse in online environments. *Language Teaching Research, 10*(2), 205–227. doi:10.1191/1362168806lr192oa

Edge, J. (2011). *The reflexive Teacher Educator in TESOL: Roots and Wings.* Routledge. doi:10.4324/9780203832899

Edwards, D., & Mercer, N. (2013). *Common Knowledge (Routledge Revivals): The Development of Understanding in the Classroom.* Routledge. doi:10.4324/9780203095287

Eisenhart, M. (1991). *Conceptual frameworks for research circa 1991: Ideas from a cultural anthropologist; implications for mathematics education researchers.* Paper presented at the Thirteenth Annual Meeting North American Paper of the International Group for the Psychology of Mathematics Education, Blacksburg, Virginia, USA.

Ek, T., Kirkegaard, C., Jonsson, H., & Nugues, P. (2011). Named Entity Recognition for Short Text Messages. *Social and Behavioral Sciences, 27*, 178–187. doi:10.1016/j.sbspro.2011.10.596

Ellison, N. B., Steinfield, C., & Lampe, C. (2007). The benefits of Facebook "friends:" Social Capital and college students' use of online social network sites. *Journal of Computer-Mediated Communication, 12*(4), 1143–1168. doi:10.1111/j.1083-6101.2007.00367.x

Estacio, R. R., & Raga, R. C. Jr. (2017). Analyzing students online learning behavior in blended courses using Moodle. *Asian Association of Open Universities Journal, 12*(1), 52–68. doi:10.1108/AAOUJ-01-2017-0016

Fahlman, S. A., Mercer, K. B., Gaskovski, P., Eastwood, A. E., & Eastwood, J. D. (2009). Does a lack of life meaning cause boredom? Results from psychometric, longitudinal, and experimental analyses. *Journal of Social and Clinical Psychology, 28*(3), 307–340. doi:10.1521/jscp.2009.28.3.307

Famularsih, S. (2020). Students' experiences in using online learning applications due to COVID-19 in English classroom. *Studies in Learning and Teaching, 1*(2), 112–121. doi:10.46627ilet.v1i2.40

Fan, Y. W., & Miao, Y. F. (2012). Effect of electronic word-of-mouth on consumer purchase intention: The perspective of gender differences. *International Journal of Electronic Business Management, 10*(3), 175.

Fardouly, J., & Vartanian, L. R. (2015). Negative comparisons about one's appearance mediate the relationship between Facebook usage and body image concerns. *Body Image, 12*, 82–88. doi:10.1016/j.bodyim.2014.10.004 PMID:25462886

Farmer, R., & Sundberg, N. D. (1986). Boredom Proneness-The Development and Correlates of a New Scale. *Journal of Personality Assessment, 50*(1), 4–17. doi:10.120715327752jpa5001_2 PMID:3723312

Farrell, T. S. C. (2019). Reflective Practice in L2 teacher education. In S. Walsh & S. Mann (Eds.), *The Routledge Handbook of English Language Teacher Education*. Routledge. doi:10.4324/9781315659824-5

Farrell, T. S. C. (2021). *TESOL Teacher Education - A Reflective Approach*. Edinburgh University Press. doi:10.1515/9781474474443

Faryadi, Q. (2017). Effectiveness of Facebook in English language learning: A case study. *Open Access Library Journal, 4*(11), 1–11. doi:10.4236/oalib.1104017

Feick, L. F., & Price, L. L. (1987). The Market Maven: A Diffuser of Marketplace Information. *Journal of Marketing, 51*(1), 83–97. doi:10.1177/002224298705100107

Filieri, R., McLeay, F., Tsui, B., & Lin, Z. (2018). Consumer perceptions of information helpfulness and determinants of purchase intention in online consumer reviews of services. *Information & Management, 55*(8), 956–970. doi:10.1016/j.im.2018.04.010

Firth, J. R. (1957). *Papers in linguistics, 1934 - 1951*. Oxford University Press.

Fishbein, M., Ajzen, I., & Belief, A. (1975). *Intention and Behavior: An introduction to theory and research*. University of Massachusetts.

Freeman, D. (2009). The scope of second language teacher education. In A. Burns & J. C. Richards (Eds.), *The Cambridge Guide to Language Teacher Education* (pp. 11–19). Cambridge University Press.

Frost, R. L., & Rickwood, D. J. (2017). A systematic review of the mental health outcomes associated with Facebook use. *Computers in Human Behavior*, *76*, 576–600. doi:10.1016/j.chb.2017.08.001

Gao, L. X., & Zhang, L. J. (2020). Teacher learning in difficult times: Examining foreign language teachers' cognitions about online teaching to tide over COVID-19. *Frontiers in Psychology*, *11*, 1–14. doi:10.3389/fpsyg.2020.549653 PMID:33071866

Gašević, D., Dawson, S., Rogers, T., & Gasevic, D. (2016). Learning analytics should not promote one size fits all: The effects of instructional conditions in predicting academic success. *The Internet and Higher Education*, *28*, 68–84. doi:10.1016/j.iheduc.2015.10.002

Gee, J. P. (2004). *Situated language and learning: A critique of traditional schooling*. Routledge.

Ghonsooli, B., Pishghadam, R., & Mahjoobi, F. (2008). The impact of collocational instruction on the writing skill of Iranian EFL learners: A case of product and process study. *Iranian EFL Jounal*, 36 - 59. https://www.researchgate.net/publication/277841864_The_impact_of_colocational_instrucion_on_the_writing_skill_of_Iranian_EFL_learners#fullTextFileContent

Gillett-Swan, J. (2017). The challenges of online learning: Supporting and engaging the isolated learner. *Journal of Learning Design*, *10*(1), 20–30. doi:10.5204/jld.v9i3.293

Giri, V. N., & Kumar, P. (2010). Assessing the Impact of Organizational Communication on Job Satisfaction and Job Performance. *Psychological Studies*, *55*(2), 137–143. doi:10.100712646-010-0013-6

Godes, D., & Mayzlin, D. (2004). Using Online Conversations to Study Word-of-Mouth Communication. *Marketing Science*, *23*(4), 545–560. doi:10.1287/mksc.1040.0071

Godwin-Jones, R. (2008). Mobile computing technologies: Lighter, faster, smarter. *Language Learning & Technology*, *12*(3), 3–9. 10125/44150

Graciyal, D. G., & Viswam, D. (2021). Social media and emotional well-being: Pursuit of happiness or pleasure. *Asia Pacific Media Educator*, *31*(1), 99–115. doi:10.1177/1326365X211003737

Grapin, S. (2019). Multimodality in the new content standards era: Implications for English learners. *TESOL Quarterly*, *53*(1), 30–55. doi:10.1002/tesq.443

Griffiths, M. D. (2013). Social networking addiction: Emerging themes and issues. *Journal of Addiction Research & Therapy*, *4*(05), 1–2. doi:10.4172/2155-6105.1000e118

Griffiths, M., & Light, B. (2008). Social networking and digital gaming media convergence: Classification and its consequences for appropriation. *Information Systems Frontiers*, *10*(4), 447–459. doi:10.100710796-008-9105-4

Guffey, M. E., & Loewy, D. (2014). *Business communication: Process and product*. Cengage Learning.

Gupta, A. (2014). E-Commerce: Role of E-Commerce in today's business. *International Journal of Computing and Corporate Research*, *4*(1), 1–8.

Gupta, P., & Harris, J. (2010). How e-WOM recommendations influence product consideration and quality of choice: A motivation to process information perspective. *Journal of Business Research*, *63*(9–10), 1041–1049. doi:10.1016/j.jbusres.2009.01.015

Hafner, C. (2014). Embedding digital literacies in English language teaching: Students' digital video projects as multimodal ensembles. *TESOL Quarterly*, *48*(4), 655–685. doi:10.1002/tesq.138

Hafner, C. (2015). Remix culture and English language teaching: The expression of learner voice in digital multimodal compositions. *TESOL Quarterly*, *49*(3), 486–509. doi:10.1002/tesq.238

Hafner, C., & Ho, W. (2020). Assessing digital multimodal composing in second language writing: Towards a process-based model. *Journal of Second Language Writing*, *47*, 100710–100714. doi:10.1016/j.jslw.2020.100710

Hafour, M. F., & Al-Rashidy, A. S. M. (2020). Storyboarding-based collaborative narratives on Google Docs: Fostering EFL learners' writing fluency, syntactic complexity, and overall performance. *The JALT CALL Journal*, *16*(3), 123–146. doi:10.29140/jaltcall.v16n3.393

Hancı-Azizoğlu, E. B., & Kavaklı, N. (2021b). Creative digital writing: A multilingual perspective. In M. Montebello (Ed.), Handbook of Research on Digital Language Pedagogies (pp. 250-266). IGI Global. doi:10.4018/978-1-7998-6745-6.ch013

Hancı-Azizoğlu, E. B., & Kavaklı, N. (2021a). Rewriting the future through rhetorical technology. In E. B. Hancı-Azizoğlu & N. Kavaklı (Eds.), *Futuristic and linguistic perspectives on teaching writing to second language students* (pp. 1–15). IGI Global. doi:10.4018/978-1-7998-6508-7.ch001

Hancock, J. T. (2004). Verbal Irony Use in Face-To-Face and Computer-Mediated Conversations. *Journal of Language and Social Psychology*, *23*(4), 447–463. doi:10.1177/0261927X04269587

Hancock, J. T., Landrigan, C., & Silver, C. (2007). Expressing emotion in text-based communication. *Proceedings of the SIGCHI Conference on Human Factors in Computing Systems*. ACM. 10.1145/1240624.1240764

Hård af Segerstad, Y. (2002). *Use and adaptation of written language to the conditions of computer-mediated communication*. [Doctoral thesis, Göteborg University].

Haron, N. N., Yasmin, H. Z., & Ibrahim, N. A. (2015). E-learning as a platform to learn English among ESL learners: Benefits and barriers. In Mahani, S., & Haliza, J. (Eds.), *Research in Language Teaching and Learning*, 79-106. UTM Press. https://www.researchgate.net/publication/306119651_ELearning _as_a_Platform_to_Learn_English_among_ESL_Learners_Benefits_ and_Barriers

Harrison-Walker, L. J. (2001). The Measurement of Word-of-Mouth Communication and an Investigation of Service Quality and Customer Commitment As Potential Antecedents. *Journal of Service Research*, *4*(1), 60–75. doi:10.1177/109467050141006

Hass, A., & Joseph, M. (2018). Investigating different options in course delivery–traditional vs online: Is there another option? *The International Journal of Information and Learning Technology*, *35*(4), 230–239. doi:10.1108/IJILT-09-2017-0096

Hayes, J. L., & King, K. W. (2014). The Social Exchange of Viral Ads: Referral and Coreferral of Ads Among College Students. *Journal of Interactive Advertising*, *14*(2), 98–109. doi:10.108 0/15252019.2014.942473

Heaton, J. B. (1988). *Writing English language tests*. Longman.

Hendler, J. A. (2023). The future of the Web. In *The Internet and Philosophy of Science* (pp. 71–83). Routledge., doi:10.4324/9781003250470

Hennig-Thurau, T., Gwinner, K. P., Walsh, G., & Gremler, D. D. (2004). Electronic word-of-mouth via consumer-opinion platforms: What motivates consumers to articulate themselves on the Internet? *Journal of Interactive Marketing*, *18*(1), 38–52. doi:10.1002/dir.10073

Hennig-Thurau, T., Walsh, G., & Walsh, G. (2003). Electronic Word-of-Mouth: Motives for and Consequences of Reading Customer Articulations on the Internet. *International Journal of Electronic Commerce*, *8*(2), 51–74. doi:10.1080/10864415.2003.11044293

Herrington, J., & Herrington, J. (2009). New technologies, new pedagogies : Mobile learning in higher education. In *World, 0*. https://ro.uow.edu.au/newtech/

Hey Tow, W. N., Dell, P., & Venable, J. (2010). Understanding information disclosure behaviour in Australian Facebook users. *Journal of Information Technology*, *25*(2), 126–136. doi:10.1057/jit.2010.18

Hlee, S., Lee, H., & Koo, C. (2018). Hospitality and Tourism Online Review Research: A Systematic Analysis and Heuristic-Systematic Model. *Sustainability*, *10*(4), 1141. doi:10.3390u10041141

Homans, G. C. (1958). Social Behavior as Exchange. *American Journal of Sociology*, *63*(6), 597–606. doi:10.1086/222355

Hsu, T. J., & Chiu, C. (2008). Lexical collocations and their relation to speaking proficiency. *Asian EFL journal*, *10*(1), 181 - 204. https://www.asian-efl-journal.com/main-editions-new/lexical-collocations-and-their-relation-to-speaking-proficiency-of-c ollege-efl-learners-in-taiwan/index.htm

Hsu, Y. J. (2010). The effect of collocation instruction on reading comprehension and vocabulary learning of Taiwanese college English majors. *Asian EFL Journal*, *12*(1), 47 - 87. https://www.asian-efl-journal.com/main-editions-new/the-effe cts-of-collocation-instruction-on-the-reading-comprehension- and-vocabulary-learning-of-college-english-majors/index.htm

Hung, B. P., & Khoa, B. T. (2021). Communication strategies for interaction in social networks: A multilingual perspective. In I. Priyadarshini & R. Sharma, Artificial Intelligence and Cybersecurity (pp. 195-208). Taylor & Francis.

Hung, B. P., Khoa, B. T., & Hejsalembrahmi, M. (2022). Qualitative research in social sciences: Data collection, data analysis, and report writing. *International Journal of Public Sector Performance Management, 9*(4), 10038439. doi:10.1504/IJPSPM.2022.10038439

Hussain, S., Ahmed, W., Jafar, R. M. S., Rabnawaz, A., & Jianzhou, Y. (2017). EWOM source credibility, perceived risk and food product customer's information adoption. *Computers in Human Behavior, 66*, 96–102. doi:10.1016/j.chb.2016.09.034

Hu, X., & Ha, L. (2015). Which form of word-of-mouth is more important to online shoppers? A comparative study of WOM use between general population and college students. *Journal of Communication and Media Research, 7*(2), 15–35.

Hu, Y., & Kim, H. J. (2018). Positive and negative eWOM motivations and hotel customers' eWOM behavior: Does personality matter? *International Journal of Hospitality Management, 75*, 27–37. doi:10.1016/j.ijhm.2018.03.004

Hwang, K., & Zhang, Q. (2018). Influence of parasocial relationship between digital celebrities and their followers on followers' purchase and electronic word-of-mouth intentions, and persuasion knowledge. *Computers in Human Behavior, 87*, 155–173. doi:10.1016/j.chb.2018.05.029

Hwee, J., Koh, L., & Chai, C. S. (2016). Teacher professional development for TPACK- 21CL: Effects on teacher ICT integration and student outcomes. *Journal of Educational Computing Research, 55*(2), 1–25. doi:10.1177%2F0735633116656848

Hyrynsalmi, S., Seppänen, M., Aarikka-Stenroos, L., Suominen, A., Järveläinen, J., & Harkke, V. (2015). Busting Myths of Electronic Word of Mouth: The Relationship between Customer Ratings and the Sales of Mobile Applications. *Journal of Theoretical and Applied Electronic Commerce Research, 10*(2), 1–18. doi:10.4067/S0718-18762015000200002

Ismagilova, E., Dwivedi, Y. K., Slade, E., & Williams, M. D. (2017). Electronic Word of Mouth (eWOM) in the Marketing Context: A State of the Art Analysis and Future Directions (SpringerBriefs in Business) (1st ed. 2017 ed.). Springer.

Isti, M., & Istikharoh, L. (2019). EFL students' attitude toward learning English. *Journal of Sains Social dan Humaniora, 3*(2), 95–105. doi:10.5539/ass.v8n2p119

Jagongo, A., & Kinyua, C. (2013). The social media and entrepreneurship growth. *International Journal of Humanities and Social Science, 3*(10), 213–227.

Jain, V. (2014). 3D model of attitude. *International Journal of Advanced Research in Management and Social Sciences, 3*(3), 1–12. https://docplayer.net/22666307-3d- model-of-attitude.html

Jaya, H. P., Wijaya, A., & Kurniawan, D. (2019). Correlation between the ability of using English collocation and academic achievements of students of faculty of teacher training and education Universitas Sriwijaya. *Holistics Journal, 11*(2), 908–917. doi:10.20319/pijss.2019.52.908917

Jewitt, C. (2008). Multimodality and literacy in school classrooms. *Review of Research in Education, 32*(1), 241–267. doi:10.3102/0091732X07310586

Jiang, H., Wang, X., & Tian, J. (2010). Second-order HMM for event extraction from short message. In *Proceedings of NLDB*, (pp. 149–156). Springer. 10.1007/978-3-642-13881-2_15

Jiang, L., & Luk, J. (2016). Multimodal composing as a learning activity in English classrooms: Inquiring into the sources of its motivational capacity. *System, 59*, 1–11. doi:10.1016/j.system.2016.04.001

Jie, Z., Puteh, M., & Hasan Sazalli, N. A. (2020). A social constructivism framing of mobile pedagogy in english language teaching in the digital era. *Indonesian Journal of Electrical Engineering and Computer Science, 20*(2), 830–836. doi:10.11591/ijeecs.v20.i2.pp830-836

Johnson, W. A. (2015). Learning to read and write. *A Companion to Ancient Education, 120*, 137.

Kabilan, M. K., Ahmad, N., & Abidin, M. J. Z. (2010). Facebook: An online environment for learning of English in institutions of higher education? *Internet and Higher Education, 13*(4), 179–187. doi:10.1016/j.iheduc.2010.07.003

Kalelioğlu, F. (2017). Using Facebook as a learning management system: Experiences of pre-service teachers. *Informatics in Education-An International Journal, 16*(1), 83–101. doi:10.15388/infedu.2017.05

Kallinikos, J., & Tempini, N. (2014). Patient data as medical facts: Social media practices as a foundation for medical knowledge creation. *Information Systems Research, 25*(4), 817–833. doi:10.1287/isre.2014.0544

Kalpidou, M., Costin, D., & Morris, J. (2011). The relationship between Facebook and the well-being of undergraduate college students. *Cyberpsychology, Behavior, and Social Networking, 14*(4), 183–189. doi:10.1089/cyber.2010.0061 PMID:21192765

Kan, A. (2009). *Statistical procedures on measurement results. In the H. Atilgan. Assessment and evaluation in education.* Ani Publications.

Kapoor, K. K., Tamilmani, K., Rana, N. P., Patil, P., Dwivedi, Y. K., & Nerur, S. (2018). Advances in social media research: Past, present and future. *Information Systems Frontiers, 20*(3), 531–558. doi:10.100710796-017-9810-y

Kavaklı Ulutaş, N., & Abuşka, A. (2022). Understanding L2 teachers engagement with digital multimodal composing (DMC) in the changing educational landscape. In E. Duruk (Ed.), *The new normal of online language education* (pp. 127–144). Eğiten Kitap.

Kavaklı, N., & Hancı-Azizoğlu, E. B. (2021). Digital storytelling: A futuristic-second-language-writing method. In E. B. Hancı-Azizoğlu & N. Kavaklı (Eds.), *Futuristic and linguistic perspectives on teaching writing to second language students* (pp. 66–83). IGI Global. doi:10.4018/978-1-7998-6508-7.ch005

Kearney, M., Schuck, S., Burden, K., & Aubusson, P. (2012). Viewing mobile learning from a pedagogical perspective. *Research in Learning Technology, 20*(1), 1–17. doi:10.3402/rlt.v20i0.14406

Kebritchi, M., Lipschuetz, A., & Santiague, L. (2017). Issues and challenges for teaching successful online courses in higher education: A literature review. *Journal of Educational Technology Systems, 46*(1), 4–29. doi:10.1177/0047239516661713

Khanchobani, A. (2012). Input enhancement and EFL learners" collocation acquisition. *International Journal of Academic Research, 4*(1), 96–101.

Khwaja, M. G., & Zaman, U. (2020). Configuring the Evolving Role of eWOM on the Consumers Information Adoption. *Journal of Open Innovation, 6*(4), 125. doi:10.3390/joitmc6040125

Kim, K., Cheong, Y., & Kim, H. (2017). User-generated product reviews on the internet: The drivers and outcomes of the perceived usefulness of product reviews. *International Journal of Advertising, 36*(2), 227–245. doi:10.1080/02650487.2015.1096100

Kingma, D. P., & Ba, J. 2014. Adam: A method for stochastic optimization. *ICLR Conference Proceedings*, (pp. 1-15). Scientific Research Publishing.

Kiousis, S. (2002). Interactivity: A concept explication. *New Media & Society, 4*(3), 355–383. doi:10.1177/146144480200400303

Klopfer, E., Squire, K., & Jenkins, H. (2002). Environmental Detectives: PDAs as a window into a virtual simulated world. *Proceedings - IEEE International Workshop on Wireless and Mobile Technologies in Education, WMTE 2002*, (pp. 95–98). IEEE. 10.1109/WMTE.2002.1039227

Kolesnikova, I. V. (2016). Combined Teaching Method: An Experimental Study. *World Journal of Education, 6*(6), 51–59. doi:10.5430/wje.v6n6p51

Kotler, P. & Armstrong, G. (2006). Principles of Marketing 11th Edition. Prentice Hall.

Kotler, P., & Zaltman, G. (1976). Targeting prospects for a new product. *Journal of Advertising Research, 16*, 7–20.

Kramer, A. D., Guillory, J. E., & Hancock, J. T. (2014). Experimental evidence of massive-scale emotional contagion through social networks. *Proceedings of the National Academy of Sciences of the United States of America, 111*(24), 8788–8790. doi:10.1073/pnas.1320040111 PMID:24889601

Kress, G. (2010). *Multimodality: A social semiotic approach to contemporary commmunication.* Routledge.

Kross, E., & Chandhok, S. (2020). How do online social networks influence people's emotional lives? In J. P. Forgas, W. D. Crano, & K. Fiedler (Eds.), Applications of social psychology: How social psychology can contribute to the solution of real-world problems (pp. 250–263). Routledge/Taylor & Francis Group. doi:10.4324/9780367816407-13

Kruk, M., & Zawodniak, J. (2018). Boredom in practical English language classes: Insights from interview data. *Interdisciplinary Views on the English Language, Literature and Culture, January.*

Kruk, M. (2016). Variations in motivation, anxiety and boredom in learning English in Second Life. *The EuroCALL Review, 24*(1), 25. doi:10.4995/eurocall.2016.5693

Kruk, M. (2022). Dynamicity of perceived willingness to communicate, motivation, boredom and anxiety in Second Life: The case of two advanced learners of English. *Computer Assisted Language Learning, 35*(1–2), 190–216. doi:10.1080/09588221.2019.1677722

Kudeshia, C., & Kumar, A. (2017). Social eWOM: Does it affect the brand attitude and purchase intention of brands? *Management Research Review, 40*(3), 310–330. doi:10.1108/MRR-07-2015-0161

Kukulska-Hulme, A., & Traxler, J. (2005). *Mobile Learning: A Handbook for Educators and Trainers.* The Open and Flexible Learning Series.

Kumar, V., Boorla, K., Meena, Y., Ramakrishnan, G., & Li, Y. F. (2018). Automating reading comprehension by generating question and answer pairs. Lecture Notes in Computer Science (Including Subseries Lecture Notes in Artificial Intelligence and Lecture Notes in Bioinformatics), 10939 LNAI, 335–348. doi:10.1007/978-3-319-93040-4_27

Kumar, V., & Raheja, G. (2012). Business to business (b2b) and business to consumer (b2c) management. *International Journal of Computers and Technology, 3*(3), 447–451.

Kuo, Y. C., Walker, A. E., Belland, B. R., & Schroder, K. E. (2013). A predictive study of student satisfaction in online education programs. *International Review of Research in Open and Distributed Learning, 14*(1), 16–39. doi:10.19173/irrodl.v14i1.1338

Kuss, D. J., & Griffiths, M. D. (2011). Online social networking and addiction a review of the psychological literature. *International Journal of Environmental Research and Public Health, 8*(9), 3528–3552. doi:10.3390/ijerph8093528 PMID:22016701

Lafferty, J., McCallum, A., & Pereira, F. (2008). *Conditional random fields: Probabilistic models for segmenting and labelling sequence data.* University of Pennsylvania. https://repository.upenn.edu/cgi/viewcontent.cgi?article=1162&context=cis_papers

Lambe, C. J., Wittmann, C. M., & Spekman, R. E. (2001). Social Exchange Theory and Research on Business-to-Business Relational Exchange. *Journal of Business-To-Business Marketing, 8*(3), 1–36. doi:10.1300/J033v08n03_01

Lamberton, C., & Stephen, A. T. (2016). A Thematic Exploration of Digital, Social Media, and Mobile Marketing: Research Evolution from 2000 to 2015 and an Agenda for Future Inquiry. *Journal of Marketing, 80*(6), 146–172. doi:10.1509/jm.15.0415

Lample, G., & Conneau, A. (2019). Cross-lingual language model pretraining. Proceedings of NeurIPS, (pp. 7059–7069).

Larson, G. W. (2011). *Instant Messaging.* Encyclopedia Britannica. https://www.britannica.com/topic/instant-messaging

Le, T. N. A. (2021). *Interaction and evaluation of teaching and learning English online: Challenges and solution.* The 9th OPEN TESOL International Conference 2021: Language education in challenging times: Designing digital transformations, Ho Chi Minh City, Vietnam. https://opentesol.ou.edu.vn/2021proceedings.html

Le, V. T. (2018). *Social media in learning English in Viet Nam* [Unpublished doctoral dissertation, The University of Canterbury].

Leander, K. M. (2009). Composing with old and new media: Toward a parallel pedagogy. In V. Carrington & M. Robinson (Eds.), *Digital literacies: Social learning and classroom practices* (pp. 147–165). Sage. doi:10.4135/9781446288238.n10

Leary, M. R., Allen, A. B., & Terry, M. L. (2011). Managing social images in naturalistic versus laboratory settings: Implications for understanding and studying self-presentation. *European Journal of Social Psychology, 41*(4), 411–421. doi:10.1002/ejsp.813

Lee, H., Min, J., & Yuan, J. (2021). The influence of eWOM on intentions for booking luxury hotels by Generation Y. *Journal of Vacation Marketing, 27*(3), 237–251. doi:10.1177/1356766720987872

Lee, J. W. (2010). Online support service quality, online learning acceptance, and student satisfaction. *The Internet Higher Education, 13*(4), 277–283. doi:10.1016/j.iheduc.2010.08.002

Lee, J., & Hong, I. B. (2021). The Influence of Situational Constraints on Consumers' Evaluation and Use of Online Reviews: A Heuristic-Systematic Model Perspective. *Journal of Theoretical and Applied Electronic Commerce Research, 16*(5), 1517–1536. doi:10.3390/jtaer16050085

Lee, J., Park, D. H., & Han, I. (2008). The effect of negative online consumer reviews on product attitude: An information processing view. *Electronic Commerce Research and Applications, 7*(3), 341–352. doi:10.1016/j.elerap.2007.05.004

Lee, M., & Youn, S. (2009). Electronic word of mouth (eWOM): How eWOM platforms influence consumer product judgement. *International Journal of Advertising, 28*(3), 473–499. doi:10.2501/S0265048709200709

Leong, C. M., Loi, A. M. W., & Woon, S. (2021). The influence of social media eWOM information on purchase intention. *Journal of Marketing Analytics.* doi:10.1057/s41270-021-00132-9

Leong, L.-Y., Hew, T.-S., Ooi, K.-B., & Lin, B. (2019). Do Electronic Word-of-Mouth and Elaboration Likelihood Model Influence Hotel Booking? *Journal of Computer Information Systems, 59*(2), 146–160. doi:10.1080/08874417.2017.1320953

Levy, S., & Gvili, Y. (2015). How Credible is E-Word of Mouth Across Digital-Marketing Channels?: The Roles of Social Capital, Information Richness, and Interactivity. *Journal of Advertising Research, 55*(1), 95–109. doi:10.2501/JAR-55-1-095-109

Lewinski, P. (2015). Effects of classrooms' architecture on academic performance in view of telic versus paratelic motivation: A review. *Frontiers in Psychology*, *6*, 746. doi:10.3389/fpsyg.2015.00746 PMID:26089812

Lewis, M. (1993). The Lexical Approach. The state of ELT and a way forward. Commercial Colour Press Lewis, M. (Eds.). (2000). *Teaching collocation: Further developments in lexical approach.* Oxford University Press.

Lewis, M. (1997). *Implementing the lexical approach: Putting theory into practice.* Language Teaching Publications.

Liao, Y.-K., Wu, W.-Y., Le, T. Q., & Phung, T. T. T. (2022). The Integration of the Technology Acceptance Model and Value-Based Adoption Model to Study the Adoption of E-Learning: The Moderating Role of e-WOM. *Sustainability*, *14*(2), 815. doi:10.3390u14020815

Li, C., & Han, Y. (2022). *Learner-internal and learner-external factors for boredom amongst Chinese university EFL students.* Applied Linguistics Review. doi:10.1515/applirev-2021-0159

Lieberman, A., & Schroeder, J. (2020). Two social lives: How differences between online and offline interaction influence social outcomes. *Current Opinion in Psychology*, *31*, 16–21. doi:10.1016/j.copsyc.2019.06.022 PMID:31386968

Li, L. (2020). Education supply chain in the era of Industry 4.0. *Systems Research and Behavioral Science*, *37*(4), 579–592. doi:10.1002res.2702

Lin, C. L., Jin, Y. Q., Zhao, Q., Yu, S.-W., & Su, Y.-S. (2021). Factors influence students' switching behavior to online learning under COVID-19 pandemic: A push–pull–mooring model perspective. *The Asia-Pacific Education Researcher*, *30*(3), 229–245. doi:10.100740299-021-00570-0

Lin, H., Tov, W., & Qiu, L. (2014). Emotional disclosure on social networking sites: The role of network structure and psychological needs. *Computers in Human Behavior*, *41*, 342–350. doi:10.1016/j.chb.2014.09.045

Li, P. H., Fu, T. J., & Ma, W. Y. (2020). Why attention? Analyze BiLSTM deficiency and its remedies in the case of NER. *Proceedings of the AAAI Conference on Artificial Intelligence*, (pp. 8236—8244). PKP Publishing. 10.1609/aaai.v34i05.6338

Litvin, S. W., Goldsmith, R. E., & Pan, B. (2008). Electronic word-of-mouth in hospitality and tourism management. *Tourism Management*, *29*(3), 458–468. doi:10.1016/j.tourman.2007.05.011

Liu, I.-F., Chen, M. C., Sun, Y. S., Wible, D., & Kuo, C.-H. (2010). Extending the TAM model to explore the factors that affect intention to use an online learning community. *Computers & Education*, *54*(2), 600–610. doi:10.1016/j.compedu.2009.09.009

Liu, Y., Alzahrani, I. R., Jaleel, R. A., & Al Sulaie, S. (2023). An efficient smart data mining framework based cloud internet of things for developing artificial intelligence of marketing information analysis. *Information Processing & Management*, *60*(1), 103121. doi:10.1016/j.ipm.2022.103121

Livingstone, S., Kirwil, L., Ponte, C., & Staksrud, E. (2014). In their own words: What bothers children online? *European Journal of Communication, 29*(3), 271–288. doi:10.1177/0267323114521045

Loranc, B., Hilliker, S. M., & Lenkaitis, C. A. (2021). Virtual exchanges in language teacher education: Facilitating reflection on teaching practice through the use of video. *TESOL Journal, 12*(2), 1–15. doi:10.1002/tesj.580

Lo, S.-K. (2008). The nonverbal communication functions of emoticons in computer-mediated communication. *Cyberpsychology & Behavior, 11*(5), 595–597. doi:10.1089/cpb.2007.0132 PMID:18817486

Lotherington, H., & Jenson, J. (2011). Teaching multimodal and digital literacy in L2 settings: New literacies, new basics, new pedagogies. *Annual Review of Applied Linguistics, 31*, 226–246. doi:10.1017/S0267190511000110

Luff, P., Heath, C., Kuzuoka, H., Hindmarsh, J., Yamazaki, K., & Oyama, S. (2003). Fractured Ecologies: Creating Environments for Collaboration. *Human-Computer Interaction, 18*(1–2), 51–84. doi:10.1207/S15327051HCI1812_3

Luo, C., Wu, J., Shi, Y., & Xu, Y. (2014). The effects of individualism–collectivism cultural orientation on eWOM information. *International Journal of Information Management, 34*(4), 446–456. doi:10.1016/j.ijinfomgt.2014.04.001

Lynch, M. W. (2018). Using conferences poster presentations as a tool for student learning and development. *Innovations in Education and Teaching International, 55*(6), 633–639.

MacIntyre, P. D., Gregersen, T., & Mercer, S. (2020). Language teachers' coping strategies during the Covid-19 conversion to online teaching: Correlations with stress, wellbeing and negative emotions. *System, 94*, 102352. doi:10.1016/j.system.2020.102352

Maguire, M., & Delahunt, B. (2017). Doing a thematic analysis: A practical, Step - by – Step guide for learning and teaching scholar. *Journal of Teaching and Learning in Higher Education, 3*, 1–13. http://ojs.aishe.org/index.php/aishe -j/article/view/335

Mahvelati, E. H. (2019). Explicit and implicit collocation teaching methods: Empirical research and issues. *Advances in Language and Literary Studies, 10*(3), 105. http://www.journals.aiac.org.au/index.php/alls/article/view/5545. doi:10.7575//aiac.alls.v.10n.3p.105

Mailizar, M., Burg, D., & Maulina, S. (2021). Examining university students' behavioural intention to use e-learning during the COVID-19 pandemic: An extended TAM model. *Education and Information Technologies, 26*(6), 7057–7077. doi:10.100710639-021-10557-5 PMID:33935579

Malone, T. W., & Lepper, M. R. (1987). Making learning fun: A taxonomy of intrinsic motivations for learning. In R. E. Snow & M. J. Farr (Eds.), *Aptitude, learning and Instruction III: Conative and affective process analysis* (pp. 223–253). Erlbaum.

Manan, N., Alias, A., & Pandian, A. (2012). Utilizing a social networking website as an ESL pedagogical tool in a blended learning environment: An exploratory study. *International Journal of Social Sciences & Education*, 2(1), 1–9. https://rb.gy/48j3wt

Mann, S., & Walsh, S. (2013). RP or 'RIP': A critical perspective on reflective practice. *Applied Linguistics Review*, 4(2), 291–315. doi:10.1515/applirev-2013-0013

Mann, S., & Walsh, S. (2017). *Reflective Practice in English Language Teaching*. Routledge. doi:10.4324/9781315733395

Martin, K. A., Leary, M. R., & Rejeski, W. J. (2000). Self-presentational concerns in older adults: Implications for health and well-being. *Basic and Applied Social Psychology*, 22(3), 169–179. doi:10.1207/S15324834BASP2203_5

Matook, S., Cummings, J., & Bala, H. (2015). Are you feeling lonely? The impact of relationship characteristics and online social network features on loneliness. *Journal of Management Information Systems*, 31(4), 278–310. doi:10.1080/07421222.2014.1001282

Matzler, K., Veider, V., & Kathan, W. (2015). Adapting to the sharing economy. Cambridge, MA, USA: Mit.

McMillan, S. J. (2002). Exploring Models of Interactivity from Multiple Research Traditions: Users, Documents, and Systems. In L. A. Lievrouw & S. Livingstone (Eds.), *Handbook of New Media: Social Shaping and Consequences of ICTs* (pp. 163–182). Sage. doi:10.4135/9781848608245.n13

McWilliam, G. (2000). Building Stronger Brands Through Online Communities. *Sloan Management Review*, 41(Spring), 43–54.

Md Yunus, M., Ang, W. S., & Hashim, H. (2021). Factors affecting teaching English as a Second Language (TESL) postgraduate students' behavioural intention for online learning during the COVID-19 pandemic. *Sustainability*, 13(6), 3524. doi:10.3390u13063524

Means, B., Toyama, Y., Murphy, R., & Baki, M. (2013). The effectiveness of online and blended learning: A meta-analysis of the empirical literature. *Teachers College Record*, 115(3), 1–47. doi:10.1177/016146811311500307

Milić, J., Ehrler, B., Molina, C., Saliba, M., & Bisquert, J. (2020). Online Meetings in Times of Global Crisis: Toward Sustainable Conferencing. *ACS Energy Letters*, 5(6), 2024–2026. doi:10.1021/acsenergylett.0c01070 PMID:34192148

Miller, M. (2012). *B2B digital marketing: Using the web to market directly to businesses*. Que publishing.

Miller, S. M. (2013). A research metasynthesis on digital video composing in classrooms: An evidence-based framework towards a pedagogy for embodied learning. *Journal of Literacy Research*, 45(4), 385–430. doi:10.1177/1086296X13504867

Miller, S., & Pennycuff, L. (2008). The Power of Story : Using Storytelling to Improve Literacy Learning. *Journal of Cross-Disciplinary Perspectives in Education*, 1(1), 36–43.

Mills, K. A. (2010). What learners 'know' through digital media production: Learning by design. *E-Learning and Digital Media, 7*(3), 223–236. doi:10.2304/elea.2010.7.3.223

Mir, I. A., & Rehman, K. U. (2013). Factors Affecting Consumer Attitudes and Intentions toward User-Generated Product Content on YouTube. *Management & Marketing Challenges for Knowledge Society, 8*(4), 637–654.

Mohammed, A., & Al-Swidi, A. (2021). The mediating role of affective commitment between corporate social responsibility and eWOM in the hospitality industry. *Journal of Sustainable Tourism, 29*(4), 570–594. doi:10.1080/09669582.2020.1818086

Moreno, M. A., Jelenchick, L. A., Egan, K. G., Cox, E., Young, H., Gannon, K. E., & Becker, T. (2011). Feeling bad on Facebook: Depression disclosures by college students on a social networking site. *Depression and Anxiety, 28*(6), 447–455. doi:10.1002/da.20805 PMID:21400639

Motteram, G. (2013). *Innovations in learning technologies for English language teaching.* British Council.

Mousavi, S. M., & Heidari Darani, L. (2018). Effect of collocations on Iranian EFL learners' writing: Attitude in focus. *Global Journal of Foreign Language Teaching, 8*(4), 131–145. doi:10.18844/gjflt.v8i4.3568

Mukahi, T., Nakamura, M., & Not, R. D. (2003). *An Empirical Study on Impacts of Computer-Mediated Communication Management on Job Satisfaction.* 7th Pacific Asia Conference on Information Systems, Adelaid, South Australia.

Mukhtar, K., Javed, K., Arooj, M., & Sethi, A. (2020). Advantages, limitations and recommendations for online learning during COVID-19 pandemic Era. *Pakistan Journal of Medical Sciences, 36*(COVID19-S4). doi:10.12669/pjms.36.COVID19-S4.2785 PMID:32582310

Müller, Y. (2008). *Collocation - A linguistic view and didactic aspects.* GRIN Verlag.

Munoz, C., & Towner, T. (2009). *Opening Facebook: How to use Facebook in the college classroom.* Paper presented at the Proceedings of society for information technology & teacher education international conference. Research Gate.

Muñoz, C. L., & Towner, T. (2011). Back to the "wall": How to use Facebook in the college classroom. *First Monday.* doi:10.5210/fm.v16i12.3513

Murphy, M., & Sashi, C. M. (2018). Communication, interactivity, and satisfaction in B2B relationships. *Industrial Marketing Management, 68,* 1–12. doi:10.1016/j.indmarman.2017.08.020

Mustafa, M. B. (2015). One size does not fit all: Students' perceptions about Edmodo at Al Ain University of Science & Technology. *Journal of Studies in Social Sciences, 13*(2), 135–160. https://core.ac.uk/download/pdf/229607465.pdf

Myers, T. A., & Crowther, J. H. (2009). Social comparison as a predictor of body dissatisfaction: A meta-analytic review. *Journal of Abnormal Psychology, 118*(4), 683–698. doi:10.1037/a0016763 PMID:19899839

Nakamura, S., Darasawang, P., and Reinders, H. (2021). The antecedents of boredom in L2 classroom learning. *System, 98,* 102469. doi:10.1016/j.system.2021.102469

Nakamura, P. M., Pereira, G., Papini, C. B., Nakamura, F. Y., & Kokubun, E. (2010). Effects of preferred and nonpreferred music on continuous cycling exercise performance. *Perceptual and Motor Skills, 110*(1), 257–264. doi:10.2466/pms.110.1.257-264 PMID:20391890

Nartiningrum, N., & Nugroho, A. (2020). Online learning amidst global pandemic: EFL students' challenges, suggestions, and needed materials. *ENGLISH FRANCA: Academic Journal of English Language and Education, 4*(2), 115–140. doi:10.29240/ef.v4i2.1494

Naveed, N., Gottron, T., Kunegis, J., & Alhadi, A. C. (2011). Bad News Travel Fast. *Proceedings of the 3rd International Web Science Conference.* ACM. 10.1145/2527031.2527052

Nayman, H., & Bavlı, B. (2022). Online Teaching of Productive Language Skills (PLS) during Emergency Remote Teaching (ERT) in EFL Classrooms: A Phenomenological Inquiry. *International Journal of Education and Literacy Studies, 10*(1), 179. doi:10.7575/aiac.ijels.v.10n.1p.179

Nemat, R. (2011). Taking a look at different types of e-commerce. *World Applied Programming, 1*(2), 100–104.

Nestle boosts Ruby chocolate range after driving viral sensation. (2019). Indianexpress.Com. https://indianexpress.com/article/business/nestle-boosts-ruby-chocolate-kitkat-viral-sensation-5634042/

Newhagen, J. E., & Rafaeli, S. (1996). Why communication researchers should study the internet: A dialogue. *Journal of Communication, 46*(1), 4–13. doi:10.1111/j.1460-2466.1996.tb01458.x

Ngah, A. H., Halim, M. R., & Aziz, N. A. (2018). The influence of electronic word of mouth on theory of *reasoned* action and the visit intention to the world monument fund site.

Ngarmwongnoi, C., Oliveira, J. S., AbedRabbo, M., & Mousavi, S. (2020). The implications of eWOM adoption on the customer journey. *Journal of Consumer Marketing, 37*(7), 749–759. doi:10.1108/JCM-10-2019-3450

Ngo, D. H. (2021). Perceptions of EFL tertiary students towards the correlation between e-learning and learning engagement during the COVID-19 pandemic. *International Journal of TESOL & Education, 1*(3). https://eoi.citefactor.org/10.11250/ijte.01.03.013

Nguyen, H. H. T. (2009). Teaching EFL writing in Vietnam: Problems and solutions -a discussion from the outlook of applied linguistics. *VNU Journal of Science, 25,* 61 - 66. https://js.vnu.edu.vn/FS/article/view/2236

Nguyen, D. Q., & Nguyen, A. T. (2022). PhoBERT: Pre-trained language models for Vietnamese. *Findings of the Association for Computational Linguistics: EMNLP, 2020,* 1037–1042. doi:10.18653/v1/2020.findings-emnlp.92

Nguyen, T. (2015). The effectiveness of online learning: Beyond no significant difference and future horizons. *Journal of Online Learning and Teaching, 11*(2), 309–319.

Nham, P. T., & Nguyen, T. T. (2013). The impact of online social networking on students'study. *VNU Journal of Education Research, 29*(1), 1−13.https://js.vnu.edu.vn/ER/article/view/486

Ni, A. Y. (2013). Comparing the effectiveness of classroom and online learning: Teaching research methods. *Journal of Public Affairs Education, 19*(2), 199–215. doi:10.1080/1523680 3.2013.12001730

Nizonkiza, D. (2017). Improving academic literacy by teaching collocations. *Stellenbosch Papers in Linguistics, 47*(0), 153 - 179. https://journals.co.za/doi/abs/10.5774/47-0-267

Number of internet users in India from 2010 to 2020, with estimates until 2040. (2022). Statista. https://www.statista.com/statistics/255146/number-of-interne t-users-in-india/

Nuseir, M. T. (2019). The impact of electronic word of mouth (e-WOM) on the online purchase intention of consumers in the Islamic countries – a case of (UAE). *Journal of Islamic Marketing, 10*(3), 759–767. doi:10.1108/JIMA-03-2018-0059

Nxumalo, L. K., & Chiweshe, N. (2019). Social enterprise digital marketing. *Strategic Marketing for Social Enterprises in Developing Nations*, 103-130. doi:10.4018/978-1-5225-7859-8.ch005

Nykiel, R. (2007). *Handbook of marketing research methodologies for hospitality and tourism.* Haworth Press. doi:10.4324/9780203448557

O'Dell, F., & McCarthy, M. (2017). *English collocations in use advanced.* Cambridge University Press.

O'Malley, J. M., & Chamot, A. U. (1990). *Learning strategies in second language acquisition.* Cambridge University Press. doi:10.1017/CBO9781139524490

Ogbonna, C. G., Ibezim, N. E., & Obi, C. A. (2019). Synchronous versus asynchronous e-learning in teaching word processing: An experimental approach. *South African Journal of Education, 39*(2), 1–15. https://hdl.handle.net/10520/EJC-168a98cd12. doi:10.15700aje.v39n2a1383

Okita, S. Y. (2012). Social interactions and learning. In Seel N. M. (Ed.), Encyclopedia of the Sciences of Learning, 182-211. Springer. doi:10.1007/978-1-4419-1428-6_1770

Orzan, G., Iconaru, C., Popescu, I. C., Orzan, M., & Macovei, O. I. (2013). PLS-based SEM analysis of apparel online buying behavior. *The importance of eWOM. Industria Textila, 64*(6), 362–367.

Oxford, R. L. (1990). *Language learning strategies: What every teacher should know.* Heinle & Heinle.

Oxford, R. L. (2011). *Teaching and researching language learning strategies.* Pearson Longman.

Pallant, J. (2016). *SPSS survival manual: A step by step guide to data analysis using IBM SPSS* (6th ed.). Allen & Unwin.

Palloff, R. M., & Pratt, K. (2013). *Lessons from the virtual classroom: The realities of online teaching*. John Wiley & Sons.

Palvia, S., Aeron, P., Gupta, P., Mahapatra, D., Parida, R., Rosner, R., & Sindhi, S. (2018). Online education: Worldwide status, challenges, trends, and implications. *Journal of Global Information Technology Management*, *21*(4), 233–241. doi:10.1080/1097198X.2018.1542262

Pantti, M., & Tikka, M. (2014). Cosmopolitan Empathy and User Generated Disaster Appeal Videos on YouTube. In T. Benski & E. Fisher (Eds.), *Internet and Emotions* (pp. 178–192). Routledge.

Parapi, J. M. O., Maesaroh, L. I., Basuki, B., & Masykuri, E. S. (2020). Virtual education: A brief overview of its role in the current educational system. *Scripta: English Department Journal*, *7*(1), 8–11. doi:10.37729cripta.v7i1.632

Pardo, L. S. (2004). What Every Teacher Needs to Know About Comprehension. *The Reading Teacher*, *58*(3), 272–280. doi:10.1598/RT.58.3.5

Park, D.-H., & Lee, J. (2008). EWOM overload and its effect on consumer behavioral intention depending on consumer involvement. *Electronic Commerce Research and Applications*, *7*(4), 386–398. doi:10.1016/j.elerap.2007.11.004

Park, D.-H., Lee, J., & Han, I. (2007). The Effect of On-Line Consumer Reviews on Consumer Purchasing Intention: The Moderating Role of Involvement. *International Journal of Electronic Commerce*, *11*(4), 125–148. doi:10.2753/JEC1086-4415110405

Park, E. L., & Choi, B. K. (2014). Transformation of classroom spaces: Traditional versus active learning classroom in colleges. *Higher Education*, *68*(5), 749–771. doi:10.100710734-014-9742-0

Park, H. S. (2000). Relationships among attitudes and subjective norms: Testing the theory of reasoned action across cultures. *Communication Studies*, *51*(2), 162–175. doi:10.1080/10510970009388516

Park, H., Choi, S., & Lee, M. (2012). Visual input enhancement, attention, grammar learning, & reading comprehension: An eye movement study. *English Teaching*, *67*(4), 241–265. http://journal.kate.or.kr/wp-content/uploads/2015/01/kate_67 _4_11.pdf. doi:10.15858/engtea.67.4.201212.241

Park, N., Lee, S., & Kim, J. H. (2012). Individuals' personal network characteristics and patterns of Facebook use: A Social Network approach. *Computers in Human Behavior*, *28*(5), 1700–1707. doi:10.1016/j.chb.2012.04.009

Partnership for 21st Century Skills. (2006). *Results that matter: 21st century skills and high school reform.* 21st Century Skills. http://www.21stcenturyskills.org/documents/RTM2006.pdf

Paul, J., Alhassan, I., Binsaif, N., & Singh, P. (2023). Digital entrepreneurship research: A systematic review. *Journal of Business Research*, *156*, 113507. doi:10.1016/j.jbusres.2022.113507

Paul, J., & Jefferson, F. (2019). A comparative analysis of student performance in an online vs. face-to-face environmental science course from 2009 to 2016. *Frontiers of Computer Science*, 7, 7. doi:10.3389/fcomp.2019.00007

Pawlak, M., Zawodniak, J., & Kruk, M. (2020). Boredom in the foreign language classroom: A micro-perspective. In Second Language Learning and Teaching. doi:10.1007/978-3-030-50769-5

Pawlak, M., Zawodniak, J., & Kruk, M. (2021). Individual trajectories of boredom in learning English as a foreign language at the university level: Insights from three students' self-reported experience. *Innovation in Language Learning and Teaching*, 15(3), 263–278. doi:10.1080/175 01229.2020.1767108

Pegrum, M. (2020). Mobile lenses on learning: Languages and literacies on the move. In *Mobile Lenses on Learning*. Languages and Literacies on the Move. doi:10.1007/978-981-15-1240-7

Pekrun, R., Goetz, T., Daniels, L. M., Stupnisky, R. H., & Perry, R. P. (2010). Boredom in Achievement Settings: Exploring Control-Value Antecedents and Performance Outcomes of a Neglected Emotion. *Journal of Educational Psychology*, 102(3), 531–549. doi:10.1037/a0019243

Pekrun, R., Hall, N. C., Goetz, T., & Perry, R. P. (2014). Boredom and academic achievement: Testing a model of reciprocal causation. *Journal of Educational Psychology*, 106(3), 696–710. doi:10.1037/a0036006

Peng, H., Su, Y. J., Chou, C., & Tsai, C. C. (2009). Ubiquitous knowledge construction: Mobile learning re-defined and a conceptual framework. *Innovations in Education and Teaching International*, 46(2), 171–183. doi:10.1080/14703290902843828

Petrescu, M., O'Leary, K., Goldring, D., & Ben Mrad, S. (2016). Incentivized reviews: Promising the moon for a few stars. *Journal of Retailing and Consumer Services*, 41, 288–295. doi:10.1016/j. jretconser.2017.04.005

Pham, M. T., Luu, T. T. U., Mai, T. H. U., Thai, T. T. T., & Ngo, T. C. T. (2022). EFL students' challenges of online courses at Van Lang University during the COVID-19 pandemic. *International Journal of TESOL & Education*, 2(2), 1–26. doi:10.54855/ijte.22221

Pham, P. Q. M. (2018). *A feature-rich Vietnamese Named-Entity Recognition Model*. Cornell University. doi:10.48550/arXiv.1803.04375

Phan, Q. P. T., Pham, N. T., & Nguyen, L. H. L. (2020). How to Drive Brand Engagement and eWOM Intention in Social Commerce: A Competitive Strategy for the Emerging Market. *Journal of Competitiveness*, 12(3), 136–155. doi:10.7441/joc.2020.03.08

Polifroni, J., Kiss, I., & Adler, M. (2010). Bootstrapping named entity extraction for the creation of mobile services. *Proceedings of the Seventh International Conference on Language Resources and Evaluation (LREC'10)*, (pp. 1515-1520). European Language Resources Association (ELRA).

Polok, K., & Harężak, J. (2018). Facebook as a beneficial tool while used in learning second language environment. *Open Access Library Journal*, 5(07), 1–13. doi:10.4236/oalib.1104732

Post, D., Carr, C., & Weigand, J. (1998). Teenagers: Mental health and psychological issues. *Primary care*, 25(1), 181–192. doi:10.1016/S0095-4543(05)70331-6 PMID:9469922

Prikhodko, O. V., Cherdymova, E. I., Lopanova, E. V., Galchenko, N. A., Ikonnikov, A. I., Mechkovskaya, O. A., & Karamova, O. V. (2020). Ways of expressing emotions in social networks: Essential features, problems and features of manifestation in internet communication. *Online Journal of Communication and Media Technologies*, 10(2). doi:10.29333/ojcmt/7931

Qader, D. S. (2018). The role of teaching lexical collocations in raising EFL learners' speaking fluency. *Journal of Literature. Language and Linguistics (Taipei)*, 46, 42–53. https://www.iiste.org/Journals/index.php/JLLL/article/view/43454

Qiu, J. (2019). A preliminary study of english mobile learning model based on constructivism. *Theory and Practice in Language Studies*, 9(9), 1167–1172. doi:10.17507/tpls.0909.13

Quan-Haase, A., & Young, A. L. (2010). Uses and gratifications of social media: A comparison of Facebook and instant messaging. *Bulletin of Science, Technology & Society*, 30(5), 350–361. doi:10.1177/0270467610380009

Rafaeli, S. (1988). Interactivity: from new media to communication. In R. P. Hawkins, J. M. Wiemann, & S. Pingree (Eds.), *Advancing communication science: Merging mass and interpersonal process* (pp. 110–134). Sage.

Ramshaw, L. A., & Marcus, M. P. (1995). Text chunking using transformation-based learning. *Computer Languages*, 11, 82–94. https://aclanthology.org/W95-0107/

Rassaei, E., & Karbor, T. (2012). The effects of three types of attention drawing techniques on the acquisition of English collocations. *International Journal of Research Studies in Language Learning*, 2(1), 15–28. doi:10.5861/ijrsll.2012.117

Reed, P., Smith, B., & Sherratt, C. (2008). A New Age of Constructivism: 'Mode Neutral.'. *E-Learning and Digital Media*, 5(3), 310–322. doi:10.2304/elea.2008.5.3.310

Resnik, P., & Dewaele, J. M. (2021). *Learner emotions, autonomy and trait emotional intelligence in "in-person" versus emergency remote English foreign language teaching in Europe.* Applied Linguistics Review., doi:10.1515/applirev-2020-0096

Rice, R. E., & Love, G. (1987). Electronic Emotion: Socioemotional Content in a Computer-Mediated Communication Network. *Communication Research*, 14(1), 85–108. doi:10.1177/009365087014001005

Richmond, W., Rader, S., & Lanier, C. (2017). The "digital divide" for rural small businesses. *Journal of Research in Marketing and Entrepreneurship*, 19(2), 94–104. doi:10.1108/JRME-02-2017-0006

Ríos, A. A., & Campos, J. L. E. (2015). The role of Facebook in foreign language learning. *Revista de Lenguas ModeRnas*, 23(23), 253–262. doi:10.15517/rlm.v0i23.22349

Ririn, D. (2020). European Journal of Education Studies. *European Journal of Education Studies*, *7*(1), 326–337. doi:10.5281/zenodo.582328

Ritzer, G., & Jurgenson, N. (2010). Production, consumption, prosumption: The nature of capitalism in the age of the digital 'prosumer'. *Journal of Consumer Culture*, *10*(1), 13–36. doi:10.1177/1469540509354673

Romiszowski, A., & Mason, R. (2004). Computer-mediated communication. In A. Romiszowski & R. Mason, Handbook of research on educational communications and technology (pp. 391-431). Lawrence Erlbaum Associates.

Rong, K., Hu, G., Lin, Y., Shi, Y., & Guo, L. (2015). Understanding business ecosystem using a 6C framework in Internet-of-Things-based sectors. *International Journal of Production Economics*, *159*, 41–55. doi:10.1016/j.ijpe.2014.09.003

Rosário, A. T., Raimundo, R. G., & Cruz, R. (2022). The Impact of Digital Technologies on Marketing and Communication in the Tourism Industry: In I. R. Management Association (Ed.), Research Anthology on Business Continuity and Navigating Times of Crisis (pp. 748–760). IGI Global. doi:10.4018/978-1-6684-4503-7.ch037

Rowsell, J., Morrell, E., & Alvermann, D. E. (2017). Confronting the digital divide: Debunking brave new world discourses. *The Reading Teacher*, *71*(2), 157–165. doi:10.1002/trtr.1603

Rubin, D. B. (1987). *Multiple imputation for nonresponse in surveys*. John Wiley & Sons Inc. doi:10.1002/9780470316696

Rufii, R. (2015). Developing module on constructivist learning strategies to promote students' independence and performance. *International Journal of Education*, *7*(1), 18. doi:10.5296/ije.v7i1.6675

S Sharma, R., Morales-Arroyo, M., & Pandey, T. (2011). The Emergence of Electronic Word-of-Mouth as a Marketing Channel for the Digital Marketplace. *Journal of Information, Information Technology, and Organizations (Years 1-3)*, *6*, 041–061. doi:10.28945/1695

Sagioglou, C., & Greitemeyer, T. (2014). Facebook's emotional consequences: Why Facebook causes a decrease in mood and why people still use it. *Computers in Human Behavior*, *35*, 359–363. doi:10.1016/j.chb.2014.03.003

Sardar, A., Manzoor, A., Shaikh, K. A., & Ali, L. (2021). An Empirical Examination of the Impact of eWOM Information on Young Consumers' Online Purchase Intention: Mediating Role of eWOM Information Adoption. *SAGE Open*, *11*(4), 21582440211052525. doi:10.1177/21582440211052547

Satici, S. A., & Uysal, R. (2015). Well-being and problematic Facebook use. *Computers in Human Behavior*, *49*, 185–190. doi:10.1016/j.chb.2015.03.005

Scheurs, J., & Dumbraveanu, R. (2014). A shift from teacher centered to learner centered approach. *learning*, *1*(2).

Schiffman, L. G., & Kanuk, L. L. (2004). Consumer Behavior, 8th International edition. Prentice Hall, Upper Saddle.

Schmitt, N. (2000). *Vocabulary in language teaching*. Cambridge University Press.

Schön, D. A. (1992). The Theory of Inquiry: Dewey's Legacy to Education. *Curriculum Inquiry*, *22*(2), 119–139. doi:10.1080/03626784.1992.11076093

Schwandt, T. A. (1994). Constructivist, interpretivist approaches to human inquiry. *Handbook of Qualitative Research, January 1994*, 118–137. http://psycnet.apa.org/psycinfo/1994-98625-006

Serra Cantallops, A., & Salvi, F. (2014). New consumer behavior: A review of research on eWOM and hotels. *International Journal of Hospitality Management*, *36*, 41–51. doi:10.1016/j.ijhm.2013.08.007

Serrano-Puche, J. (2016). Internet and emotions: New trends in an emerging field of research. *Comunicar*, *24*(46), 19–26. doi:10.3916/C46-2016-02

Seuren, L. M., Wherton, J., Greenhalgh, T., & Shaw, S. E. (2021). Whose turn is it anyway? Latency and the organization of turn-taking in video-mediated interaction. *Journal of Pragmatics*, *172*, 63–78. doi:10.1016/j.pragma.2020.11.005 PMID:33519050

Shaker, A. K., Mostafa, R. H. A., & Elseidi, R. I. (2021). Predicting intention to follow online restaurant community advice: A trust-integrated technology acceptance model. *European Journal of Management and Business Economics*. doi:10.1108/EJMBE-01-2021-0036

Sheldon, P., & Bryant, K. (2016). Instagram: Motives for its use and relationship to narcissism and contextual age. *Computers in Human Behavior*, *58*, 89–97. doi:10.1016/j.chb.2015.12.059

Shen, Z. (2021). A persuasive eWOM model for increasing consumer engagement on social media: Evidence from Irish fashion micro-influencers. *Journal of Research in Interactive Marketing*, *15*(2), 181–199. doi:10.1108/JRIM-10-2019-0161

Short, J., Williams, E., & Christie, B. (1976). *The social psychology of telecommunication*. Wiley.

Siegel, M. (2012). New times for multimodality? Confronting the accountability culture. *Journal of Adolescent & Adult Literacy*, *55*(8), 671–681. doi:10.1002/JAAL.00082

Siik, S. (2006). *The teaching of lexical collocations and its effects on the quality of essays and knowledge of collocations among students of program Persedian Ijazah Sarjana Muda of Instut Perguruan Batu Lintang, Kuchiing*. [Unpublished Master Thesis, University Teknologi Malasia, Malaysia].

Singh, V., & Thurman, A. (2019). How many ways can we define online learning? A Systematic literature review of definitions of online learning (1988-2018). *American Journal of Distance Education*, *33*(4), 289–306. doi:10.1080/08923647.2019.1663082

Sirimanna, U. I., & Gunawardana, T. S. L. W. (2020). Impact of computer mediated communication systems on job satisfaction: Employees in the transmission division of ceylon electricity board (pp. 379-402). *The 9th International Conference on Management and Economics.* Sri Lanka.

Smith, A. N., Fischer, E., & Yongjian, C. (2012). How Does Brand-related User-generated Content Differ across YouTube, Facebook, and Twitter? *Journal of Interactive Marketing, 26*(2), 102–113. doi:10.1016/j.intmar.2012.01.002

Smits, M. T., & Mogos, S. (2013). The impact of social media on business performance. *ECIS 2013 Completed Research.* 125. https://aisel.aisnet.org/ecis2013_cr/125

Smythe, S., & Neufeld, P. (2010). 'Podcast time': Negotiating digital literacies and communities of learning in a middle years ELL classroom. *Journal of Adolescent & Adult Literacy, 53*(6), 488–496. doi:10.1598/JAAL.53.6.5

Soini, A., & Eräranta, K. (2023). Collaborative construction of the closet (in and out): The affordance of interactivity and gay and lesbian employees' identity work online. *Organization, 30*(1), 21–41. doi:10.1177/13505084221115833

Sokolova, K., & Kefi, H. (2020). Instagram and YouTube bloggers promote it, why should I buy? How credibility and parasocial interaction influence purchase intentions. *Journal of Retailing and Consumer Services, 53*, 101742. doi:10.1016/j.jretconser.2019.01.011

Song, B. L., Liew, C. Y., Sia, J. Y., & Gopal, K. (2021). Electronic word-of-mouth in travel social networking sites and young consumers' purchase intentions: An extended information adoption model. *Young Consumers, 22*(4), 521–538. doi:10.1108/YC-03-2021-1288

Spagnoletti, P., Resca, A., & Sæbø, Ø. (2015). Design for social media engagement: Insights from elderly care assistance. *The Journal of Strategic Information Systems, 24*(2), 128–145. doi:10.1016/j.jsis.2015.04.002

Sproull, L., & Kiesler, S. (1986). Reducing social context cues: Electronic Mail in organizational communication. *Management Science, 32*(11), 1492–1512. doi:10.1287/mnsc.32.11.1492

Steffes, E. M., & Burgee, L. E. (2009). Social ties and online word of mouth. *Internet Research, 19*(1), 42–59. doi:10.1108/10662240910927812

Steinhoff, L., & Palmatier, R. W. (2021). Commentary: Opportunities and challenges of technology in relationship marketing. *Australasian Marketing Journal, 29*(2), 111–117. doi:10.1016/j.ausmj.2020.07.003

Stewart, B. E. (2023). The problem of the web: Can we prioritize both participatory practices and privacy. *Contemporary Educational Technology, 15*(1), ep402. doi:10.30935/cedtech/12668

Stieglitz, S., & Dang-Xuan, L. (2013). Emotions and information diffusion in social media—Sentiment of microblogs and sharing behavior. *Journal of Management Information Systems, 29*(4), 217–248. doi:10.2753/MIS0742-1222290408

Stimpson, A. J., & Cummings, M. L. (2014). Assessing intervention timing in computer-based education using machine learning algorithms. *IEEE Access: Practical Innovations, Open Solutions, 2*, 78–87. doi:10.1109/ACCESS.2014.2303071

Stockwell, G. (2007). Vocabulary on the move: Investigating an intelligent mobile phone-based vocabulary tutor. *Computer Assisted Language Learning, 20*(4), 365–383. doi:10.1080/09588220701745817

Sun, A., & Chen, X. (2016). Online education and its effective practice: A research review. *Journal of Information Technology Education, 15*, 157–190. doi:10.28945/3502

Sundaram, D. S., Mitra, K., & Webster, C. (1998). Word of-Mouth Communications: A Motivational Analysis. *Advances in Consumer Research. Association for Consumer Research (U. S.), 25*, 527–531.

Sykes, J. M. (2005). Synchronous CMC and pragmatic development: Effects of oral and written chat. *CALICO Journal, 22*(3), 399–431. doi:10.1558/cj.v22i3.399-431

Tam, K. Y., & Ho, S. Y. (2005). Web Personalization as a Persuasion Strategy: An Elaboration Likelihood Model Perspective. *Information Systems Research, 16*(3), 271–291. doi:10.1287/isre.1050.0058

Tan, J. P. L., & McWilliam, E. (2009). From literacy to multiliteracies: Diverse learners and pedagogical practice. *Pedagogies, 4*(3), 213–225. doi:10.1080/15544800903076119

TeachThought. (n.d.). *100 ways to use Facebook in education by categories*. TeachThought. http://www.teachthought.com/technology/100-ways-to-use-facebook-in-education-by-category/

Thao, N., & Shurong, T. (2020). Is It Possible for "Electronic Word-of-Mouth" and "User-Generated Content" to be Used Interchangeably? *Journal of Marketing and Consumer Research*. doi:10.7176/JMCR/65-04

Thompson, D., Mackenzie, I. G., Leuthold, H., & Filik, R. (2016). Emotional responses to irony and emoticons in written language: Evidence from EDA and facial EMG. *Psychophysiology, 53*(7), 1054–1062. doi:10.1111/psyp.12642 PMID:26989844

Thurlow, C., & Poff, M. (2011). Text messaging. C. Susan, D. S. Herring, & V. Tuija Virtanen (Eds.), Handbook of the Pragmatics of CMC. Mouton de Gruyter

Tiago, M. T. P. M. B., & Veríssimo, J. M. C. (2014). Digital marketing and social media: Why bother? *Business Horizons, 57*(6), 703–708. doi:10.1016/j.bushor.2014.07.002

Timperley, H. (2011). *Realizing the Power of Professional Learning*. Open University Press.

Todorov, A., Chaiken, S., & Henderson, M. D. (2002). The heuristic-systematic model of social information processing. *The persuasion handbook: Developments in theory and practice, 23*, 195-211.

Tóth, Z., Mrad, M., Itani, O. S., Luo, J., & Liu, M. J. (2022). B2B eWOM on Alibaba: Signaling through online reviews in platform-based social exchange. *Industrial Marketing Management, 104*, 226–240. doi:10.1016/j.indmarman.2022.04.019

Tour, E. (2015). Digital mindsets: Teachers' technology use in personal life and teaching. *Language Learning & Technology, 19*, 124–139. http://llt.msu.edu/issues/october2015/tour.pdf

Tran, H. N., & Bui, H. P. (2022). Causes of and coping strategies for boredom in L2 classroom: A case in Vietnam. *Language Related Research*. https://lrr.modares.ac.ir/article-14-62169-fa.html (In Press).

Tran, N. H., & Bui, H. P.(2022) *Causes of and Coping Strategies for Boredom in Language Classrooms: A Case in Vietnam,* 0-196.

Tran, T. Q., & Ngo, D. X. (2020). *Attitudes towards Facebook-based activities for English language learning among non-English majors.* Proceedings of the International conference 2020: Language for global competence: Finding authentic voices and embracing meaningful practices (pp. 624–643). Ho Chi Minh City: Publishing House of Economics.

Tran, T. P., & Nguyen, T. T. A. (2022). Online education at Saigon University during the COVID-19 pandemic: A survey on non-English major college students' attitudes towards learning English. *AsiaCALL Online Journal, 13*(2), 1–20. doi:10.54855/acoj.221321

Tran, T. Q., & Duong, H. (2021). Tertiary non-English majors' attitudes towards autonomous technology-based language learning. *Advances in Social Science, Education and Humanities Research, 533*, 141–148. doi:10.2991/assehr.k.210226.018

Tran, T. Q., & Tran, T. N. P. (2020). Attitudes toward the use of project-based learning: A case study of Vietnamese High school students. *Journal of Language and Education, 6*(3), 140–152. doi:10.17323/jle.2020.10109

Trisviana, W., Afriazi, R., & Hati, G. M. (2019). The correlation between students' mastery on lexical collocation and their reading comprehension. *Journal of English Education and Teaching, 3*(1), 53–65. doi:10.33369/jeet.3.1.53-65

Trusov, M., Bucklin, R. E., & Pauwels, K. (2009). Effects of Word-of-Mouth versus Traditional Marketing: Findings from an Internet Social Networking Site. *Journal of Marketing, 73*(5), 90–102. doi:10.1509/jmkg.73.5.90

Tularam, G. A. (2018). Traditional vs Non-traditional Teaching and Learning Strategies-the case of E-learning! *International Journal for Mathematics Teaching and Learning, 19*(1), 129–158.

Turban, E., Outland, J., King, D., Lee, J. K., Liang, T. P., & Turban, D. C. (2018). *Electronic commerce 2018: a managerial and social networks perspective.* Springer International Publishing., doi:10.1007/978-3-319-58715-8

Turel, O., & Serenko, A. (2012). The benefits and dangers of enjoyment with social networking websites. *European Journal of Information Systems, 21*(5), 512–528. doi:10.1057/ejis.2012.1

Tze, V., Daniels, L. M., & Klassen, R. M. (2016). Evaluating the relationship between boredom and academic outcomes: A meta-analysis. *Educational Psychology Review, 28*(1), 119–144. doi:10.100710648-015-9301-y PMID:28458499

Udell, C., & Woodill, G. (2014). Mastering Mobile Learning. In C. Udell & G. Woodill (Eds.), *Mastering Mobile Learning.* John Wiley & Sons, Inc., doi:10.1002/9781119036883

Ulla, M. B., & Perales, W. F. (2021). Facebook as an integrated online learning support application during the COVID19 pandemic: Thai university students' experiences and perspectives. *Heliyon, 7*(11), 1–8. doi:10.1016/j.heliyon.2021.e08317 PMID:34746477

Urokova, S. B. (2020). Advantages and disadvantages of online education. *ISJ Theoretical & Applied Science, 9*(89), 34–37. doi:10.15863/TAS.2020.09.89.9

Uther, M. (2019). Mobile learning—Trends and practices. *Education Sciences, 9*(1), 33. doi:10.3390/educsci9010033

Valenzuela, S., Park, N., & Kee, K. F. (2009). Is there social capital in a social network site? Facebook use and college students' life satisfaction, trust, and participation. *Journal of Computer-Mediated Communication, 14*(4), 875–901. doi:10.1111/j.1083-6101.2009.01474.x

Valkenburg, P. M., Peter, J., & Schouten, A. P. (2006). Friend networking sites and their relationship to adolescents' well-being and social self-esteem. *Cyberpsychology & Behavior, 9*(5), 584–590. doi:10.1089/cpb.2006.9.584 PMID:17034326

Valk, J. H., Rashid, A. T., & Elder, L. (2010). Using mobile phones to improve educational outcomes: An analysis of evidence from Asia. *International Review of Research in Open and Distance Learning, 11*(1), 117–140. doi:10.19173/irrodl.v11i1.794

Vishwanath, A. (2014). Diffusion of deception in social media: Social contagion effects and its antecedents. *Information Systems Frontiers, 17*(6), 1353–1367. doi:10.100710796-014-9509-2

Vogel-Walcutt, J. J., Fiorella, L., Carper, T., & Schatz, S. (2012). The definition, assessment, and mitigation of state boredom within educational settings: A comprehensive review. *Educational Psychology Review, 24*(1), 89–111. doi:10.100710648-011-9182-7

Vroeginday, B. J. (2005). *Traditional vs. online education: A comparative analysis of learner outcomes.* Fielding Graduate University.

Vu, D. V., & Peters, E. (2021). Incidental learning of collocations from meaningful input. *Studies in Second Language Acquisition,* 1 -23. doi:10.1017/S0272263121000462

Vu, N. N. (2016). Mobile Learning in Language Teaching Context of Vietnam: an Evaluation of Students' Readiness. *Journal of Science, HCMC University of Education, 7*(85), 16–27. https://www.vjol.info/index.php/sphcm/article/viewFile/24861/21273

Vu, N. N., Hung, B. P., Van, N. T. T., & Lien, N. T. H. (2021) Theoretical and Instructional Aspects of Using Multimedia Resources in Language Education: A Cognitive View. In: Kumar R., Sharma R., Pattnaik P.K. (eds) Multimedia Technologies in the Internet of Things Environment, (pp. 165-194). Springer. doi:10.1007/978-981-16-3828-2_9

Wallace, M. J. (1991). *Training Foreign Language Teachers - A reflective approach.* Cambridge University Press.

Walsh, S., & Mann, S. (2015). Doing reflective practice: A data-led way forward. *ELT Journal, 69*(4), 351–362. doi:10.1093/elt/ccv018

Walther, J. B., & D'Addario, K. P. (2001). The impacts of emoticons on message interpretation in computer-mediated communication. *Social Science Computer Review, 19*(3), 324–347. doi:10.1177/089443930101900307

Wang, C., & Chen, C. (2013). Effects of Facebook tutoring on learning English as a second language. *IADIS International Conference e-Learning*, 135–142. https://files.eric.ed.gov/fulltext/ED562299.pdf

Wang, J. T., & Good, R. L. (2007, November). The repetition of collocations in EFL textbooks: A corpus study [Paper presentation]. *The Sixteenth International Symposium and Book Fair on English Teaching, Taipei.* https://eric.ed.gov/?id=ED502758

Wang, L., Wang, Z., Wang, X., & Zhao, Y. (2022). Assessing word-of-mouth reputation of influencers on B2C live streaming platforms: The role of the characteristics of information source. *Asia Pacific Journal of Marketing and Logistics, 34*(7), 1544–1570. doi:10.1108/APJML-03-2021-0197

Wang, Q., Woo, H. L., Quek, C. L., Yang, Y., & Liu, M. (2012). Using the Facebook group as a learning management system: An exploratory study. *British Journal of Educational Technology, 43*(3), 428–438. doi:10.1111/j.1467-8535.2011.01195.x

Wardani, D. K., Martono, T., Pratomo, L. C., Rusydi, D. S., & Kusuma, D. H. (2018). Online learning in higher education to encourage critical thinking skills in the 21st century. *International Journal of Educational Research Review, 4*(2), 146–153. doi:10.24331/ijere.517973

Waterloo, S. F., Baumgartner, S. E., Peter, J., & Valkenburg, P. M. (2018). Norms of online expressions of emotion: Comparing Facebook, Twitter, Instagram, and WhatsApp. *New Media & Society, 20*(5), 1813–1831. doi:10.1177/1461444817707349 PMID:30581358

Webb, K., Mann, S., & Shafie, K. A. (2022). Using Computer-Mediated Cooperative Development in a Virtual Reflective Environment Among English Language Teachers. In Z. Tajeddin & A. Watanabe (Eds.), *Teacher Reflection: Policies, Practices and Impacts* (pp. 224–237). Multilingual Matters. doi:10.21832/9781788921022-019

Wedell, M. (2017). Teacher education planning handbook: Working together to support teachers' continuing professional development. In M. Wedell (Ed.), *Teacher Education Planning Handbook: Working together to support teachers' continuing professional development.* British Council.

Wei, Y. (1999). *Teaching Collocations for Productive Vocabulary Development*. Paper presented at the Annual Meeting of the Teachers of English to Speakers of Other Languages. New York, USA.

Weisfeld-Spolter, S., Sussan, F., & Gould, S. (2014). An integrative approach to eWOM and marketing communications. *Corporate Communications*, *19*(3), 260–274. doi:10.1108/CCIJ-03-2013-0015

Wenden, A. (1991). *Learner strategies for learner autonomy*. Prentice Hall.

Westgate, E. C., & Wilson, T. D. (2018). Boring thoughts and bored minds: The MAC model of boredom and cognitive engagement. *Psychological Review*, *125*(5), 689–713. doi:10.1037/rev0000097 PMID:29963873

Wilkins, D. A. (1972). *Linguistics in Language Teaching*. Edward Arnold.

Wilkinson, D., & Birmingham, P. (2003). *Using research instruments: A guide for researchers*. Psychology Press.

Williamson, B., Potter, J., & Eynon, R. (2019). New research problems and agendas in learning, media and technology: The editors' wishlist. *Learning, Media and Technology*, *44*(2), 87–91. doi:10.1080/17439884.2019.1614953

Wilson, A. A., Chaves, K., & Anders, P. L. (2012). "From the Koran and Family Guy": Expressions of identity in English learners' digital podcasts. *Journal of Adolescent & Adult Literacy*, *55*(5), 374–384. doi:10.1002/JAAL.00046

Wixson, K. K. (2017). An interactive view of reading comprehension: Implications for assessment. *Language, Speech, and Hearing Services in Schools*, *48*(2), 77–83. doi:10.1044/2017_LSHSS-16-0030 PMID:28395296

Wolny, J., & Mueller, C. (2013). Analysis of fashion consumers' motives to engage in electronic word-of-mouth communication through social media platforms. *Journal of Marketing Management*, *29*(5–6), 562–583. doi:10.1080/0267257X.2013.778324

Xia, M., Huang, Y., Duan, W., & Whinston, A. (2009). Ballot box communication in online communities. *Communications of the ACM*, *52*(9), 138–142. doi:10.1145/1562164.1562199

Xiao, M., Wang, R., & Chan-Olmsted, S. (2018a). Factors affecting YouTube influencer marketing credibility: A heuristic-systematic model. *Journal of Media Business Studies*, *15*(3), 188–213. doi:10.1080/16522354.2018.1501146

Xiong, G., & Bharadwaj, S. (2014). Prerelease Buzz Evolution Patterns and New Product Performance. *Marketing Science*, *33*(3), 401–421. doi:10.1287/mksc.2013.0828

Yang, S.-B., Shin, S.-H., Joun, Y., & Koo, C. (2017). Exploring the comparative importance of online hotel reviews' heuristic attributes in review helpfulness: A conjoint analysis approach. *Journal of Travel & Tourism Marketing*, *34*(7), 963–985. doi:10.1080/10548408.2016.1251872

Yan, L., Peng, J., & Tan, Y. (2015). Network dynamics: How can we find patients like us? *Information Systems Research, 26*(3), 496–512. doi:10.1287/isre.2015.0585

Yan, L., & Tan, Y. (2014). Feeling blue? Go online: An empirical study of social support among patients. *Information Systems Research, 25*(4), 690–709. doi:10.1287/isre.2014.0538

Yi, Y., & Choi, J. (2015). Teachers' views of multimodal practices in K-12 classrooms: Voices from teachers in the United States. *TESOL Quarterly, 29*(4), 838–847. https://www.jstor.org/stable/43893789. doi:10.1002/tesq.219

Yi, Y., Shin, D., & Cimasko, T. (2020). Special issue: Multimodal composing in multilingual learning and teaching contexts. *Journal of Second Language Writing, 47*, 100717–100716. doi:10.1016/j.jslw.2020.100717

Yüksel, H. F. (2016). Factors affecting purchase intention in YouTube videos. *The Journal of Knowledge Economy & Knowledge Management, 11*(2), 33–47.

Yurovskiy, V. (2014). Pros and cons of internet marketing. *Research Paper*, 1-12.

Yusuf, A. S., Che Hussin, A. R., & Busalim, A. H. (2018). Influence of e-WOM engagement on consumer purchase intention in social commerce. *Journal of Services Marketing, 32*(4), 493–504. doi:10.1108/JSM-01-2017-0031

Zaferanieh, E., & Behrooznia, S. (2011). On the impacts of four collocation instructional methods: Web - based concordancing vs. traditional method, explicit vs. implicit instruction. *Studies in Literature and Language, 3*(3), 120–126. http://cscanada.net/index.php/sll/article/view/j.sll.1923156320110303.110

Zaichkowsky, J. L. (1985). Measuring the Involvement Construct. *The Journal of Consumer Research, 12*(3), 341. doi:10.1086/208520

Zaki, A. A., & Md Yunus, M. (2015). Potential of mobile learning in teaching of ESL academic writing. *English Language Teaching, 8*(6), 11–19. doi:10.5539/elt.v8n6p11

Zawodniak, J., Kruk, M., & Pawlak, M. (2021). Boredom as an Aversive Emotion Experienced by English Majors. *RELC Journal*. doi:10.1177/0033688220973732

Zhang, K. Z. K., Zhao, S. J., Cheung, C. M. K., & Lee, M. K. O. (2014). Examining the influence of online reviews on consumers' decision-making: A heuristic–systematic model. *Decision Support Systems, 67*, 78–89. doi:10.1016/j.dss.2014.08.005

Zhang, M., Akoto, M., & Li, M. (2021). Digital multimodal composing in post-secondary L2 settings: A review of the empirical landscape. [CALL]. *Computer Assisted Language Learning*, 1–28. doi:10.1080/09588221.2021.1942068

Zhang, P. (2021). Understanding digital learning behaviors: Moderating roles of goal setting behavior and social pressure in large-scale open online courses. *Frontiers in Psychology, 12*, 783610. doi:10.3389/fpsyg.2021.783610 PMID:34899535

Zhou, W., & Duan, W. (2013). How Does the Distribution of Word-of-Mouth Across Websites Affect Online Retail Sales? SSRN *Electronic Journal*. doi:10.2139/ssrn.2396072

Zhu, D. H., Ye, Z. Q., & Chang, Y. P. (2017). Understanding the textual content of online customer reviews in B2C websites: A cross-cultural comparison between the U.S. and China. *Computers in Human Behavior*, *76*, 483–493. doi:10.1016/j.chb.2017.07.045

Zia, A. (2020). Exploring factors influencing online classes due to social distancing in COVID-19 pandemic: A business students perspective. *International Journal of Information and Learning Technology*, *37*(4), 197–211. doi:10.1108/IJILT-05-2020-0089

About the Contributors

Hung Bui is working as a lecturer and researcher at School of Foreign Languages, University of Economics Ho Chi Minh City, Vietnam. Hung also serves as an editor in several journals indexed in Scopus and Web of Science. His broad research interests have stretched across different aspects of second/ foreign language (L2) education and the use of computer-mediated communication (CMC) in L2 teaching and learning. His recently published works have mainly concentrated on applications of cognitive linguistics in L2 acquisition, sociocultural theory in L2 acquisition, teacher and student cognition, social interaction in L2 classrooms, L2 classroom assessment, teaching English for specific purposes, and computer-assisted language teaching and learning. Influenced by educational, linguistic, and psychological perspectives, his endeavors, mainly published in leading journals in the fields of language education, applied linguistics, and educational psychology, have been stimulating interesting discussions. Serving as the keynote and plenary speaker in many national and international conferences in the world, Hung has had opportunities to spread his knowledge and research interests to students, colleagues, and novice researchers.

***

Aleyna Abuşka received her Bachelor's degree in English Language Teaching at Izmir Demokrasi University, and is currently an MA student at the Department of Foreign Language Education at the same university. During her 4-year university education, she took part in many educational projects as a participant and worked as a project assistant. She is working as an editorial assistant for online academic journals. She has presented her studies at education congresses and conferences. She has book chapters published by international publishers. Her research interests include language teacher education, language teaching, young learners and second language acquisition.

Le Nguyen Nhu Anh is currently a lecturer of English Department, Ho Chi Minh City University of Education. He got his BA degree in Teaching English as a Foreign Language from Ho Chi Minh City University of Education, and his MA in Applied Linguistics from Curtin University. He is the chief author of a textbook for Vietnamese grade 7 students and a co-author of several textbooks for other grades. He is also an expert in ICT in education and has been providing trainings in different subjects for in-service teachers organized by the National Foreign Languages Project. His research interests are instructional design, applied linguistics, and ICT integration in language teaching.

Hong Quan Bui holds a Ph.D. in Educational Psychology and is a renowned author in the fields of communication, emotions, and happiness. He is currently a lecturer at Ho Chi Minh City University of Education in Vietnam. His research focuses on young people's emotions in social media communication. He believes that conveying emotions on social media platforms can be challenging due to the primarily text-based nature of the medium and the limitations of nonverbal cues. It is essential to explore the effects of social media on users' emotions and provide implications for social media communication improvement.

Khoa Do is a young lecturer at the National College of Education, Ho Chi Minh city. He is also a teacher trainer and academic manager at IELTS Vietop. He was a 2021-2022 Hornby Scholar, with an MA TESOL at University of Warwick, UK. His thematic areas of interest include Computer-assisted language learning (CALL), Reflective practice (RP) for second language teacher education, corpus-informed language teaching and material and curriculum development.

Nguyen Dung received a Ph.D. degree from Hanoi Medical University, Hanoi, Vietnam, in 2009. He became an Associate Professor of Internal Medicine in 2013. From 2019 to the present, he has been a Rector of Thai Nguyen University of Medicine and Pharmacy, Thai Nguyen University, Vietnam. His research interests include education, medicine, pharmacy, and medical communication.

Tham My Duong is currently a Vice Dean of the Faculty of English at Ho Chi Minh City University of Economics and Finance, Vietnam. She has worked with EFL students at both undergraduate and postgraduate levels in Vietnam and abroad. Her academic areas of interest predominantly lie in content and language integrated learning (CLIL), English language teaching (ELT), discourse analysis, language skills, language learning strategies, learner autonomy, project-based learning (PBL), and task-based language teaching (TBLT).

Ayushi Gupta is currently pursuing her PhD from Indian Institute of Foreign Trade, New Delhi, India. She received her Master's degree in Commerce from University of Delhi, New Delhi. Her research interests mainly lie in the areas of gamification, electronic word of mouth and consumer behavior.

Thien Hang is currently a developer at Mobile World Investment Corporation and a lecturer at Mindx Technology School. He received a Bachelor of Engineering in Software Engineering from University of Information Technology, Vietnam National University. He worked as a researcher at Soongsil University, Korea in 2019. His work focuses on Automated communication systems and fault prediction systems.

Chau Thi Hoang Hoa, PhD. is an EFL teacher, vice director of International Collaboration Office, Director of Trà Vinh King Sejong Institute, Tra Vinh University, Vietnam. She finished her master program in TESOL from Can Tho University in 2009 and her doctoral degree from Hue University of Foreign Languages in 2021. Her research interests are teaching EFF in general education and integrating cultures into teaching English as a foreign language (EFL).

Nurdan Kavakli Ulutaş received her Ph.D. degree in English Language Teaching at Hacettepe University. She is currently a full-time academic at the department of English Language Teaching at Izmir Demokrasi University. She divides her loyalties between teaching undergraduate and graduate classes, and academic research. She has book chapters, and articles published in national and international academic journals. She has coordinated or participated in the steering committees of several national and international education projects. Her research interests include language teacher education, language testing and assessment, and language attrition.

I**Nguyen Thi Hong Lien** is a lecturer of English at Hoa Sen University, Vietnam. She teaches courses in language skills and business English. Her research interests include technology in language teaching, teaching English to speakers of other languages and content, and language integrated learning.

Ayman Nassif is a highly motivated educator, researcher and consultant engineer. His teaching and research experience spans solid mechanics, materials, structural engineering at all undergraduate and postgraduate levels. His teaching expertise includes fire structural engineering, thermo-mechanical FE modelling, concrete materials and technology. His industrial experience includes structural design and supervision of construction of buildings, roads, bridges, telecommunication towers and irrigation systems. His experience was gained in Egypt, Finland, UK, Switzerland and Vietnam.

Dang Thanh Tam is a teacher of English at Chu Van An High School, Vietnam. He teaches courses in language skills for K12 students. His research interests include second language acquisition and educational technology.

Pham Thang received a Ph.D. degree from the University of Social Sciences and Humanities, Hanoi, Vietnam, in 2019. From 2014 to the present, he has been a lecturer at University of Sciences, Thai Nguyen University, Vietnam. In 2020, he became Head of the Department at the Faculty of Journalism and Communication, University of Sciences, Thai Nguyen University, Vietnam. His areas of research include the use of new technology in journalism, media, education, computer-mediated communication, and social media.

Thao Quoc Tran is currently a lecturer of English language at the Faculty of English language, Ho Chi Minh City University of Technology (HUTECH), Vietnam. His research areas are discourse analysis, English language skills, teacher professional development and instructional design model.

Ta Thi Nguyet Trang received a Ph.D. degree from the University of Social Sciences and Humanities, Hanoi, Vietnam, in 2020. From 2018 to the present, she has been a lecturer at Department of Economics and Management, International School, Thai Nguyen University, Vietnam. Her areas of research include international studies, marketing communication, new media, education, economics, and management.

Le Vo holds a PhD in TEFL/TESOL from the University of Portsmouth, U.K. She is particularly interested in materials design, using technology in English teaching and how English language training can be improved with IT support, classroom research, teacher education, global Englishes and English language communicative competence and intercultural communication required of graduates at the workplace.

Tho Vo is an English lecturer at the University of Economics Ho Chi Minh City (Vietnam). He obtained his Ph.D in Education at Victoria University of Wellington (New Zealand) where he worked on the teachers' and students' use of digital technologies in the English-medium context of Vietnamese higher education. His research interests are in the areas of technology in language education and English-medium education.

Nguyen Ngoc Vu is currently Director of Training and Applied Research Institute (TARI), HUFLIT University and chairman of STESOL founded by Association of Vietnam Universities and Colleges. With training experience from more than 25 countries, his main expertise is building digital transformation competency and providing consultation services to higher education institutions and businesses across Vietnam. He won Vietnam Technology Innovation Award in 2012 and got recognized as Microsoft Innovative Educator (MIE) Master Trainer in 2014. His research interests include Computer Assisted Language Learning, Cognitive Linguistics, Educational Technology, and ELT Methodology.

Index

4English Mobile app 67-69, 71-73, 77-83

A

Affect Terms 188, 195-196, 199, 201, 210
Asynchronous 1, 6-8, 13, 16, 18, 87, 90, 108-109, 112, 132, 158, 258
Asynchronous CMC 16, 258
Attitude 24, 80-81, 91-92, 104, 111, 113, 125-126, 129, 132-133, 137, 140, 147-148, 151, 195, 227, 229, 233, 239, 248

B

Behavior 52, 91, 104-107, 113, 130, 132, 186, 190, 202-207, 209-210, 225-226, 229, 233, 237-238, 241-247, 250-251, 253-255
Benefit 9-10, 21, 25, 108, 136, 148, 190, 216
Blended Learning 13, 35, 87, 90, 129
Bonus points 51, 58-60, 66
Boredom 51-66, 91, 207
Business 11, 39-41, 43, 84, 99, 107, 114, 128, 164, 190, 211-224, 243-247, 250-251, 253-254, 258, 273
Business-to-Business (B2B) 211, 215-216, 224

C

Challenge 21, 45, 58, 108, 134, 145, 157
CMC 16, 18-19, 29, 111, 127, 156, 183, 188-190, 192-193, 196, 200-201, 236, 258-259, 275
Collaborative Learning 7, 35, 44, 87

communication 1-2, 6-7, 9, 12, 16-22, 28-30, 36, 46, 56-59, 61-62, 72, 78, 83, 91, 102, 110-112, 114, 128, 131, 156-157, 172, 179, 186-190, 195-196, 200-210, 212-222, 224-225, 227-229, 232, 236, 238-242, 244-247, 250-253, 256, 258-259, 261, 269, 272-275
Communicator 228, 231, 255
Computer assisted language learning 29, 47, 63, 67, 85
Computer-Mediated Communication (CMC) 16, 18, 111, 189, 236, 258
Consumer 28, 130, 214-215, 222, 225-228, 230, 234-238, 240-246, 248-255
Consumer Behaviour 225, 244
Consumer Generated 225, 229
Cooperative Development (CD) 157, 187
Credible 226, 236, 249, 255
Customer Satisfaction 244, 256, 259, 265, 276

D

Digital Literacy 18, 20, 27, 30, 49
Digital multimodal composing 17-18, 26-27, 29
DMC 17-25, 27, 29-30

E

Educational technology 50, 69, 105, 131, 223
EFL 26, 29, 51-53, 56, 58, 63-65, 67-69, 88-89, 93, 103-104, 106-108, 110-112, 126, 129, 132-133, 137, 139,

147, 149-153

E-learning 2, 6, 9, 12-16, 28, 43, 69, 85-87, 89-92, 104-107, 131, 249

Electronic 2, 19, 42, 87, 94, 97-100, 102, 189, 206-207, 213, 216, 223-229, 242-255

Electronic Word of Mouth 225-228, 243, 247-248, 250, 254-255

EMI 31, 33-36, 40, 42, 44-45, 48-50

Emoticons 188, 190, 195, 197, 199-201, 203, 205, 207-208, 210, 269-270

English Language Face-to-Face 51

English majors 65, 107, 110-112, 114-127, 132, 150

English medium instruction (EMI) 49-50

eWOM 225-235, 237-245, 247-252, 254

Explicit Instruction 134-135, 137, 145-146, 148, 154

F

Facebook 39, 42-43, 93, 98, 111-115, 117-118, 120-123, 125-132, 188, 190-194, 196-197, 199-209, 219, 251

Facebook-Based Language Learning Activities 113, 132

Facebook-Based Language Learning Strategies 113, 132

G

Games 21, 51, 56, 58-61, 66, 68, 71, 87

Gamification 87

I

ICT 30, 45, 47, 105, 128, 213-214

L

L2 18, 20-21, 23-25, 27, 29-30, 53-54, 64, 152, 184, 207

Language Teaching 13, 18, 25-26, 47, 61, 65, 84, 86, 88, 104-105, 110-112, 114, 131-132, 134, 149-154, 183-185

Learning Management System 89, 110-111, 129, 131-132, 209

Lexical Collocation 133-140, 142-149, 151, 154

M

Market Maven 232, 246, 255

Marketing 43, 130, 190, 193, 202, 206, 215, 218, 221-225, 227-229, 233, 236, 238, 240-254, 276

Mobile Learning 35, 69-70, 83-87

Multimodality 19-20, 24, 26-28, 30

N

Named Entity Recognition (NER) 256-257, 260-261, 271

O

Online Communication 62, 189, 211, 213-220, 224, 239

Online Education 1-4, 6-16, 52, 54, 62, 90, 104, 107-108, 134, 137

Online Learning 1-2, 7, 10, 12-13, 15, 43, 50, 52-55, 61, 63, 66, 87-99, 101-102, 104-107, 109, 111, 113, 131-132, 134

Online Teaching 1, 5, 9, 13, 15-16, 52, 56-57, 61-62, 64-65, 88-91, 93, 103, 105, 128, 133, 135, 138, 142, 145, 148, 154, 273

P

Perception 53, 91, 94, 109, 146, 177, 239

Practice 14, 17-18, 21-22, 28, 41, 50, 58, 60-62, 69, 72, 79-83, 85, 88, 93-95, 98, 100, 109, 131, 150, 152, 155-157, 162, 164, 179, 183-187, 221, 252

Prosumer 22, 28, 30

Purchase Intention 227, 229, 232-233, 239, 243-244, 246, 248-251, 253, 255

Q

Quality 2, 4, 9-11, 54, 62, 71, 92, 94, 98-99, 103, 105, 109-112, 127, 133-134, 142, 146-147, 151, 221, 226, 233, 235-236,

244, 246, 259
quantitative Research 66

R

Reading Comprehension 68, 80, 82, 84, 86, 112, 150-151
Reflective Practice 50, 155-157, 162, 183-187
Research Question 36, 56, 142, 145, 196-197, 200

S

Second Language Teacher Education 156-157, 184, 186-187
Self-Regulated Learning (SRL) 32-33, 50
Social media 108, 127-129, 188-204, 206-209, 214, 218-219, 221-224, 229, 232-233, 235, 238-240, 248-249, 251-252, 254, 275
Strategy 9, 32, 36, 87, 110, 114-115, 125, 195-197, 215, 218, 250, 252
Students' Attitude 104, 111, 129, 137, 140
Synchronous 1, 6-8, 13, 16, 18, 87, 90, 109, 112, 132, 258-259, 275
Synchronous CMC 16, 258, 275

T

Technology 1-3, 5, 9, 11-19, 21-22, 24-26, 28, 30, 34-36, 42, 45, 47, 50-51, 57-58, 60-61, 67, 69, 84, 88, 90-92, 98, 101-102, 104-107, 109, 129-131, 183, 190, 192, 204, 206, 209, 212-213, 216-217, 219-220, 222-225, 232, 235-236, 238,

240, 242-243, 249, 251-252, 259, 274
Technology-Assisted 31, 50
Technology-Assisted Self-Regulated Learning 31
Text Message 276
Text Messages 166, 202, 256, 258-261, 263, 266, 269, 273
Traditional Teaching 1-4, 6, 10-12, 15-16

U

User 87, 104, 194, 198, 201-202, 206, 210, 214, 227, 232-235, 239, 258-259, 272
User Interface Design 194, 198, 201, 210

V

Verbosity 188, 195-197, 199-201, 210
Video-Mediated Communication 186-187
Video-Mediated Cooperative Development 155-156, 158-159, 184
Vietnamese higher education 31, 46, 48
Vocabulary Use 133-136, 139-140, 142-149, 154

W

Writing Performance 133-134, 136-137, 139-140, 142-149, 154

Z

Zalo 93, 100, 188, 190-192, 196-201, 209, 261

Printed in the United States
by Baker & Taylor Publisher Services